Land Perspectives: People, Tenure, Planning, Tools, Space, and Health

Land Perspectives: People, Tenure, Planning, Tools, Space, and Health

Editors

Uchendu Eugene Chigbu
Ruishan Chen
Chao Ye

MDPI • Basel • Beijing • Wuhan • Barcelona • Belgrade • Manchester • Tokyo • Cluj • Tianjin

Editors

Uchendu Eugene Chigbu
Namibia University of Science and Technology
Namibia

Ruishan Chen
Shanghai Jiaotong University
China

Chao Ye
East China Normal University
China

Editorial Office
MDPI
St. Alban-Anlage 66
4052 Basel, Switzerland

This is a reprint of articles from the Special Issue published online in the open access journal *Land* (ISSN 2073-445X) (available at: https://www.mdpi.com/journal/land/special_issues/Land_Perspectives).

For citation purposes, cite each article independently as indicated on the article page online and as indicated below:

LastName, A.A.; LastName, B.B.; LastName, C.C. Article Title. *Journal Name* **Year**, *Volume Number*, Page Range.

ISBN 978-3-0365-3884-6 (Hbk)
ISBN 978-3-0365-3883-9 (PDF)

Cover image courtesy of Uchendu Eugene Chigbu

Contents

About the Editors . . . . **ix**

Preface to "Land Perspectives: People, Tenure, Planning, Tools, Space, and Health" . . . . . . **xi**

Uchendu Eugene Chigbu, Ruishan Chen and Chao Ye
Land Perspectives: People, Tenure, Planning, Tools, Space, and Health
Reprinted from: *Land* **2022**, *11*, 296, doi:10.3390/land11020296 1

Shewakena Aytenfisu Abab, Feyera Senbeta Wakjira and Tamirat Tefera Negash
Determinants of the Land Registration Information System Operational Success: Empirical Evidence from Ethiopia
Reprinted from: *Land* **2021**, *10*, 1394, doi:10.3390/land10121394 7

Jiawen Zhou and Jing Xiong
Resource Opportunity in China's Market Transition and Governance: Time Factor in Urban Housing Inequality
Reprinted from: *Land* **2021**, *10*, 1331, doi:10.3390/land10121331 33

Sayeh Kassaw Agegnehu, Tilahun Dires, Worku Nega and Reinfried Mansberger
Land Tenure Disputes and Resolution Mechanisms: Evidence from Peri- Urban and Nearby Rural Kebeles of Debre Markos Town, Ethiopia
Reprinted from: *Land* **2021**, *10*, 1071, doi:10.3390/land10101071 53

Bonoua Faye and Guoming Du
Agricultural Land Transition in the "Groundnut Basin" of Senegal: 2009 to 2018
Reprinted from: *Land* **2021**, *10*, 996, doi:10.3390/land10100996 75

Dongyang Yang, Chao Ye and Jianhua Xu
Land-Use Change and Health Risks in the Process of Urbanization: A Spatiotemporal Interpretation of a Typical Case in Changzhou, China
Reprinted from: *Land* **2021**, *10*, 820, doi:10.3390/land10080820 93

Armands Auzins and Uchendu Eugene Chigbu
Values-Led Planning Approach in Spatial Development: A Methodology
Reprinted from: *Land* **2021**, *10*, 461, doi:10.3390/land10050461 107

Solomon Dargie Chekole, Walter Timo de Vries, Pamela Durán-Díaz and Gebeyehu Belay Shibeshi
Analyzing the Effects of Institutional Merger: Case of Cadastral Information Registration and Landholding Right Providing Institutions in Ethiopia
Reprinted from: *Land* **2021**, *10*, 404, doi:10.3390/land10040404 127

Marco Hölzel and Walter Timo de Vries
Digitization as a Driver fur Rural Development—An Indicative Description of German Coworking Space Users
Reprinted from: *Land* **2021**, *10*, 326, doi:10.3390/land10030326 145

Mingxing Chen, Yuan Zhou, Xinrong Huang and Chao Ye
The Integration of New-Type Urbanization and Rural Revitalization Strategies in China: Origin, Reality and Future Trends
Reprinted from: *Land* **2021**, *10*, 207, doi:10.3390/land10020207 167

Festus A. Asaaga
Building on "Traditional" Land Dispute Resolution Mechanisms in Rural Ghana: Adaptive or Anachronistic?
Reprinted from: *Land* **2021**, *10*, 143, doi:10.3390/land10020143 . . . . **183**

Jaime De Pablo Valenciano, Juan Milán-García, Juan Uribe-Toril and María Angustias Guerrero-Villalba
Rural Development from a Gender Perspective: The Case of Women Farmers in Southern Spain
Reprinted from: *Land* **2021**, *10*, 75, doi:10.3390/land10010075 . . . . **201**

Solomon Dargie Chekole, Walter Timo de Vries, Pamela Durán-Díaz and Gebeyehu Belay Shibeshi
Performance Evaluation of the Urban Cadastral System in Addis Ababa, Ethiopia
Reprinted from: *Land* **2020**, *9*, 505, doi:10.3390/land9120505 . . . . **215**

Walter Dachaga and Walter Timo de Vries
Land Tenure Security and Health Nexus: A Conceptual Framework for Navigating the Connections between Land Tenure Security and Health
Reprinted from: *Land* **2021**, *10*, 257, doi:10.3390/land10030257 . . . . **229**

Guilherme Silva Fracarolli
Global Markets, Local Issues: The Hegemonic Process of Agri-Food Construction to Present Challenges
Reprinted from: *Land* **2021**, *10*, 1182, doi:10.3390/land10111182 . . . . **251**

About the Editors

Uchendu Eugene Chigbu (PhD) is Associate Professor at the Namibia University of Science and Technology. His works fall within the interface between Social Sciences and Geodesy (or Social Geodesy). He is a multi-disciplinary consultant with diverse experience across a broad spectrum of land management specializations. He has multiple professional affiliations but is most active with the International Federation of Surveyors (FIG) and the Global Land Tool Network (GLTN). He is the Co-Chair of the Research in the GLTN. His most recent book, entitled *Land Governance and Gender: The Tenure–Gender Nexus in Land Management and Land Policy*, was published in 2022. He serves on the Editorial Boards of the journals *Land Use Policy* (Elsevier) and *Local Development & Society* (Routledge).

Ruishan Chen (PhD) is Professor at the Shanghai Jiaotong Univeristy. He researches land use change, human–nature interactions, and disaster risk reduction. He has worked on the International Platform on Biodiversity and Ecosystem Services (IPBES) Land Degradation and Restoration Assessment and is a lead author of *Transformative Change Assessment* (a thematic assessment of the underlying causes of biodiversity loss, determinants of transformative change and options for achieving the 2050 vision for biodiversity). He is also Director of the Urban Climate Change Research Network (UCCRN) East Asian hub and has worked on several UCCRN assessments, for example, COVID, Climate Change and Cities and Nature-Based Solutions for Cities.

Chao Ye (PhD) is Professor at the School of Geographic Sciences, East China Normal University. He was selected as a young scholar of the Changjiang Scholars Program of the Ministry of Education and is now the chief expert of the Major Project of the National Social Science Foundation of China. Chao's research interests are in urbanization, urban and rural governance, sustainability science, and geographical thought, with particular focus on interdisciplinary research in economy, society, geography, and culture of urban and rural areas. He has over 100 publications in such high-impact journals as *Science Bulletin, Habitat International, Journal of Rural Studies*, and *Land Use Policy*. He currently serves as the Associate Editor of *Environment, Development and Sustainability* and serves on the editorial board and as a reviewer for numerous international journals.

Preface to "Land Perspectives: People, Tenure, Planning, Tools, Space, and Health"

Efficient land (including water and forest) administration practices and spatial enablement is required to achieve numerous items of various global development agendas—e.g., related to land degradation neutrality, New Urban Agenda, COP21, Sustainable Development Goals (SDGs), and COVID-19 or coronavirus pandemic challenges. Achieving these goals requires understanding how land administration practices and spatial decisions can impact people, property tenure, and health or wellbeing. Also helpful is creating a peaceful environment by eliminating social conflicts caused by poor land administration practices. Hence, there is a need to probe natural resources administration theories and tools. Good land administration and spatial enablement help improve people's living conditions in urban, peri-urban, and rural areas. They protect people's land rights (including individuals, communities, and the state) through good governance principles and practices. This makes research concerning land administration practices and geographic (spatial) sciences—whether in developed or developing countries—essential to developing tools or methods for securing natural resource rights for people. In the time of COVID-19, understanding the land and health or wellbeing nexus is also crucial for adequate living conditions for people in living urban, peri-urban, and rural areas. This Special Issue comprises 15 articles (including the editorial) that present insights on theories and practices on land administration and geographic (spatial) sciences in the context of the land/water/forest–people–health–wellbeing nexus.

The urban-to-rural dimension of land studies is crucial because it embodies streams of knowledge that support rural-urban co-governance and development. The focus of the Special Issue is essential for understanding the influence of planning on people through its tools, space, tenure, and health. It also provides a multi-faceted lens of these issues—from urban, peri-urban, and rural lenses. These topical issues are covered at various scales (local, national, and global). The Special Issue was curated to investigate three critical questions. First, it answers the question: What exactly do people, tenure, planning, tools, space, and health imply? Second, it responds to the question: What relationships do people, tenure, planning, tools, space, and health share? Third, it uses empirical and literature evidence to present the land perspectives of these issues worldwide. We encourage all interested individuals or groups—including academics and practitioners within the discipline of land administration and geographic (spatial) sciences—to read the collection of articles in this book. These articles represent the latest research from around the world.

We would like to express our sincere thanks to some of the agencies that provided both financial and in-kind support that led to the success of this Special Issue. We would particularly like to express our gratitude to the Major Program of National Social Science Foundation of China (19ZDA086), the Hanns Seidel Foundation in Shadong, China, and the Global Land Tool Network at the UN-Habitat in Nairobi, Kenya.

Uchendu Eugene Chigbu, Ruishan Chen, and Chao Ye
Editors

land

MDPI

Editorial

Land Perspectives: People, Tenure, Planning, Tools, Space, and Health

Uchendu Eugene Chigbu [1,*], Ruishan Chen [2] and Chao Ye [3]

1 Department of Land and Property Sciences, Faculty of Engineering and Spatial Sciences, Namibia University of Science and Technology, Windhoek 9000, Namibia
2 School of Design, Shanghai Jiao Tong University, Shanghai 200040, China; rschen@sjtu.edu.cn
3 School of Geographic Sciences, East China Normal University, Shanghai 200062, China; cye@geo.ecnu.edu.cn
* Correspondence: echigbu@nust.na; Tel.: +264-61-207-2470

Citation: Chigbu, U.E.; Chen, R.; Ye, C. Land Perspectives: People, Tenure, Planning, Tools, Space, and Health. *Land* **2022**, *11*, 296. https://doi.org/10.3390/land11020296

Received: 28 January 2022
Accepted: 8 February 2022
Published: 15 February 2022

The global agendas on land and related issues (including agendas on land degradation neutrality, New Urban Agenda, climate change, United Nations' decades on ecosystem restoration, and Sustainable Development Goals) are vulnerable to being neglected due to the current global focus on eradicating the COVID-19 or coronavirus pandemic. The actions needed to reposition these agendas on a realistic path to sustainability require an understanding and the application of land perspectives in the various development policies being implemented in different national and local contexts. Land perspectives entail the potential ways (both differences and similarities)—geographies, situations, viewpoints, and approaches—necessary for (re)engaging land issues as core factors of socioeconomic and environmental developments. This is why the importance of people, tenure, planning, tools, space, and health should be critical themes for investigation by scholars wherever they may be around the world and in whatever discipline they may come from, hence the relevance of this Special Issue.

This Special Issue (SI) of the journal *Land* is entitled Land Perspectives: People, Tenure, Planning, Tools, Space, and Health. The focus of the SI is crucial for understanding the influence of planning on people through its shapes tools, space, tenure and health. Conceptualized from a land perspective, the SI provides a broad way of investigating land management, land administration, land governance and policy issues. It also provides a multi-faceted lens of these issues—from urban, peri-urban, and rural lenses. The urban-to-rural dimension of land studies is crucial because it embodies streams of knowledge that support rural–urban co-governance and development. The SI covers these topical issues at various scales (local, national, and global). It was curated to investigate three critical questions. First, it answers the question: What exactly do people, tenure, planning, tools, space, and health imply? Second, it responds to the question: What relationships do people, tenure, planning, tools, space, and health share? Third, it uses empirical and literature evidence to present the land perspectives of these issues from all over the world. The first question posed above can be answered by synthesizing all the articles published in this SI.

- People: Land and people have a relationship which is an eternal topic of geography and many other disciplines. On the one hand, people live on and use land in all aspects of their livelihood and existence on earth. On the other hand, land (the base of the environmental system) requires human activities to enjoy various forms of ecosystem services. Therefore, the people aspect of this SI recognizes that experts are interested in administering, managing, or developing land resources because of people. Humans in their various communities (people) are not just the actors in improving socioeconomic living conditions. They are the reason or motivation for engaging in multiple aspects of development (i.e., social, economic, and environmental development). Some articles in this SI have considered research exploring land-and-people relations from spatial, social, and geographical angles.

- Tenure: Land tenure stands at the heart of the development in rural and urban areas. This is because owning, using, accessing privileges, and exercising land rights are crucial dimensions of wealth creation. How people use and exercise rights over land has a tremendous influence on the direction of their development. Land tenure (including its associated property rights, historical, land-use, tenure security, institutions and political dimensions) is necessary for grasping the how-to aspect of improving the living conditions of people who own, use and exercise various rights on land. This SI provides a platform for developing theoretical and practical knowledge on improving tenure security by collating expert ideas and experiences from multiple scholars from different parts of the world. The authors of articles in this SI have considered case studies that unravel transferable experiences across the globe to ensure cross-regional knowledge building.
- Planning: Planning and planners face a critical question under the ongoing COVID-19 pandemic. How can modern urban planning improve the people's wellbeing and health. This requires that planners return planning to its health and wellbeing roots. Answering this question places more responsibilities on the planning profession and their role in the land and health nexus of resolving contemporary problems. Some articles submitted to this SI considered planning issues beyond their traditional boundaries and delved into all aspects of planning that connects to land, including health matters.
- Tools: Land is not only the physical earth with its above-and-below resources. It serves as a tool or a practical way to intervene in land administration problems. In developing country contexts, the need for practical tools that respond to country-specific conditions is necessary to facilitate the management of spaces in urban and rural areas. Articles with a methodological focus on land-use approaches (and organization of natural resources) are essential for this SI. Such articles can suggest practical ways for improving the challenges people face.
- Space: The spatial dimension of land studies is a crucial aspect of science that consistently demands renewed research attention. This is because there is a tendency for experts to focus on spatial planning while leaving out the development aspect of how people adapt to space use. The need for a more inclusive process in spatial planning should ensure that inclusive development becomes the outcome. Scholarly works that engage in bottom-up decision making that are mediative and based on consensus building are worth investigating. This SI attracted articles that spatially analyzed environmental scenarios for socio-spatial justice in human societies.
- Health: The land–health nexus of research existed long before now. However, investigations into the outbreak of COVID-19 or coronavirus are rapidly evolving. This has led to investigations into land–health relationships and how land issues influence the quality of life of people and communities. Some articles in this SI probed the health and wellbeing dimensions of land management and land administration to tease out how land uses, and the exercise of land rights (or lack of it) influence individual and community wellbeing.

The issue of relationships is essential because, by dwelling on this, the SI contributes to advancing the borders of cross-, inter- and multi-disciplinarity across all genres of studies connected to land. The question "What relationships do people, tenure, planning, tools, space, and health share?" is also answerable by analyzing the evidence from all articles published in the SI. We use Figure 1 to illustrate the relationship between *people, tenure, planning, tools, space, and health (from a land perspective).*

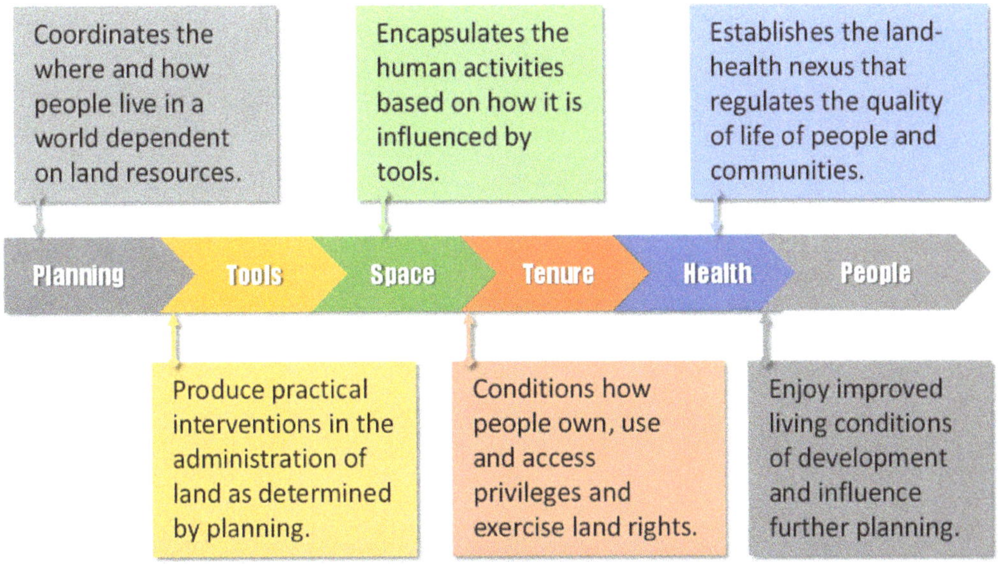

Figure 1. The relationship between people, tenure, planning, tools, space, and health (from land perspective) (authors' illustration).

As shown in Figure 1, planning is the starting point of influence for development. It coordinates where and how people live in a world dependent on land resources. It leads to creating tools that produce practical interventions in land administration as determined by planning. Tools then shape space to encapsulate human activities based on how tools influence it. Space influences tenure (the condition concerning how people own, use and access privileges and exercise land rights). Tenure influences health (positively or negatively). It establishes the land–health nexus that regulates the quality of life of people and communities. Finally, enjoy all forms of improved living conditions or development options. Noteworthy is that people go on to influence further planning for their future. This relationship indicates that planning and people are critical factors in development. Land is the resource for development.

To understand how land serves as a resource for development, it is necessary to dwell on the third (and final) question. In this regard, we present snapshots of the articles published in the SI and their key findings below.

Article 1: Land Tenure Security and Health Nexus: A Conceptual Framework for Navigating the Connections between Land Tenure Security and Health by Dachaga and de Vries [1]. This article used evidence from existing literature to show that land tenure security can influence health outcomes via four pathways—infrastructure access, environmental justice, psycho-ontological security, and social cohesion.

Article 2: *Resource Opportunity in China's Market Transition and Governance: Time Factor in Urban Housing Inequality* by Zhou and Xiong [2]. In this study, the authors investigated the influence of real-estate purchase factors (such as time, organization, human capital, and political capital) on real estate value and the appreciation of real estate in China. They found that time influences the prior possession of resources in the early stage of market transformation.

Article 3: *Agricultural Land Transition in the "Groundnut Basin" of Senegal: 2009 to 2018* by Faye and Du [3]. This article reveals the transition features of agricultural land use in the Groundnut Basin of Senegal from 2009 to 2018, especially the impact of urbanization on agricultural land and the viewpoint of farmland spatiotemporal evolution.

Article 4: *Values-Led Planning Approach in Spatial Development: A Methodology* by Auzins and Chigbu [4]. This study proposes a methodology for introducing a values-led planning approach in spatial development. It presents and discusses the essential elements required to design methods for values-focused planning.

Article 5: *Building on "Traditional" Land Dispute Resolution Mechanisms in Rural Ghana: Adaptive or Anachronistic?* by Asaaga [5]. This research explores the importance of traditional dispute resolution institutions in land-related disputes in southcentral and western Ghana. It highlights practical ways to incorporate traditional dispute resolution in Ghana's overall land governance setup and elsewhere in sub-Saharan Africa.

Article 6: *Global Markets, Local Issues: The Hegemonic Process of Agri-Food Construction to Present Challenges* by Fracarolli [6]. This article uses dialectics to analyze the historical process of agrarian systems according to their complexity, origins and effects of hegemonic interests in the agri-food markets. It shows that markets evolve from different trade types as the capitalist system also evolve, changing the mechanics of trade and functions of food production.

Article 7: *Performance Evaluation of the Urban Cadastral System in Addis Ababa, Ethiopia* by Chekole et al. [7]. This study evaluated the performance of the urban cadastral system of Addis Ababa (Ethiopia) based on the European Foundation for Quality Management (EFQM) excellence model. They found that the most bottlenecks to organizational achievement are the strategic plan, quality of leadership, bureaucratic processes, and supply of resources.

Article 8: *Rural Development from a Gender Perspective: The Case of Women Farmers in Southern Spain* by Valenciano et al. [8]. This article probed the land-based women's working conditions in Spain. It found that women workers in the fruit- and vegetable-handling sector are satisfied with their jobs. It presents a local development model for increasing women's empowerment in the land-based labor market.

Article 9: *The Integration of New-Type Urbanization and Rural Revitalization Strategies in China: Origin, Reality and Future Trends* by Chen et al. [9]. This article reviews the classic theories and cognition of the research on urban–rural relations at home and abroad. It outlines the stage evolution characteristics of urban–rural relations in China.

Article 10: *Digitization as a Driver fur Rural Development—An Indicative Description of German Coworking Space Users* by Hölzel and de Vries [10]. The research investigated the conditions of users of coworking spaces in Germany. It found that the choice of working in rural coworking spaces draws on benefits and opportunities for its users in the aspects of avoiding social isolation, separating private and professional life and reducing commuting challenges.

Article 11: Land-Use Change and Health Risks in the Process of Urbanization: A Spatiotemporal Interpretation of a Typical Case in Changzhou, China by Yang et al. [11]. This study established the relationship between urban land-use changes and health in the context of Changzhou, China.

Article 12: Land Tenure Disputes and Resolution Mechanisms: Evidence from Peri-Urban and Nearby Rural Kebeles of Debre Markos Town, Ethiopia by Agegnehu et al. [12]. This study analyzes the nature, types, and causes of land tenure disputes and the resolution mechanisms in peri-urban and nearby rural areas of Debre Markos town in Ethiopia.

Article 13: Analyzing the Effects of Institutional Merger: Case of Cadastral Information Registration and Landholding Right Providing Institutions in Ethiopia by Chekole et al. [13]. This research is based on a survey conducted with the directors of the two institutions and their employees to determine how to reduce the effects of data duplication and provide one-window services (among other factors) to improve efficiency in the Ethiopian land markets

Article 14: *Determinants of the Land Registration Information System Operational Success: Empirical Evidence from Ethiopia* by Abab et al. [14]. This research assessed the most extensive digitalization program for rural land registration in Africa. It revealed that system quality, information quality, service quality, and perceived usefulness of the program have positively and significantly influenced the acceptance and actual use of the system.

Articles published in this SI can become multi-disciplinary reference material for in-class and on-field learning in land studies. The SI contains 14 positively evaluated (peer-reviewed) articles as listed above. Each article presents 14 lessons learned from across Africa, Europe and Asia. All scholars within the land profession—whether in the business, geography, sociology, area studies, anthropology, planning, engineering and the built environment disciplines—are encouraged to read, use and apply these lessons in their different roles in the land sector.

Author Contributions: Conceptualization, U.E.C.; writing—original draft preparation, U.E.C.; writing—review and editing, U.E.C., C.Y. and R.C.; visualization, U.E.C. All authors have read and agreed to the published version of the manuscript.

Funding: This research received no external funding.

Institutional Review Board Statement: Not applicable.

Informed Consent Statement: Not applicable.

Data Availability Statement: Not applicable.

Acknowledgments: We would like to acknowledge the institutional affiliation of the Guest Co-editors of this Special Issue: The Department of Land and Property Sciences at the Namibia University of Science and Technology; The School of Design, Shanghai Jiao Tong University, Shanghai; and the School of Geographic Sciences, East China Normal University. Special thanks go to the authors (both for papers accepted and rejected), and the reviewers of the articles published in this Special Issue. Most important is that we would love to recognize the efforts of all those people from communities worldwide who provided the data used in writing the articles published in this Special Issue.

Conflicts of Interest: The authors declare no conflict of interest.

References

1. Dachaga, W.; de Vries, W.T. Land Tenure Security and Health Nexus: A Conceptual Framework for Navigating the Connections between Land Tenure Security and Health. *Land* **2021**, *10*, 257. [CrossRef]
2. Zhou, J.; Xiong, J. Resource Opportunity in China's Market Transition and Governance: Time Factor in Urban Housing Inequality. *Land* **2021**, *10*, 1331. [CrossRef]
3. Faye, B.; Du, G. Agricultural Land Transition in the "Groundnut Basin" of Senegal: 2009 to 2018. *Land* **2021**, *10*, 996. [CrossRef]
4. Auzins, A.; Chigbu, U.E. Values-Led Planning Approach in Spatial Development: A Methodology. *Land* **2021**, *10*, 461. [CrossRef]
5. Asaaga, F.A. Building on "Traditional" Land Dispute Resolution Mechanisms in Rural Ghana: Adaptive or Anachronistic? *Land* **2021**, *10*, 143. [CrossRef]
6. Fracarolli, G.S. Global Markets, Local Issues: The Hegemonic Process of Agri-Food Construction to Present Challenges. *Land* **2021**, *10*, 1182. [CrossRef]
7. Chekole, S.D.; de Vries, W.T.; Durán-Díaz, P.; Shibeshi, G.B. Performance Evaluation of the Urban Cadastral System in Addis Ababa, Ethiopia. *Land* **2020**, *9*, 505. [CrossRef]
8. De Pablo Valenciano, J.; Milán-García, J.; Uribe-Toril, J.; Guerrero-Villalba, M.A. Rural Development from a Gender Perspective: The Case of Women Farmers in Southern Spain. *Land* **2021**, *10*, 75. [CrossRef]
9. Chen, M.; Zhou, Y.; Huang, X.; Ye, C. The Integration of New-Type Urbanization and Rural Revitalization Strategies in China: Origin, Reality and Future Trends. *Land* **2021**, *10*, 207. [CrossRef]
10. Hölzel, M.; de Vries, W.T. Digitization as a Driver fur Rural Development—An Indicative Description of German Coworking Space Users. *Land* **2021**, *10*, 326. [CrossRef]
11. Yang, D.; Ye, C.; Xu, J. Land-Use Change and Health Risks in the Process of Urbanization: A Spatiotemporal Interpretation of a Typical Case in Changzhou, China. *Land* **2021**, *10*, 820. [CrossRef]
12. Agegnehu, S.K.; Dires, T.; Nega, W.; Mansberger, R. Land Tenure Disputes and Resolution Mechanisms: Evidence from Peri-Urban and Nearby Rural Kebeles of Debre Markos Town, Ethiopia. *Land* **2021**, *10*, 1071. [CrossRef]
13. Chekole, S.D.; de Vries, W.T.; Durán-Díaz, P.; Shibeshi, G.B. Analyzing the Effects of Institutional Merger: Case of Cadastral Information Registration and Landholding Right Providing Institutions in Ethiopia. *Land* **2021**, *10*, 404. [CrossRef]
14. Abab, S.A.; Wakjira, F.S.; Negash, T.T. Determinants of the Land Registration Information System Operational Success: Empirical Evidence from Ethiopia. *Land* **2021**, *10*, 1394. [CrossRef]

land

MDPI

Article

Determinants of the Land Registration Information System Operational Success: Empirical Evidence from Ethiopia

Shewakena Aytenfisu Abab *, Feyera Senbeta Wakjira and Tamirat Tefera Negash

Center for Environment and Development, College of Development Studies, Addis Ababa University, Addis Ababa P.O. Box 1176, Ethiopia; feyera.senbeta@aau.edu.et (F.S.W.); tamirat.tefera@aau.edu.et (T.T.N.)
* Correspondence: shewakena.aytenfisu@aau.edu.et

Abstract: Ethiopia has embarked on one of the largest digitalization programs for rural land registration in Africa. The program is called the national rural land administration information system (NRLAIS). Over the past couple of years, NRLAIS was rolled-out and made operational in over 180 woredas (districts). There is, however, limited empirical evidence on whether and to what extent NRLAIS has been successful. This study explores the factors that influence the acceptance and actual use of NRLAIS to gauge its operational success in Ethiopia. Data were collected both from primary and secondary sources using surveys, key informant interviews, and a literature review. Survey data were collected from 201 staff of 50 woreda land administration offices in three regional states (Amhara, Oromia, and SNNP) and analyzed using a structural equation model. The results revealed that system quality, information quality, service quality, and perceived usefulness of NRLAIS have positively and significantly influenced the acceptance and actual use of the system. However, perceived ease of use has an insignificant influence. The predictive relevance of the research model is significant and indicates substantial operational success of NRLAIS. The quick acceptance and use of NRLAIS will likely improve service delivery, promote data integration, and strengthen informed decision-making. The study recommends strengthening behavioral changes of the land administration experts through two enhanced service quality measures—technical and operational capacity to a robust and sustainable digitalization. Policymakers could leverage operational success to upgrade the NRLAIS into a unified national land registration information system that bridges the urban–rural land governance divide.

Keywords: land register; digitalization; standardization; tenure security; service delivery; governance; structural equation model

Citation: Abab, S.A.; Wakjira, F.S.; Negash, T.T. Determinants of the Land Registration Information System Operational Success: Empirical Evidence from Ethiopia. *Land* **2021**, *10*, 1394. https://doi.org/10.3390/land10121394

Academic Editors: Uchendu Eugene Chigbu, Ruishan Chen and Chao Ye

Received: 28 October 2021
Accepted: 10 December 2021
Published: 16 December 2021

1. Introduction

Land is a key natural resource, means of livelihood, and development asset for many societies in Ethiopia. The relationship between people and land also bisects almost all sustainable development goals (SDGs) [1,2]. The SDGs and other recent global initiatives have renewed and increased the need to improve land tenure to address a multitude of development challenges [3]. Secure land tenure has regularly been prioritized by policymakers to ensure long-term land-based investment, agricultural productivity, as well as to address changing climate risks, biodiversity loss, poverty reduction, food security, and spur sustainable growth [3–6].

Traditionally, land administration systems were created to record information about property ownership, rights, and boundaries, and other attributes of real properties [7]. However, depending on the context, the strengthening of land rights can take a variety of forms, from documenting customary uses to formalizing individual rights [3]. In other instances, existing customary land tenure systems may be sufficient to ensure land tenure security [8]. Land registration and information system programs and procedures to implement these programs are among the major reasons for failure or success to achieve stronger tenure security [6]. Thus, land registration and information system organization should be part of studies on land tenure and investment in land and productivity [9].

Land information systems (LIS) in this paper are understood as a set of land information technologies (hardware, software, infrastructure, and equipment), personnel, data, rules-based procedures, and organizational structures. The existence of good and well-functioning land information infrastructure is key to answering the fundamental land development and management decision-making questions (i.e., why, who, what, where, when, and how) [10]. The LIS should be complete, reliable, uniform, up to date, and sustainable [11]. Accurate and up-to-date land administration data are also understood as occurring between land information system establishment and maintenance phases [12]. Hence, LIS needs to reflect the reality on the ground and transform itself into a dynamic and sustainable innovation hub and meet service expectations [13,14].

However, it is worth noting that the overall establishment, effectiveness, and maintenance of an integrated geospatial land information system depends on governance, technology, and people factors [15,16]. The governance factor further decomposes to the establishment of appropriate policy, legal framework, and governance structures, while the technological aspects include the data, innovation, and standards. The people dimension is concerned with capacity building, education, partnership, communication, and engagement [2]. Particularly, understanding factors influencing the intention of personnel to accept and use the system is important for the digital transformation and effective land administration service deliveries [17].

In the land registration and land tenure discourse, the 1970s and 1980s marked a move from a paper-based and manual type of data management and process to digital and automated systems in developed economies [1]. Many attempts have been made to set up titling systems in numerous African countries in the same period [18]. However, the efforts have largely failed to achieve the expected transformation, owing to colonial imported rules and systems that disregard local context, among other reasons [19,20]. In most cases, land registries failed to provide authoritative records of titles and transactions and quickly became outdated due to poor planning, lack of capacity, and the flouting and manipulation of law by officials and elites [21].

Despite low coverage of land registration in Africa (only 10%), the surge of land titling programs following the recent reform of land policy and legal framework, such as in Ethiopia and Rwanda, have shown positive trends and outcomes [19,22]. The reform provides a rapid scale-up across the continent that improves access to land and tenure security [19]. These approaches take different forms, ranging from fit-for-purpose to pro-poor land recordation approaches [22–24]. Advancements in information technology have also facilitated the speedy acquisition, storage, dissemination, and application of data related to tenure, use, value, as well as the development of land and other natural resource governance [25]. Demands for a more customer-oriented focus is also one of the drivers for the automation of the land administration systems [1,15]. In this regard, a first step in introducing a new information technology (IT)-enabled land administration system is the determination of the user needs [7]. Hence, if LIS is to be successful, it will be designed to fulfill the requirements of its end-users [14]. Evidence shows that some African countries such as Rwanda and Ethiopia have launched modernization initiatives for their land information systems [13]. These countries accelerate and securely register land titles into a functional land registration information system [26,27].

In Ethiopia, at the heart of the land administration reform is the digitalization of the manual land register. Recognizing the manual land register shortcomings related to the maintenance, security, accessibility, and integration of land information at different administration levels, the Ministry of Agriculture (MoA) has opted for digitalization. Between 2015 and 2017, the MoA has developed and successfully piloted the national rural land administration information system (NRLAIS) in the highland regions [28]. NRLAIS is a web-based system developed on open-source licensed software and based on the land administration domain model (LADM). Hence, NRLAIS is the programmatic approach of the MoA to address the shortcomings of the semi-manual land register, improve the standardization of service delivery, and promote informed policymaking [28].

While the transition to the NRLAIS is at an initial stage, its operational success has not been researched based on explanation and prediction theories. There is little knowledge about what causes user acceptance and actual use of the NRLAIS in the land administration domain. It is also not know why and how the woreda land administration experts choose to discontinue the use of the semi-manual land register that they are using currently. There are a few similar studies in Africa such as Zeng and Cleon [17] on the implementation and development of land information systems (LIS) in Liberia, which adopted the diffusion paradigm. To the best of the authors' knowledge, this is the first attempt in the country. Taking the woreda land administration experts as a unit of analysis, the study seeks to understand the behavior of the woreda land administration experts towards the acceptance and actual use of the NRLAIS and document its operational success. The proposed and empirically tested model of this study reveals a strong construct validity and predictive power. The model captures multiple aspects of each variable, which is a change from much of the measurement of LIS success model constructs that focus on only one aspect of the construct. The study also highlights the needed policy and strategic actions to achieve robust and sustainable digitalization.

2. Literature Review

2.1. Land Tenure and Registration in Ethiopia

Dominated by the agricultural economy, the available scholarly literature documents that widespread tenure insecurity hinders long-term land-based investment in rural Ethiopia [28]. Land tenure insecurity contributes to unprecedented environmental degradation, ecosystem depletion, biodiversity loss, decrease in productivity, and food insecurity [28–31]. These development challenges have been compounded and amplified by climate change and disaster risks that threaten the sustainability of productive landscapes and livelihood resilience [32]. This section briefly highlights the historical account of land tenure and the land certification program and its implication to NRLAIS development in Ethiopia.

During the last century, the land tenure history of Ethiopia has experienced extensive changes. Ethiopia has a long legacy of state intervention in land tenure relations that influence local tenure regimes throughout different political discourses [33]. Hence, the creation and recording of land rights by the national state has been a development theme since the 1960s in the contemporary land tenure history of Ethiopia [34]. The land tenure registration innovations before 1960 have generally been swept away by subsequent changes, but they still have relevance as the model of tenure reform. Before the 1974 revolution, the land tenure systems of Ethiopia were grounded in historically shaped, local institutions, complex and varied across the regions [31]. The military socialist regime's redistributive land reform of 1975 ensured that rural farming households received access to land through only usufruct rights, while ensuring state ownership. Notably, this reform legacy not only weakened the remaining customary institutions but also swept away the overall imperial land governance systems [35]. The current land registration system is highly affected by the 1975 radical land reform of the military socialist regime (1974 to 1991).

In post-socialist Ethiopia, tenure insecurity is linked to a history of limited empowerment of smallholder farmers and significant control by the state in determining access to and control over land resources [35]. During the Ethiopia Revolutionary Democratic Front (EPRDF)-led government (1991 to 2018), state ownership was maintained and enshrined in the 1995 constitution. The 1995 constitution of the federal democratic republic of Ethiopia Article 40 inherited the state land ownership and usufruct rights for landholders from the military socialist regime. The governance structure also changed from a centralized socialistic arrangement to market-led decentralization [36]. Article 52 of the 1995 constitution gives the regional governments the right to administer land and other natural resources following the federal laws. As a result, the land tenure system is evolving differently in rural and urban areas [37]. Different federal proclamations govern its development, and reforms have been progressing at different speeds across the country [38,39]. Moreover, there are essentially two parallel land registration and information system infrastructures—one for rural and one for urban. This rural–urban divide costs the country hugely in terms of policy, institutional, technical, operational, and human resource challenges for integrated and transparent land administration and resource governance systems that foster sustainable development [29].

In rural Ethiopia, the theme of this research, the government has been implementing a progressive two-stage land registration and certification program since 1998 [30]. The first-level landholding certification (FLLC) program that started in 1998 has been claimed by the Government of Ethiopia (GoE) as a policy response of improving tenure security to reverse land degradation, food insecurity, and poverty reduction [40]. Till 2010, the FLLC, which claimed to be cheap and fast, has mainly been financed by regional states but without spatial data of parcels [33,34], whereas the second level landholding certification (SLLC) is being coordinated by the Federal Government in collaboration with the regional states and has attracted strong technical and financial support from international development partners [41]. The latter approach introduced parcel-level cadastral mapping and the transition of the manual registry into a harmonized computerized LIS [42]. Since 2013, Ethiopia has continued investing in the SLLC program to cover over 50 million rural parcels and improve tenure security and land administration service delivery [40,42]. According to MoA [43], between 2013 and 2021, over 21 million rural parcels have been demarcated and mapped, of which close to 18 million parcels have been issued with SLLC. The demarcation and mapping cover about 42 percent of the estimated 50 million parcels found in the highland parts of the country. In Ethiopia, the household-level positive impact of these massive land certification programs has been well studied by different scholars [33,40,41].

2.2. Land Information System in Ethiopia

In Ethiopia, the land registration information system (LRIS) follows the rural–urban cadastral divide. NRLAIS was developed for rural land and the cadaster and real property registration system (CRPRS) for urban land. The systems are being administered by two different agencies. NRLAIS is operated by the rural land administration agencies under the guidance of MoA, while CRPRS is operated by the Urban Land and Cadaster Chief Executive Officer under the guidance of the Ministry of Urban and Infrastructure Development (MoUID). The focus of this paper is NRLAIS serving rural land administration.

In 2010, MoA developed its information system/information communication technology (IS/ICT) and software development strategy for the first time [28]. The strategy was developed following analysis of the requirements of a harmonized land administration system that is suitable for the adoption and implementation of a unified LIS in Ethiopia. According to MoA [28], this IS/ICT strategy provides a single overarching requisite framework embracing both urban and rural lands for the safe and secure maintenance and updating of land records. However, following legal mandates MoA revised its IS/ICT and software development and implementation strategy twice, i.e., in 2012 and, later, in 2017, with a focus on rural land.

As part of the standardization of the rural land administration system, the development of NRLAIS was well established based on the analysis and business reengineering of the four existing organizational structures (federal, regional, zonal, and woreda) [28]. NRLAIS is a web-based system developed on open-source licensed software and based on the land administration domain model (LADM). The system utilizes a modular technology stack and meets the functional and legal requirements for registering rural landholding rights in all the non-pastoral and highland regional states of Ethiopia. The definition of technical specifications for the development of NRLAIS considered the requirements of the functional and legal framework at both the federal and regional levels, including inheritance, gift, exchange, divorce, rent, and encumbrances.

Between 2015 and 2017, supported by the Finnish government-financed project Responsible and Innovative Land Administration in Ethiopia (REILA), MoA developed and piloted NRLAIS. A production version was delivered with an operational acceptance report (OAR) to the then MoA in March 2018 [44]. Between 2016 and 2017, NRLAIS has undergone due diligence processes through a series of pilot testing and upgrades. Since late 2018, the MoA initiated the rollout of NRLAIS into more regions and woredas. The NRLAIS roll-out was divided into two phases of two years each, starting with a comparative trial of four-to-six months that may reveal necessary improvements and changes, followed by a one-and-a-half year period for full-scale roll-out at national level. Since 2020, the second phase of the rollout has continued at increasing speed, addressing sustainability factors such as capacity building and upgrading of the software with emerging functionality requirements [28,43].

According to MoA [28], up to 400 woredas covering 25 million parcels of land records are targeted to establish a functional and operational NRLAIS by 2024. NRLAIS is operational in over 180 woredas of Amhara, Beneshangul Gumuz, Oromia, Southern Nations Nationalities and Peoples (SNNP), and Tigray regional states as of November 2021 [43]. By mid-2021, about 113 woredas with NRLAIS have been verified by third party or independent verifying agency for the system being made operational. During the same period, the information from approximately 11 million parcels has been migrated into the system, 5 to 13 subsequent land transactions were updated per day per woreda, and over 102,000 transactions per year were updated in total. NRLAIS is the largest distributed LIS in Ethiopia and currently operates in 6 regional states, 37 Zones, and over 180 woredas. However, the wider area network (WAN) that connect woredas to zonal, regional to federal/central servers for an online data replication and information flow is yet to deployed due to underdeveloped network infrastructure in the country.

NRLAIS will provide security, transparency, service quality, and continuous maintenance of land records, with enhanced data management functionality and usability at the woreda level in an effective, spatially integrated, and sustainable manner [28]. NRLAIS is considered the key strategic component within the land administration modernization endeavor and an integral part of standardization in the country. Under this context, NRLAIS forms the framework and defines the role of stakeholders and their relationship among personnel, technology, and standard procedures. It also serves as a legitimate bearer of land information including the socio-spatial aspects of landholdings and users' interests in land and natural resources.

3. Theoretical Base

The theoretical base of this study is the DeLone and McLean Information System success model (D&M IS success model) of 2003 and the theory of technology acceptance model (TAM) of Davis and modified by Venkatesh et al. [45–48]. An information system (IS) is developed using information technology (IT) to enhance the performance of individuals and organizations. However, the adoption of an IS is influenced by people, the organization, and other environmental factors [49]. Measurement of information system success is both complex and elusive [50]. Knowledge advancement in IT and related practices currently verifies that the right practice is the main factor of technology and knowledge success regarding diffusion and assimilation of IT innovations [17,51]. Researchers have derived several models to explain what makes some IS successful. For instance, Davis adapted the theory of planned behavior (TPB) and developed the technology acceptance model (TAM), which explains why some IS are more accepted by users than others [52,53]. Acceptance, however, is not equivalent to success, although acceptance of an information system is a necessary precondition to success [49]. In recent years, intention-based models were one important lines of research that employed behavioral intention to predict usage [46]. In turn, this focuses on the identification of the determinants of intention, such as attitudes, social influences, and facilitating conditions [54].

Since its invention, the D&M IS success model and the TAM model have been applied across several IS domain research contexts but are rare in the LRIS context [55,56]. DeLone and MacLean reviewed the existing definitions of IS success and their corresponding measures and classified them into six major categories [57]. Thus, they created a multidimensional measuring model with interdependencies between the different success categories [48,58,59]. Ten years after the publication of their first model, DeLone and McLean [60] proposed an updated IS success model based on the evaluation of many other contributions to it. The updated D&M IS success models of 2003 consist of six interrelated dimensions that include (1) system quality, (2) information quality, (3) intention to use, (4) user satisfaction, (5) individual impact, and (6) organizational impact [49]. Each of these variables is a composite of numerous and diverse constructs and measures. The practical application of the D&M model is naturally dependent on the organizational context [49,60].

On the other hand, TAM is an adaptation of the TPB to the field of IS. TAM later advanced to TAM 2, which incorporated additional theoretical constructs spanning social influence processes (SIP) and cognitive instrumental processes [61]. TAM posits that perceived usefulness and perceived ease of use determine an individual intention to use an IS, while the intention to use serves as a mediator of the actual use of a system. Perceived usefulness is also seen as being directly impacted by perceived ease of use [45,47]. Both TAM and TPB have strong behavioral elements, which assume that when someone forms an intention to act, they will be free to act without limitations. In practice, constraints such as limited ability, time, environmental or organizational limits, and unconscious habits will limit the freedom to act [49].

However, much of the existing theory in related research areas has not been articulated in a manner that lends it to rigorous testing to land tenure IS [62]. For instance, after more than a decade of modeling and building international consensus, the land administration domain model (LADM) only became a formal international standard(ISO 19152) in 2012 [15]. LADM offers a very generic spatial representation model, and it is becoming a common language in establishing geospatial referenced cadastral and land information systems [1,13,15,63]. On the other hand, a recent study by Biraro et al. [13] summarizes parameters and indicators to be taken to account when updating a LIS in the context of the land administration domain. In addition, Bennett et al. [1] systematically reviewed land administration system maintenance and indicated pathways for future research. Although all these models have different approaches, they have commonality in that they provide variables to be considered while evaluating or investigating a LIS development, implementation, and maintenance. This helps the authors to develop the proposed research model and empirically test it to gauge the acceptance and actual use of NRLAIS and predict its operational success in Ethiopia.

Accordingly, six latent variables identified include system quality, information quality, service quality, perceived ease of use, perceived usefulness, and intention to use as a latent construct to determine acceptance and actual use of NRLAIS. The actual use behavior of the woreda land administration experts is considered a proxy predictor of NRLAIS operational success. First, the authors posit that the land administration experts at the woreda level, as internal system end-users/operators, need to accept NRLAIS. However, acceptance alone is not enough for continued use. In addition, as part of the service quality measure, the NRLAIS requires continuous maintenance of the system infrastructure and updating of the land records. In addition, the staff needs continuous competency and skill upgrading as a critical element of the system to operate and render land administration services on a daily basis. This will be explored through the identification by land administration staff of the perceived factors that determine the acceptance and actual use of NRLAIS as a proxy predictor to its operational success.

The land administration experts at the woreda land administration offices are taken as a unit of analysis for this study, because they are the key players and responsible for daily business service delivery. NRLAIS is the main and reliable source of information of the land administration experts for decision-making for service delivery per organizational rules and standards. The woreda land administration experts are the most experienced in creating, describing, defining, and altering the human to parcels of land relationship to legal interests (rights, restrictions, and responsibilities). Hence, the acceptance and actual use of NRLAIS at the woreda land administration offices would have the utmost policy and operational relevance for the success and sustainability of the land administration system in the country.

On this basis, this study proposed the definition of the identified variable constructs, their measurement, and their hypothetical relationships to each variable, which are presented as follows.

System Quality (SYQU)-System quality relies on user needs and overall performance, as perceived by users [49] and as specified in the system technical requirements and development. System quality measures the technical success aspects of NRLAIS. High-level system quality would serve users with useful perception for doing their daily business effectively and under a secured condition that is easy to use and learn. Hence, the main measurement items identified are ease of use, usefulness, and ease of learning [49]. Thus, the following is hypothesized.

Hypothesis 1 (H1). *System quality has a positive and significant influence on actual use of NRLAIS (a) on service quality, (b) on information quality, and (c) on perceived ease of use of NRLAIS.*

Information Quality (INQU)-INQU is the desirable characteristics of the system outputs, such as outcome reports [60]. All types of generation of information by application of information technology cannot be used for decision making. The INQU represents the success of a land registration information system (LRIS) as the information aid to make appropriate business decisions. Seven attributes are identified, of which five are considered in this study, including availability, usability, accuracy, relevance, understandability, format, and ease of access or retrieval [49]. Information quality is often a dimension of end-user satisfaction instruments. INQU is measured as a component of user satisfaction, since it is often not distinguished as a unique construct. While this holds true, the authors contend and rather support the 2003 modified D&M success model construct, which embedded information quality as an independent construct. Hence, the following is hypothesized.

Hypothesis 2 (H2). *Information quality has a positive and significant influence on actual use of NRLAIS (a) on perceived usefulness and (b) on perceived ease of use of NRLAIS.*

Service Quality (SRQU)-According to Petter et al. [49], service quality refers to the quality of the institutional support that system users receive from the IS department and support personnel. Hence, service quality is considered an important organizational dimension that determines individual performance. A specific service quality improvement depends on the status of the measurable service quality attributes, which include five dimensions—tangibles, reliability, responsiveness, assurance, and empathy [64,65]. Similarly, in this study, the SRQU measures are an internal service provider dimension and are rendered to the woreda NRLAIS users by regional and federal land institutions. The measurement attributes include reliability, availability or assurance, and empathy of support staff. Reliability includes the ability to perform the promised service dependably and accurately. Assurance includes the knowledge and courtesy of an IT (technical) and operational support staff and their ability to inspire trust and confidence to the woreda land administration experts. Empathy, on the other hand, includes the caring and individualized attention the IT and support staff provides its woreda land administration experts. In addition, SRQU significantly affects information quality, perceived ease of use, perceived usefulness, and intention to use. Hence, the following is hypothesized.

Hypothesis 3 (H3). *Service quality has a positive and significant influence on actual use of NRALIS (a) on perceived ease of use, (b) on information quality, and (c) on perceived usefulness of NRLAIS.*

Perceived Ease of Use (PEOU)-The extent to which individuals believe that using part of a system does not require much effort is known as perceived ease of use [47]. TAM is considered a flexible model, as it includes variables that explain technology acceptance. Perceived ease of use has a direct impact on behavioral intention and on perceived usefulness [45]. Hence, the following is hypothesized.

Hypothesis 4 (H4). *Perceived ease of use has a positive and significant effect on actual use of NRLAIS (a) on perceived usefulness of NRLAIS.*

Perceived Usefulness-Davis [47] proposed that certain factors such as perceived usefulness, attitude, and perceived ease of use can be the components of TAM. TAM defines individual positive or negative reactions towards a certain thing, which are referred to as attitudes. However, the perspectives of individuals of a certain system being useful to them through influencing their performance are called perceived usefulness [49]. Innovative technologies' acceptance or adoption can be determined and explained by perceived usefulness. TAM further discovered usefulness as one of the noteworthy perceptions leading to intention to adopt new systems. TAM contends that actual system use is an indicator of IS success and is associated with the ultimate impact rendered from IS [49,66]. Hence, in this study, PRUS and PEOU represent the perceived behavior of the woreda land administration experts towards acceptance and actual use of the system for the daily operational success of NRLAIS.

Hence, the following was hypothesized.

Hypothesis 5 (H5). *Perceived usefulness has a positive and significant effect on actual use of NRLAIS.*

System Actual Use (SYAU)-Petter et al. [49] reviewed the updated DeLone and McLean 2003 IS success model and added service quality as a new dimension, grouping all "impact" measures into a single "net benefit" [49]. The construction of "system actual use" and "intent to use" are still considered in this model as an important measure. Hence, the authors considered system actual use (SYAU) in the proposed research model. The research model is graphically represented in Figure 1 below.

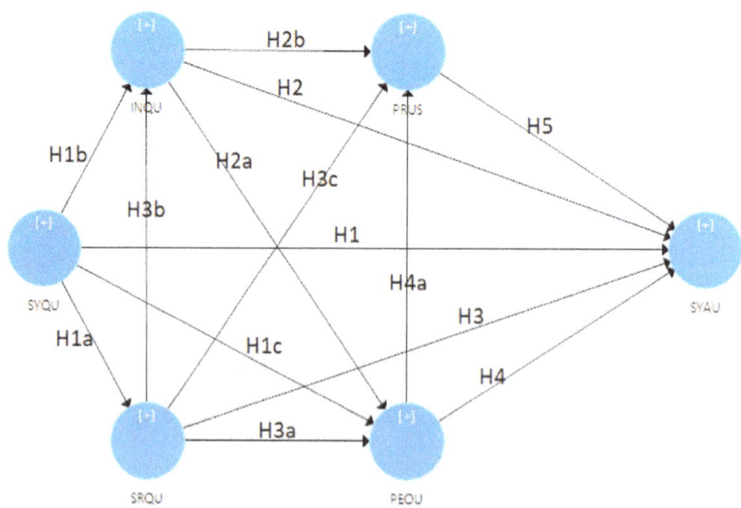

Figure 1. Research model with latent variables and hypotheses construct adapted from information system success model and technology acceptance model.

4. Materials and Methods

The paper draws on primary and secondary data. The primary data were collected from the woreda land administration experts through surveys and key informant interviews of national-level senior experts. Key informant interviews were also employed to collect information on NRLAIS development, support and maintenance services, and operational deployment at the federal and regional land institutions. In addition, the secondary data were collected through a review of scientific literature and policy and program documents.

A partial least square structural equation model (PLS-SEM) was used for the data analysis, which integrates several different multivariate techniques into one model-fitting framework. Smart-PLS software version 3.0 was used to process the data analysis related to the coefficient of interaction terms. PLS-SEM includes confirmatory factor analysis, path analysis and partial least square to impute relationships between latent variables [67]. SEM was used to test the structural relationships between the 14 hypotheses and the actual use of NRLAIS. Cronbach Alpha analysis was performed to examine the consistency of data, and the value of Cronbach alpha should be greater than 0.7. SEM combines the usage of latent (unobserved) variables that represent the concept of theory and data from measures (indicators or manifest variables). The manifest variables are used as input for statistical analysis that provides evidence about relationships among latent variables. Descriptive statistics are also employed to analyze the results of survey data. Figures, tables, and maps are mainly utilized to present the results and findings of the study.

Questionnaires were formulated to collect the professional perceptions and experience of the woreda land administration experts on the technological (SYQU), organizational (INQU and SRQU), and behavioral aspects (PEOU and PRUS) of NRLAIS. The survey consists of three main sections. The first section comprises ten questions on the demographics of the woreda land administration experts. The second section consists of five questions on the NRLAIS use experience of the experts. The third section includes 29 questions related to the measurement variables and their respective indicator items. All the measurement variable indicators are formative and adapted from various earlier related studies [49,55]. The respondents were asked to state their opinions using a seven-point Likert scale from strongly disagree (1) to strongly agree (7).

Five to six items were initially formulated to develop pilot survey questionnaires for direct measures. The formulation of these pilot questionnaires aimed to assess each of the theory's major constructs: system quality, service quality, information quality, perceived ease of use, perceived usefulness, intention to use, and actual use behavior. Seven-point bipolar adjective scales were employed. The pilot questionnaires also included measures of background factors and other variables, including demographic characteristics, professional experience, and system use. The pilot questionnaires were distributed to 30 land administration experts convened in a national workshop in February 2021. The experts came from national and regional level land administration institutions that had been supporting the roll-out of NRLAIS at the woreda land administration offices. The results of the pilot questionnaires also allowed the authors to evaluate the validity and consistency of each item and utility of the background measures. Based on these inputs, necessary adjustment was made, and the standard questionnaires to be used in the main study were produced.

4.1. Study Site

The study covers 50 sample woredas of three regional states (Amhara, Oromia, and SNNP) in Ethiopia. These three regional states hold over 80 percent of the total population and close to half of the country's landmass [68]. Geographically, most of the study areas are located in the central highlands of the country and some along the south-central parts of the Rift Valley, characterized by high population density and diverse land uses. The male (50.1%) population is slightly higher than the female (49.9%). Over 70% of the population in the study areas is under age 30, which aligns with the overall national age breakdown [68], showing a high density of younger people in the study areas. Agriculture, forestry, and livestock raising contributed directly or indirectly to the livelihoods of most of the population in the study areas. Climate change is leading to above average temperatures and greater rainfall variability, with a pronounced effect on agricultural productivity and the suitability of major crops in the study areas [32].

According to official figures from the Ethiopian Central Statistics Agency (CSA), the urbanization rate is growing at an average rate of 5.2 percent per year since 2018. If these trends continue, the urban population is projected to reach 50 million by 2034 [69]. Natural increase rather than rural-to-urban migration was the main driver of urban population growth up to 2018, with rural-to-urban migration being the main driver since 2018 [70]. As population density increases, combined with continued land fragmentation, large cohorts of young people will increasingly become functionally landless. This fuels intense land use competitions and conversions of rural land to built environments. This is becoming a serious land governance issue, particularly in the urban–rural frontiers of most Ethiopian cities [71]. Currently, access to land continues to be difficult due to increasing land scarcity and the total area of landholding per household diminishing over time in the study areas [72]. The average number of rural land parcels per study woreda is 83,000. The woreda land administration offices had an average annual subsequent land transactions turnover of 1 percent, mainly through inheritance, donation, and land rentals. The average size of parcels involved in these transactions was about a quarter of a hectare [43].

According to the MoA [43], as of November 2021, over 180 woredas had established NRLAIS and made it operational. Woredas with operational NRLAIS are found in Amhara (61 woredas), Oromia (68 woredas), and SNNP (56 woredas) regional states (see Figure 2 for a map of the study woredas). The study excluded the inaccessible Tigray regional state due to the ongoing armed conflict and instability. According to the same source, information of about 11 million parcels has been registered in NRLAIS, covering close to 6 million hectares.

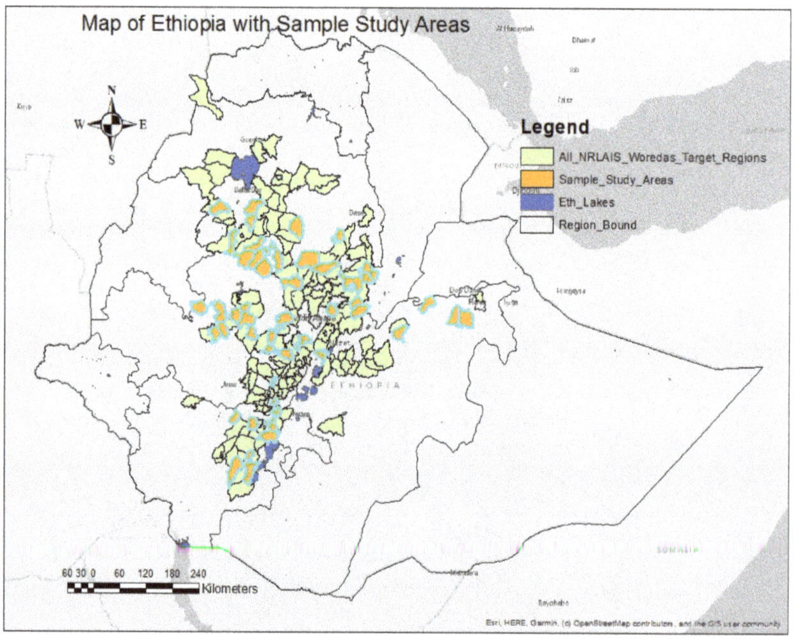

Figure 2. Study Site Map. Data Source: Ministry of Agriculture, November 2021.

4.2. Sampling Method

To determine the sample size required for a study that uses a structural equation model (SEM), the authors applied the Soper [73] online free statistic calculator, which calculates prior sample sizes for structural equation models. This sampling calculator considers the number of observed and latent variables in the model, the anticipated effect size, and the desired probability and statistical power levels. Accordingly, the model of this study contains 29 observed variables and six construct latent variables. The model considers the anticipated effect size of medium (0.3), the desired probability level of 0.05, and desired statistical power level of variables of 0.8. Hence, the minimum initial sample size to detect effect was determined to be 161, the minimum sample size for the model structure was 100, and the recommended minimum sample size was 161. The random selection process was stratified by regions in proportion to each region's number of woredas covered with operational NRLAIS.

4.3. Sample Size

As a result of these sample size requirements, 50 woredas were selected randomly (17 from Amhara, 18 from Oromia, and 15 from SNNP). There are four to six land administration experts per woreda on average who operate the NRLAIS, which means about 450 experts in total. In this study, land administration expert is used as a common name for people working on land administration matters in the woreda land administration offices with different titles, including land registration experts, cadastral surveyors, geospatial and land information management experts, land law and compliant management experts, land transaction experts, team leaders, etc.

From the 50 sampled woredas, about 275 land administration experts were targeted and invited to respond to the self-administered quantitative survey questionnaire by email and the Telegram social media platform. The survey data collection was conducted between April and May 2021. This virtual method of data collection was preferred due to restrictions on movement to field sites caused by the COVID-19 pandemic outbreak and the state of emergency in some of the study areas following social instabilities, particularly in the northern parts of the country. Telephone follow-up calls were also employed to clarify question items to respondents and enhance the quality of the survey data. Of the 220 filled and returned questionnaires, 19 were incomplete. The result shows a 73% success rate of properly completed questionnaires. Depending on the study design model selected, the sample and effect size of the survey data were found satisfactory.

5. Results

5.1. Characteristics of Respondents

The sample respondents included 157 (78%) males and 44 (22%) females (Figure 3). Despite their low numbers, the presence of women land administration professionals in the woreda land administration offices would help the policy reform move towards gender-sensitive land tenure security. About 188 (87%) of the respondents were between 21 and 40 years of age. In addition, 82% and 18% had completed their bachelor's degrees and diplomas, respectively. This also indicates that the woreda land administration offices are filled with relatively young and degreed land administration professionals. This would likely foster innovative technology acceptance in the rural land administration sector and facilitate the establishment of the NRLAIS at woreda level. The range of disciplines the university graduates had studied were very broad, including surveying (17%), ICT and computer science (16%), agriculture (14%), natural resource management (14%), geography (12%), land administration (8%), economics (7%), and others (13%). The diversity of the disciplines would also reflect the multi-disciplinary nature of the land administration domain. However, the number of existing land administration professionals with land law and economics backgrounds appeared low.

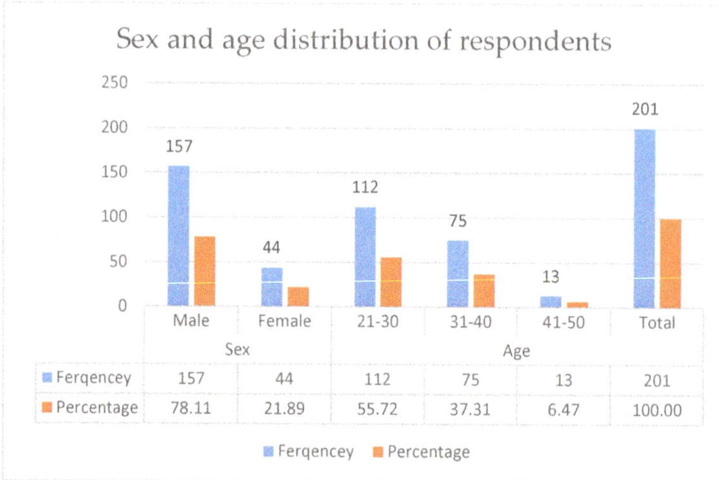

Figure 3. Sex and age distribution of respondents.

In terms of work experience, respondents were asked how long they had been working for their respective woreda land administration offices. The experience levels ranged from less than a year (8%), to between one and seven years (70%), to over seven years (22%). The result revealed that about 61% of the respondents had worked for over 5 years, while 25% had worked for between 3 and 5 years. Only 24% of the respondents had worked less than 3 years in their respective woreda land administration offices. During data collection, on average, 83% of the land administration expert positions were filled. Despite frequent staff turnover reported during the key informant interviews as a key challenge for NRLAIS deployment, the survey result revealed a substantial level of staff retention in the woreda land administration offices. However, this does not mean that the reported land administration experts' turnover did not affect NRLAIS roll-out and activation.

In terms of system use, 67.2% of the respondents had NRLAIS use experience between six months and one year. In addition, 26.2% of the respondents had used NRLAIS for over one year and less than two years. Only 1.5% of the respondents had used NRLAIS for over three years, which probably indicates respondents from the pilot woreda of NRLAIS (Figure 4). This system use experience revealed that all respondents have had adequate familiarity with NRLAIS functional and operational issues.

Respondents were also asked how many minutes or hours per day they spent working on NRLAIS to discharge their service delivery related to land transaction management. As presented in Figure 5, 47% of the respondents spent between six to eight hours. Moreover, 31% of the respondents spent between four to six hours of office hours using NRLAIS to process land transaction management and service delivery. This indicates that about 78% of respondents use NRLAIS for over half of typical office hours. In addition, this shows that NRLAIS is being used as a source and maintains land record information at woreda land administration offices.

Similarly, the respondents were asked how many times a day on average they log into the system. The question measures the frequency of system login as a proxy indicator of access security awareness and rule compliance. The survey shows that about 45% of respondents answered that they log in over ten times per workday. About 30% of respondents log in between six and ten times per workday. Only 2% of respondents log in once per day and process the land transaction management tasks assigned to them on the system. This result seems congruent with the average number of land transactions processed (5 to 13 per day) at the woreda land administration offices. This, in turn, implied relatively good compliance with standard producers and rules by the woreda land administration offices. However, the 2% of responses indicating only one log in per day seemed to reflect a misunderstanding of the question, as NRLAIS has a session time-out functionality.

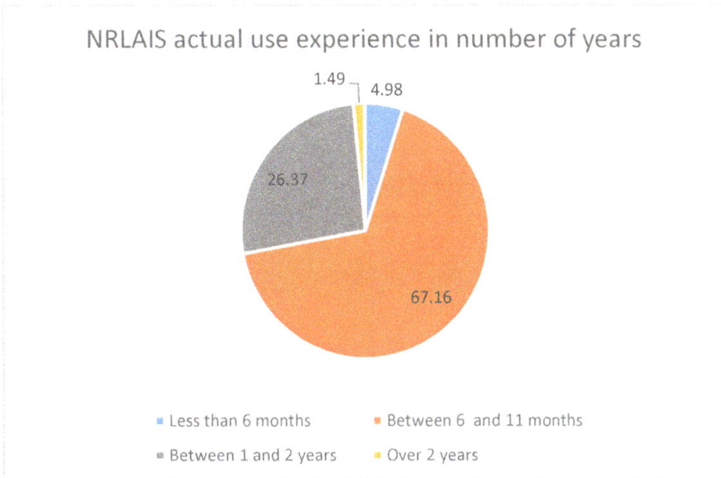

Figure 4. Experience of the respondents' actual usage of the NRLAIS in their daily official business discharge.

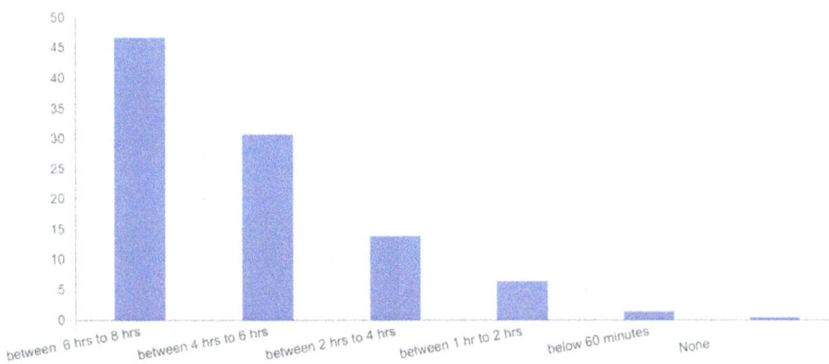

Figure 5. Time spent to process land transaction management and service delivery using NRLAIS.

According to the technical specification of NRLAIS, user-specific roles and user administration have been defined through the business processes. NRLAIS also supports user role definition, assignment, auditing, and reporting with separate management trees according to the administrative structure (federal, region, zone, and woreda) and within the same hierarchy of the woreda land administration offices [74]. To this end, respondents were asked which access privileges they were assigned as internal system users or operators. As presented in Figure 6, the respondents answered that about 44% held an expert role, 31% an officer role, 21% a supervisor role, and 3% a system administrator specific role. However, the 3% responses most likely misunderstand the question, since system administration specific roles are assigned at federal, regional, and sometimes zonal or mobile IT support teams only.

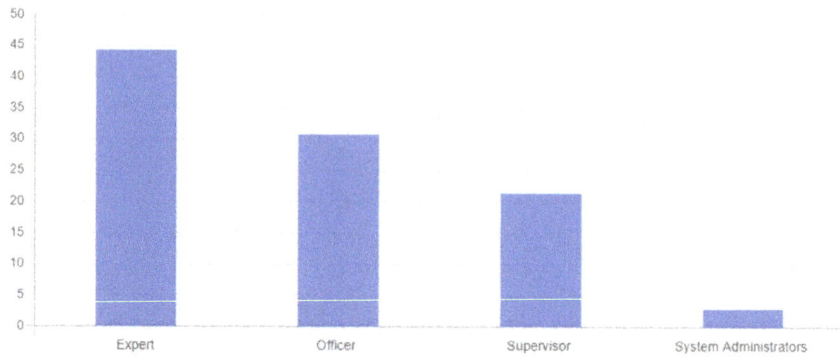

Figure 6. Users' specific assigned role of respondents.

Similarly, according to the architecture design of NRLAIS, the woreda is the primary level at which the system functions. The technical requirement also specifies five subsystems at the woreda level, which affect only parts of the system while processing transactions management, including web information, cadastral maintenance, property registration, document management, and process subsystems [28]. The processing subsystem handles the main operations to register rights, while the cadaster maintenance subsystem handles the management of spatial features (parcels, maps, points, boundaries, etc.). Both the document management subsystem and the processing subsystem have been customized for specific rural land registration processes. Respondents were asked which subsystems they have been using the most while processing land transaction management and service delivery in priority orders. The result revealed that all respondents were used the web information subsystem, as this is the user interface to log in to access the other subsystems. This is followed by the property registration subsystem (48%), cadastral maintenance subsystem (35%), and process subsystem and document management subsystems (17%).

5.2. Validity and Reliability

The first test conducted in this study was testing the validity and reliability of the outer model. The outer model testing was performed through a process of algorithm iteration, a parameter of measurement model that includes convergence validity, discriminant validity, composite reliability, and Cronbach's alpha. Validity and reliability ensure that the multiple indications of each latent variable in the measurement model converge to measure a single construct and hence develop legitimacy, defined as the level to which things used to measure can calculate the idea they meant to quantify [75]. All items used to measure the construct should pile essentially to their constructs rather than different builds. As for the component analysis, it ensures that items are designated to their constructs, as they express high loading on them that stands out from several constructs [76]. The measurement model assessment of vertical collinearity is presented in Table 1. This shows the subjective independence of every indicator on its latent variable using cross-loading criteria.

The individual item reliability was evaluated by examining the loading and cross-loadings of indicators on their respective construct. According to Fornell and Larcker's criteria [77], a reliability score of Cronbach alpha 0.6 is considered minimally acceptable, with 0.70 preferred (50% of the explained variance). The theory also recommends that an indicator loading having a value of less than 0.40 should be removed from the model. Hence, this study found three indicator items with less than or equal to 0.4 outer loading. As per the rules, the indicator items removed from the model include INQU4 (format), PEOU4 (clear and understandable), and SYQU4 (risk of losing data).

Table 1. Indicator item cross loading.

Variables	Code	INQU	PEOU	PRUS	SRQU	SYAU	SYQU
Information Quality	INQU1	0.698	0.507	0.432	0.453	0.458	0.571
	INQU2	0.862	0.569	0.578	0.571	0.654	0.672
	INQU3	0.890	0.716	0.556	0.608	0.602	0.741
	INQU5	0.743	0.599	0.501	0.577	0.568	0.508
Perceived Ease of Use	PEOU1	0.588	0.796	0.462	0.560	0.426	0.560
	PEOU2	0.608	0.863	0.578	0.641	0.507	0.593
	PEOU3	0.512	0.724	0.622	0.588	0.519	0.428
	PEOU5	0.718	0.867	0.507	0.699	0.631	0.626
Perceived Usefulness	PRUS1	0.452	0.402	0.737	0.411	0.424	0.404
	PRUS2	0.412	0.434	0.680	0.472	0.340	0.309
	PRUS3	0.288	0.350	0.591	0.249	0.184	0.302
	PRUS4	0.489	0.483	0.807	0.432	0.451	0.441
	PRUS5	0.469	0.502	0.801	0.489	0.458	0.429
	PRUS6	0.640	0.663	0.773	0.624	0.576	0.484
Service Quality	SRQU1	0.498	0.615	0.433	0.756	0.458	0.405
	SRQU2	0.371	0.455	0.322	0.668	0.266	0.296
	SRQU3	0.570	0.613	0.482	0.806	0.489	0.482
	SRQU4	0.482	0.553	0.509	0.730	0.526	0.422
	SRQU5	0.585	0.570	0.541	0.721	0.570	0.499
System Actual Use	SYAU1	0.648	0.599	0.506	0.696	0.750	0.545
	SYAU2	0.491	0.502	0.431	0.398	0.733	0.465
	SYAU3	0.475	0.391	0.347	0.364	0.718	0.467
	SYAU4	0.417	0.318	0.375	0.315	0.725	0.347
System Quality	SYQU1	0.737	0.643	0.535	0.551	0.626	0.843
	SYQU2	0.462	0.423	0.316	0.327	0.430	0.786
	SYQU3	0.662	0.561	0.449	0.504	0.477	0.819

The composite reliability index and the average variance extracted (AVE) were applied to assess the internal consistency and convergent validity [77]. According to the rule, the square root of the AVE of a particular construct should also be greater than its correlation with other constructs. Generally, the AVE should be higher than 0.5. Table 2 shows the internal consistency of each construct. In the measurement model, the study used Cronbach's alpha and composite reliability (CR) to test the reliability of the constructs. The study found that all the CRs were higher than the recommended value of 0.700, ranging from 0.822 to 0.887. The Cronbach's alpha of each construct exceeded the recommended 0.700 threshold, which is 0.720 to 0.831 in the current study. Hence, convergence validity was acceptable, because the average variance extracted (AVE) was over 0.500.

Discriminant validity concerns the uniqueness of a construct, whether the phenomenon captured by a construct is unique and not reflected in the model by the other construct [75]. The subjective independence can help reduce the presence of multicollinearity amongst the latent variables, denoting that the average variance extracted (AVE) of a latent variable should be higher than the squared correlations between the latent variable and all other variables [77]. Discriminant validity was assessed by the Fornell–Larcker criterion [77]. Table 3 shows that the square-root of AVE for the construct in the diagonal was greater than the inner-construct correlation, which ranges from 0.732 to 0.816. The test result of the current study may therefore imply the strong reliability of all the items.

Table 2. Reliability and Validity.

	Cronbach's Alpha	rho_A	Composite Reliability	Average Variance Extracted (AVE)
INQU	0.811	0.827	0.877	0.644
PEOU	0.829	0.837	0.887	0.664
PRUS	0.831	0.856	0.875	0.541
SRQU	0.791	0.799	0.856	0.544
SYAU	0.720	0.730	0.822	0.535
SYQU	0.754	0.776	0.857	0.666

Table 3. Fornell–Larcker Criterion. Note: Value in diagonal represent the Square-root of AVE.

	INQU	PEOU	PRUS	SRQU	SYAU	SYQU
INQU	0.802					
PEOU	0.749	0.815				
PRUS	0.648	0.664	0.736			
SRQU	0.692	0.767	0.632	0.737		
SYAU	0.715	0.645	0.581	0.644	0.732	
SYQU	0.782	0.681	0.548	0.582	0.639	0.816

5.3. Structural Model (Inner Model) Analysis

The structural model (Figure 7) reflects the path hypothesized in the research framework. Using the bootstrap resampling technique (5000 resamplings), the path coefficient was then tested to investigate the significance of the hypothesis. The t-value > 1.96 is significant at $p < 0.05$, and t-value > 2.58 is significant at $p < 0.01$ [67]. A structural model is also assessed based on the R^2, Q^2, and significance of paths. The goodness of the model is determined by the strength of each structural path, determined by the R^2 value for the dependent variable; the value for R^2 should be equal to or over 0.1 [67]. The values of R^2 in PLS are interpreted similarly to those obtained from multiple regression analysis. It was considered that R^2 values of 0.75, 0.50, and 0.25 are substantial, moderate, and weak, respectively [67], and evaluated subsequently. Hence, the predictive capability is established.

In this study, information quality accounted for 69.6 percent of the variance in explaining perceived usefulness and perceived ease of use. Meanwhile, service quality accounted for 33.9 percent of the variance explaining perceived usefulness and perceived ease of use. Likewise, perceived ease of use accounted for 69.5 percent, and perceived usefulness accounted for 51.2 percent of the variance explaining SYAU. Finally, the current model explained 57.5 of the variance in acceptance and actual use of NRLAIS, which provided substantial explanatory power and predictive capability (see Figure 7).

Furthermore, the model fit was assessed using SRMR. The value of SRMR was 0.100, equal to the required value of 0.10, indicating acceptable model fit [78]. Further assessment of the goodness of fit hypotheses were tested to ascertain the significance of the relationship (Table 4).

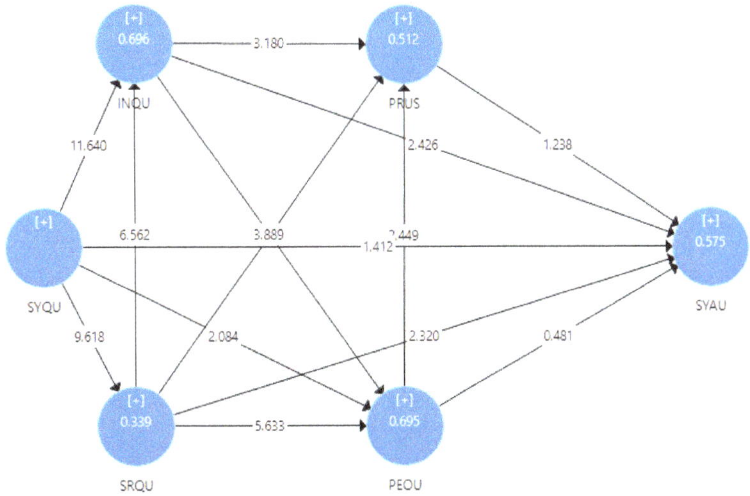

Figure 7. Measurement and structure equation model results of the research model.

Table 4. Model fit analysis of the proposed model.

	Original Sample (O)	Sample Mean (M)	95%	99%	Original Sample (O)	Sample Mean (M)	95%	99%
Saturated Model	0.100	0.059	0.070	0.074	3.520	1.241	1.727	1.944
Estimated Model	0.100	0.060	0.072	0.075	3.521	1.298	1.808	1.963

The studied 5000 resamples also generate 95% confidence intervals, and hypotheses testing results are summarized in Table 5. A confidence interval not equal to zero indicates a significant relationship. Further, Q^2 establishes the predictive relevance of the endogenous constructs. A Q^2 above 0 shows that the model has predictive relevance. The result shows that there is significance in the prediction of the construct in the research model. Therefore, except for H4 and H5, the rest of the hypotheses have positive and significant impact on the acceptance, which is a proxy predictor to NRLAIS operational success.

H1 evaluates whether SYQU has a positive and significant effect on the acceptance and SYAU of NRLAIS. The result revealed that a positive and significant impact on SYAU ($\beta = 0.639$, $t = 10.019$, $p = 0.000$), on the SRQU (H1a: $\beta = 0.583$, $t = 9.879$, $p = 0.000$), on INQU (H1b: $\beta = 0.583$, $t = 9.879$, $p = 0.000$), and on the PEOU (H1c: $\beta = 0.194$, $t = 2.128$, $p = 0.034$). Hence, H1 was fully supported. The result shows (Figure 7 and Table 5) that 63.9% of NRLAIS acceptance and actual usage was explained by the system quality variables with indicators including easy to learn (SYQU1: 84.3%), easy to use (SYQU2: 78.6%), and useful for doing daily business effectively under secured conditions (SYQU3: 81.9%).

Secondly, the result shows that (Figure 7 and Table 5) information quality is found to have a positive and significant influence on acceptance and actual use of NRLAIS (H2). The result revealed that the relation between INQU and SYAU ($\beta = 0.436$, $t = 3.600$, $p = 0.000$) (2a) on perceived usefulness, i.e., INQU -> PRUS ($\beta = 0.283$, $t = 3.347$, $p = 0.001$), and (2b) on perceived ease of use, i.e., INQU -> PEOU ($\beta = 0.275$, $t = 3.818$, $p = 0.000$), has a positive and significant influence. Hence, H2 was supported. Respondents admitted that information quality of NRLAIS has a positive and significant influence on acceptance and actual use. When the information generated by NRLAIS is accurate, useable, relevant, reliable, and understandable, the woreda land administration experts consider the information system useful and valuable. This result revealed that information availability (69.8%) is the main characteristic of NRLAIS to ensure a necessary level of acceptance for land administration to provide seamless services that do not halt business. In addition, information usability (86.2%), information accuracy (89%), information relevance, and understandability. (74.3%) are influencing factors that determine information quality.

Table 5. Mean, STDEV, T-Values, *p*-Values, R^2, and Q^2.

	β	STDEV	T Tatistics	*p* Values	2.50%	97.50%
INQU -> PEOU	0.275	0.072	3.818	0.000	0.154	0.420
INQU -> PRUS	0.283	0.085	3.347	0.001	0.105	0.427
INQU -> SYAU	0.436	0.121	3.600	0.000	0.182	0.655
PEOU -> PRUS	0.291	0.114	2.556	0.011	0.083	0.508
PEOU -> SYAU	0.086	0.101	0.852	0.395	−0.111	0.294
PRUS -> SYAU	0.111	0.086	1.285	0.199	−0.039	0.290
SRQU -> INQU	0.356	0.053	6.681	0.000	0.246	0.453
SRQU -> PEOU	0.465	0.084	5.537	0.000	0.285	0.615
SRQU -> PRUS	0.213	0.096	2.222	0.027	0.022	0.401
SRQU -> SYAU	0.208	0.101	2.057	0.040	0.027	0.416
SYQU -> INQU	0.575	0.048	12.019	0.000	0.477	0.665
SYQU -> PEOU	0.194	0.091	2.128	0.034	−0.006	0.355
SYQU -> SRQU	0.583	0.059	9.879	0.000	0.475	0.693
SYQU -> SYAU	0.639	0.064	10.019	0.000	0.503	0.752
	R^2	Q^2				
INQU	0.697	0.434				
PEOU	0.695	0.451				
PRUS	0.512	0.252				
SRQU	0.340	0.175				
SYAU	0.567	0.258				
SYQU	0.000					

Thirdly, service quality has a positive and significant influence on acceptance and actual use of NRLAIS (H3) (3a) on the perceived ease of use, (3b) on the information quality, and (3c) on the perceived usefulness of NRLAIS. The result revealed that a positive and significant impact on SYAU (β = 0.208, t = 2.057, *p* = 0.040); (3a) on the perceived ease of use, i.e., SRQU -> PEOU (β = 0.465, t = 5.537, *p* = 0.000); (3b) on the information quality, i.e., SRQU -> INQU (β = 0.575, t = 12.019, *p* = 0.000); and (3c) on the perceived usefulness, i.e., SRQU -> PRUS (β = 0.213, t = 2.222, *p* = 0.027). Hence, H3 was fully supported. The SRQU construct is one of the most influential service quality measurement instruments and is widely used in many applications, and the development of service quality affects IS success factor. The result revealed that the technical support service provided by the federal and regional land institutions is significantly affecting the acceptance and actual usage of NRLAIS at the woreda offices of land administration. The result revealed that 20.8% of the acceptance and actual use of NRLAIS were explained by the quality of support services (Figure 7). This includes the knowledge and skill transfer made through classroom training and during data migration, on-the-spot support by a mobile IT team, and remote helpdesk support. Such support services were found relevant to the job performance and awareness regarding compliance with actual system use by the woreda land administration experts.

Conversely, H4 evaluates whether PEOU has a positive and significant effect on SYAU of NRALIS. The result revealed that PEOU has a positive but an insignificant impact on acceptance and SYAU (β = 0.086, t = 0.852, *p* = 0.395), and (4a) PEOU has a positive and significant effect on perceived usefulness, i.e., PEOU -> PRUS (β = 0.291, t = 2.556, p = 0.011). Hence, H4 was not supported, while H4a was fully supported.

Similarly, H5 evaluates whether PRUS has a positive and significant effect on acceptance and SYAU of NRLAIS. The result revealed that PRUS has a positive but insignificant impact on acceptance and SYAU (β = 0.111, t = 1.285, p = 0.199). Hence, H5 was not supported.

5.4. Mediation Analysis

Mediation analysis was performed to assess the mediating (INQU, SRQU, PEOU and PRUS) role of system actual use outcomes. The results revealed an insignificant ($p > 0.05$) mediating role of PRUS (H3a: β = 0.024, t = 2.000, p = 0.322) and a partially significant (p = 0.046) mediating role of PEOU (H3c: β = 0.040, t = 2.000, p = 0.046). INQU was found to significantly mediate the relationship between SRQU and SYAU of NRLAIS (see Table 6).

Table 6. Total, specific, and indirect effect between the independent variables and dependent variable.

	Total Effect	T	Sig	Direct Effect	Sig	Specific Indirect Effect	Effect	T	Sig
SR -> SYAU	0.464	7.397	0.000	0.208	0.040	SRQU -> INQU -> SYAU	0.251	3.665	0.000
						SRQU -> PEOU -> SYAU	0.040	2.000	0.046
						SRQU -> PRUS -> SYAU	0.024	2.000	0.322

6. Discussion

Land administration governance in Ethiopia is complex and requires the use and integration of an innovative and robust set of land registration information technologies that meet the social, economic, and environmental goals of tenure security and service delivery. Innovative information technologies and systems can be used to help build a quicker, accessible, affordable, and more reliable LIS. This, in turn, provides landholders, communities, business firms, and the general public with a clearer sense of tenure security, particularly to women, by providing evidence of recognized and enforceable land and resource rights. Landholders with secure tenure rights will be incentivized to make long-term land-based investment that improve welfare and environmental outcomes.

This study demonstrated that the theoretical application of the research model that integrated selected variables from the IS success and TAM models to explain the acceptance and actual usage of NRLAIS in Ethiopia; acceptance and usage serve as proxy predictors of NRLAIS' operational success. The paper explored the determining factors and their relationships with technical (SYQU), organizational (SRQU and INQU), and the behavioral (PEOU and PRUS) aspects of the woreda land administration experts that influence the acceptance and actual usage of NRLAIS in Ethiopia. These constructs, along with measuring the LRIS' operational success, pave the way to scientifically research land administration system digitalization, particularly in developing economies. The research suggests that the IS success model integrated with TAM with selected variables was applicable for explaining LRIS acceptance and actual usage as a proxy predictor for operational success. This study does not include the organizational net benefit in its latent variable constructs measured by productivity, competitiveness, and management improvements. The discussion of the results is presented in detail as follows.

Firstly, based on the structural model, the study examined the relationship between the system quality and acceptance and actual use of NRLAIS. The results indicate that there is a positive and significant relationship between the system quality and the acceptance and actual use of NRLAIS (H1) on SRQU (1a), on INQU (1b), and on PEOU (1c). Based on that, it can be inferred that the woreda land administration experts relate this system quality with the acceptance and actual use of NRLAIS. The system quality constructs reflect the technical aspects of NRLAIS to its acceptance and actual use. These are that it is easy to learn, easy to use, and useful to do daily business related to land transaction management and service delivery at the woreda land administration offices. Malik et al. [55] and Hamdan and Al-Hajra [79] observed that a system's level of association had a positive influence on the perceived ease of use and the perceived usefulness of the system under investigation [55,56]. Moreover, the system quality would also affect the information quality and the service quality of the organization under study, and this, in turn, affects the acceptance and actual usage of the system.

As described in the background section, the technical requirement review provides an understanding of the different components of NRLAIS as an automated and streamlined land registration information system for rural land. The functional and legal aspects of the rural land administration are key requirements of NRLAIS. This includes data capturing, data management, visualization, and workflow management and reporting. The NRLAIS system development process has engaged the regional and federal level land administration professionals in terms of the definition, verification, and approval of the functional and legal requirements as part of its quality assurance system. Hence, the multi-tiered application architecture makes NRLAIS possible to use at the different administrative levels with differentiated functionality and user interface. The login function and session time-out are also suited to secure the user's access to the system and the subsystems before any transaction begins. Furthermore, the classified roles of system users (officer, expert, and supervisor) represent read-only, data entry, and approval, and change secure system access as primary actors in the existing administrative roles in the institutions that operate the NRLAIS.

Regarding information quality, the results of this study are consistent with what was found by Nugroho and Chang et al. [48,80]. NRLAIS also manages information regarding land transaction performance at the woreda level, such as the type and patterns of land transactions made. Updating the land record is one of the key functions of the woreda land administration offices. This affects the efficiency and effectiveness of the land administration system and staff job performance related to all land transactions. Relevant and accurate land information affects operational business decisions of transactions to the regional governments, to federal ministries, and to the general public. Since land relationships change frequently, the information in NRLAIS needs to be updated and maintained. Timely and accurate management of information ensured through well-maintained cadastral and land use related information should reflect the reality on the ground. Complete and up-to-date land information will support expedited business decisions made by the woreda land administration offices, thereby enhancing legitimacy and trustworthiness of the woreda land administration offices. Further to this, the availability of such geospatial land information lays a foundation for the national spatial data infrastructure to flourish and be accessible to all concerned in order to inform strategic and policy reform.

Moreover, NRLAIS has a modular design that allows the system to be deployed at several administrative levels. The modular stack and the web-based server design of NRLAIS enable the transfer of data and information from lower to higher administration levels to include zone, region, and federal levels that facilitate easier deployment. However, this data and information flow is yet to happen due to low and weak telecom network infrastructure coverage in the country for access to strong bandwidth Internet services.

With regard to service quality, the result of the current study is consistent with the findings of Al Fraihat [81], Nugroho [48], and Malik et al. [55]. These studies found that the effective role or support service of the technical staff, (i.e., service quality) is positively related to the eventual use of the system [81]. Competency of the support staff, vendor support, and availability of training affected acceptance and use of IS [82]. The current study also revealed that the technical support service is significantly affecting the acceptance and actual usage of NRLAIS. However, an IT-enabled LRIS at this decentralized scale needs a solid IT management approach, which is dependable, available, and has good empathy of support staff. NRLAIS with sufficient quality affects the type and intensity of technical support in the transition operation from manual to digital service delivery.

The quality of support service, in turn, affects adherence to standardized methods and procedures for service delivery in the acceptance and actual use of NRLAIS.

The transition and service operation is the highest priority for NRLAIS to function and be operational at the woreda level. In the context of the establishment of NRLAIS, it is essential to have a proper knowledge transfer from the developer to an in-house or outsourced local IT company, which carries out the ongoing system maintenance, upgrading, and operational support services. The woreda land administration experts' knowledge and skill acquired through training, experience gained during data migration, self-practice, and helpdesk support are also critical success factors. Therefore, the support service quality should be strong, regular, dependable, and available when needed at the woreda level.

Fourthly, the analysis revealed that perceived ease of use has a positive but insignificant effect on the acceptance and actual use of NRLAIS. However, perceived ease of use has a positive and significant effect on the perceived usefulness of NRLAIS (H4a). Hence, H4 was not supported, while H4a was supported. Malik et al. [55,56] found that perceived usefulness is influenced by an understanding of the information quality; [83] the current study analysis also revealed similar findings (H2b). Machdar and Malik et al. [55,56,84] found that the quality of information positively affects perceived usefulness and ease of use, and perceived ease of use positively affects perceived usefulness [84]. Several studies have found strong relationships between perceived usefulness and self-reported use [85], extent of use [86], or dependence on an information system [87]. Empirical studies in various contexts have confirmed that the post-usage perception of usefulness has a strong association with actual use [88]. This holds true in the current study too, since the hypothetical relationships between the perceived ease of use effect on acceptance and actual use was found to be significant. This may be due to the fact that the survey collected post-usage perceptions rather than measuring intention to use prior-actual system use. Therefore, the behavioral antecedences of perceived ease of use and actual use should not be underestimated to meet operational success, though perceived ease of use has shown a weak relationship to actual use in the analysis.

The result of this study revealed that 69.5% of the acceptance and actual use of NRLAIS is explained by the perceived ease of use construct (Figure 7). The quality of information affects the perceived ease of use (27.5%), as well as the perceived usefulness (28.3%) of the woreda land administration experts. This, in turn, affects the acceptance and actual use of NRLAIS to maintain the land records of subsequent land transactions and to make effective and efficient business decisions in daily service delivery.

7. Conclusions

Investigation of the general antecedents of what causes users' acceptance and use of innovative information technology in the land administration domain is critical for digital transformation and improve service delivery. The research has attempted to understand the influencing factors of the acceptance and actual use of NRLAIS in Ethiopia to gauge and predict its operational success. The research found that, despite the fact that the NRLAIS program of Ethiopia is at the initial stage of establishment, the speed and scale of its implementation have implications for its acceptance and actual use. The authors argue that investigating what factors affect this transition is worth studying in order to document the experience and generate a knowledge base for similar initiatives in other countries.

The research demonstrated that a system must be developed that understands the functional, technical, legal, and administrative requirements. These factors significantly affect the acceptance and actual use of the system and better predict its operational success. Similarly, the organizational aspects related to support services and information quality determine the behavioral attributes of staff to accept and use NRLAIS and rely on it to make business decisions and deliver services on a daily basis.

Moreover, twelve of the fourteen constructs (hypotheses) in this study were found to be significant in terms of affecting the acceptance and actual use of NRLAIS. The findings rendered a theoretical and practical knowledge base about land administration information digitalization and operational success. Practically, the findings of this study help the country to make strategic and policy decisions on the planning, implementation, and use of a land registration information system for sustainable land resource management and governance systems. Secondly, the study demonstrated that, while the service quality and information quality constructs to the acceptance and actual use of NRLAIS deserve an independent construct, they are also playing a mediating role between system quality and actual use of the system. The proposed model empirically revealed strong construct validity, in that it captures multiple aspects of each variable, which is a change from much of the measurement of LIS success model constructs that focus on only one aspect of the construct.

Historically, the land administration system in Ethiopia has evolved in the urban–rural cadastral and land registration system divide. This costs the country a great deal in terms of economic, social, and environmental management and governance systems. With the depth of functional and legal requirements compliance, NRLAIS is instrumental in strengthening secure tenure rights in rural jurisdictions. It also demonstrates that standardization would pave the way for the development of a unified land administration information system in Ethiopia that embraces both urban and rural land tenure.

Despite the study's contribution to policy, research, and practice, the authors suggest that a similar future research undertaking would be recommendable in other developing countries and should include other variables that are not considered in this study. The method applied in this study adds to the knowledge base and replicability of the proposed model under the land administration information domain. Further, future studies should include the organizational net benefit in their latent variable constructs as measured by productivity, competitiveness, and management improvements. Finally, the research did not consider the success of NRLAIS from the perspective of an external system user, such as financial institutions, courts, businesses, and smallholder farmers, which would be worth studying in a future research undertaking.

Author Contributions: Conceptualization, S.A.A., F.S.W. and T.T.N.; methodology, S.A.A.; software, S.A.A.; validation, S.A.A., F.S.W. and T.T.N.; formal analysis, S.A.A.; investigation, S.A.A.; resources, S.A.A., F.S.W. and T.T.N.; data curation, S.A.A.; writing—original draft preparation, S.A.A.; writing—review and editing, S.A.A., F.S.W. and T.T.N.; visualization, S.A.A.; supervision, F.S.W. and T.T.N.; project administration, Kelly Robbins; funding acquisition, S.A.A., F.S.W. and T.T.N. All authors have read and agreed to the published version of the manuscript.

Funding: The research was supported by the Partnerships for Enhanced Engagement in Research (PEER) Program, financed by the U.S. Agency for International Development (USAID) and administered by the U.S. National Academy of Sciences (NAS) under cooperative agreement AID-OAA-A-11-00012. The authors highly acknowledge and appreciate this generous financial support, without which this research could not have been realized.

Institutional Review Board Statement: All respondents of survey questionnaire and key informant interviewees are not mentioned by name. Results are aggregated and cannot be tracked back to individual person.

Informed Consent Statement: All person involved in the study participated voluntarily and agreed the study results derived from their responses.

Data Availability Statement: Not applicable.

Acknowledgments: Our heartfelt gratitude goes to Dilu Shaleka, Dean of College of Development Studies of Addis Ababa University and his finance team who have been supportive to all administrative requests. We would also like to extend our thanks to Kelly Robbins, Senior Program Officer, of the National Academies of Sciences, Engineering, and Medicine (NAS), who administers the PEER grant funds, for her unwavering support to our research work. We owe our full-hearted thanks to Yohannes Redda from Ministry of Agriculture of the Responsible and Innovative Land Administration Project (REILA II), Dawit Woldemariam from Oromia Land Administration and Use Bureau, Demisachew Agegnehu from Amhara Region Bureau of Land, and Henok Yosef from

SNNP regional state, who have supported the data collection process. Without their close support the remote data collection would not have been successful. Our thanks also goes to all the study woreda land administration experts who set aside time to answer the research questions under an unprecedented operating context in the country due to the COVID-19 outbreak restrictions, social instability, and security conditions in the country.

Conflicts of Interest: The authors declare no conflict of interest.

References

1. Bennett, R.M.; Unger, E.M.; Lemmen, C.; Dijkstra, P. Land administration maintenance: A review of the persistent problem and emerging fit-for-purpose solutions. *Land* **2021**, *10*, 509. [CrossRef]
2. Framework for Effective Land Administration Expert Group on Land Administration and Management United Nations Committee of Experts on Global Geospatial Information Management (UN-GGIM) Content. 2020. Available online: https://ggim.un.org/meetings/GGIM-committee/10th-Session/documents/E-C.20-2020-29-Add_2-Framework-for-Effective-Land-Administration.pdf (accessed on 23 September 2021).
3. Masuda, Y.J.; Kelly, A.C.; Robinson, B.E.; Holland, M.B.; Bedford, C.; Childress, M.; Game, E.T.; Ginsburg, C.; Hilhorst, T.; Lawry, S.W.; et al. How do practitioners characterize land tenure security? *Conserv. Sci. Pract.* **2020**, *2*, e186. [CrossRef]
4. Byamugisha, F.F.K. Experiences and development impacts of securing land rights at scale in developing countries: Case studies of China and Vietnam. *Land* **2021**, *10*, 176. [CrossRef]
5. Mitchell, D.; Barth, B.; Ho, S.; Sait, M.S.; McEvoy, D. The benefits of fit-for-purpose land administration for urban community resilience in a time of climate change and COVID-19 pandemic. *Land* **2021**, *10*, 563. [CrossRef]
6. Singirankabo, U.A.; Ertsen, M.W. Relations between land tenure security and agricultural productivity: Exploring the effect of land registration. *Land* **2020**, *9*, 138. [CrossRef]
7. Krigsholm, P.; Riekkinen, K.; Ståhle, P. The changing uses of cadastral information: A user-driven case study. *Land* **2018**, *7*, 83. [CrossRef]
8. McLain, R.; Lawry, S.; Guariguata, M.R.; Reed, J. Toward a tenure-responsive approach to forest landscape restoration: A proposed tenure diagnostic for assessing restoration opportunities. *Land Use Policy* **2021**, *104*, 103748. [CrossRef]
9. Lawry, S.; Samii, C.; Hall, R.; Leopold, A.; Hornby, D.; Mtero, F. The impact of land property rights interventions on investment and agricultural productivity in developing countries: A systematic review. *J. Dev. Eff.* **2017**, *9*, 61–81. [CrossRef]
10. Schwartz, M.W.; Cook, C.N.; Pressey, R.L.; Pullin, A.S.; Runge, M.C.; Salafsky, N.; Sutherland, W.J.; Williamson, M.A. Decision Support Frameworks and Tools for Conservation. *Conserv. Lett.* **2018**, *11*, e12385. [CrossRef]
11. Ingram, G.K.; Hong, Y. *Property Rights and Land Policies*; Lincoln Institute of Land Policy: Cambridge, MA, USA, 2009; ISBN 9781558441880.
12. Jing, Y.; Bennett, R.; Zevenbergen, J. Up-to-dateness in land administration: Setting the record straight. *Coordinates* **2014**, *10*, 37–42.
13. Biraro, M.; Zevenbergen, J.; Alemie, B.K. Good practices in updating land information systems that used unconventional approaches in systematic land registration. *Land* **2021**, *10*, 437. [CrossRef]
14. Aydinoglu, A.C.; Bovkir, R. Generic land registry and cadastre data model supporting interoperability based on international standards for Turkey. *Land Use Policy* **2017**, *68*, 59–71. [CrossRef]
15. Unger, E.M.; Bennett, R.M.; Lemmen, C.; Zevenbergen, J. LADM for sustainable development: An exploratory study on the application of domain-specific data models to support the SDGs. *Land Use Policy* **2021**, *108*, 105499. [CrossRef]
16. UN-GGIM. Framework for Effective Land Administration A Reference for Developing, Reforming, Renewing, Strengthening or Modernizing Land Administration and Management Systems Expert Group on Land Administration and Management United Nations Committee of Experts on Global Geospatial Information Management (UN-GGIM). 2019. Available online: https://ggim.un.org/meetings/GGIM-committee/9th-Session/documents/E_C.20_2020_10_Add_1_LAM_background.pdf. (accessed on 20 January 2021).
17. Zeng, Z.; Cleon, C.B. Factors affecting the adoption of a land information system: An empirical analysis in Liberia. *Land Use Policy* **2018**, *73*, 353–362. [CrossRef]
18. Bolwig, S.; Cold-Ravnkilde, S.M.; Rasmussen, K. *Achieving Sustainable Natural Resource Management in the Sahel after the Era of Desertification Markets, Property Rights, Decentralisation and Climate Change*; Danish Institute for International Studies: København, Denmark, 2009; ISBN 9788776053086.
19. Byamugisha, F. *Securing Land Tenure and Easing Access to Land*; African Center for Economic Transformation: Accra, Ghana, 2016.
20. Peters, P.E. *Challenges in Land Tenure and Land Reform in Africa: An Anthropological Perspective*; CID Working Paper No. 141; Center for International Development at Harvard University: Cambridge, MA, USA, 2007.
21. Byamugisha, F.F.K. *Agricultural Land Redistribution and Land Administration in Sub-Saharan Africa: Case Studies of Recent Reforms*; Direction in Development Agriculture and Rural Development; World Bank: Washington, DC, USA, 2014.
22. Simbizi, M.C.D.; Bennett, R.M.; Zevenbergen, J. Land tenure security: Revisiting and refining the concept for Sub-Saharan Africa's rural poor. *Land Use Policy* **2014**, *36*, 231–238. [CrossRef]
23. Enemark, S.; Mclaren, R.; Lemmen, C.; Antonio, D.; Gitau, J.; De Zeeuw, K.; Dijkstra, P.; Quinlan, V.; Freccia, S. *Fit-For-Purpose Land Administration: Guiding Principles for Country Implementation*; Global Land Tool Network: Nairobi, Kenya, 2016.

24. Zevenbergen, J.; Augustinus, C.; Antonio, D.; Bennett, R. Pro-poor land administration: Principles for recording the land rights of the underrepresented. *Land Use Policy* **2013**, *31*, 595–604. [CrossRef]
25. Mitchell, D.P.; Zevenbergen, J.A. Toward Land Administration Systems to Support Climate Change Mitigation Payments. *Land Tenure J.* 2011, pp. 57–79. Available online: http://www.fao.org/nr/tenure/land-tenure-journal/index.php/LTJ/article/view/33. (accessed on 23 September 2021).
26. James, K.; Geosystems, L. Improving Rwanda Land Administration Information Systems. *Geomat. Indaba 2016-Stream* **2016**, *1*.
27. Cochrane, L.; Hadis, S. Functionality of the land certification program in Ethiopia: Exploratory evaluation of the processes of updating certificates. *Land* **2019**, *8*, 149. [CrossRef]
28. Ministry of Agriculture. *Upgraded Information System/IT Strategy for National Rural Land Administration Information System (NRLAIS), 2017–2021. Federal Democratic Republic of Ethiopia Ministry of Agriculture and Natural Resources, Rural Land Administration and Use Directorate Supported by Ministry for Foreign Affairs of Finland*; Responsible and Innovative Land Administration in Ethiopia (REILA II): Addis Ababa, Ethiopia, 2017.
29. Wubie, A.M.; de Vries, W.T.; Alemie, B.K. Synthesizing the dilemmas and prospects for a peri-urban land use management framework: Evidence from Ethiopia. *Land Use Policy* **2021**, *100*, 105122. [CrossRef]
30. Lemmen, C.; Zevenbergen, J.; Lengoiboni, M.; Deininger, K.; States, U.; Burns, T. First experiences with High Resolution Imagery Based Adjudication Approach for Social Tenure Domain Model in Ethiopia. In Proceedings of the FIG—World Bank Conference: Land Governance in Support of the Millennium Development Goals, Responding to New Challenges, Washington, DC, USA, 9–10 March 2009.
31. Bruce, J.W.; Hoben, A.; Rahmato, D. *After the Derg: An Assessment of Rural Land Tenure Issues in Ethiopia*; Land Tenure Center, University of Wisconsin-Madison: Madison, WI, USA, 1994.
32. Shimeles, A.; Verdier-Chouchane, A.; Boly, A. *Building a Resilient and Sustainable Agriculture in Sub-Saharan Africa*; Springer International Publishing: Cham, Switzerland, 2018; ISBN 9783319762227.
33. Bezu, S.; Holden, S. Demand for second-stage land certification in Ethiopia: Evidence from household panel data. *Land Use Policy* **2014**, *41*, 193–205. [CrossRef]
34. Belay, G. Toolbox for the Development of Cadastral and Registration Proclamation for Second Level Certification Program in Ethiopia. *Int. J. Sci. Basic Appl. Res.* **2014**, *13*, 244–259.
35. Crewett, W.; Korf, B. Ethiopia: Reforming land tenure. *Rev. Afr. Polit. Econ.* **2008**, *35*, 203–220. [CrossRef]
36. Dessalegn, R. Land rights and tenure security: Rural land registration in Ethiopia. In *Legalising Land Rights: Local Practices, State Responses and Tenure Security in Africa, Asia and Latin America*; Leiden University Press: Leiden, The Netherlands, 2009.
37. Wabelo, T.S. Legal and Institutional Frameworks Regulating Rural Land Governance in Ethiopia: Towards a Comparative Analysis on the Best Practices of Other African Countries. *Beijing Law Rev.* **2020**, *11*, 64–98. [CrossRef]
38. Adam, A.G.; Birhanu, T.A. Decentralised Rural Land Administration in Ethiopia. *J. Land Rural Stud.* **2018**, *6*, 34–39. [CrossRef]
39. Belachew, M.; Aytenfisu, S. Facing the challenges in building sustainable land administration capacity in Ethiopia. In Proceedings of the FIG Congress 2010 Facing the Challenges—Building the Capacity, Sydney, Australia, 11–16 April 2010.
40. Deininger, K.; Ali, D.A.; Alemu, T. *Impacts of Land Certification on Tenure Security, Investment, and Land Markets: Evidence from Ethiopia*; World Bank: Washington, DC, USA, 2008; Available online: https://openknowledge.worldbank.org/handle/10986/6897 (accessed on 8 December 2021).
41. Persha, L.; Greif, A.; Huntington, H. Assessing the Impact of Second-Level Land Certification in Ethiopia. In Proceedings of the 2017 World Bank Conference on Land and Poverty, Washington, DC, USA, 20–24 March 2017.
42. Chekole, S.D.; de Vries, W.T.; Durán-Díaz, P.; Shibeshi, G.B. Analyzing the effects of institutional merger: Case of cadastral information registration and landholding right providing institutions in ethiopia. *Land* **2021**, *10*, 404. [CrossRef]
43. MoA National Rural Land Administration Information System (NRLAIS) Implementation Progress Report, Addis Ababa. 2021.
44. DOrsolya, K. Operational Acceptance Test Report: National Rural Land Administration Information System (NRLAIS) Software Operational Acceptance Test Report ver. 1.0 Responsible and Innivative Land Administration in Ethiopia (REILA), Addis Ababa, Ethiopia. 2018.
45. Dečman, M. Factors that increase active participation by higher education students, and predict the acceptance and use of classroom response systems. *Int. J. High. Educ.* **2020**, *9*, 84. [CrossRef]
46. Oye, N.D.; A.Iahad, N.; Ab.Rahim, N. The history of UTAUT model and its impact on ICT acceptance and usage by academicians. *Educ. Inf. Technol.* **2014**, *19*, 251–270. [CrossRef]
47. Wang, H.-Y.; Wang, S.-H. User acceptance of mobile internet based on the Unified Theory of Acceptance and Use of Technology: Investigating the determinants and gender differences. *Soc. Behav. Personal. Int. J.* **2010**, *38*, 415–426. [CrossRef]
48. Nugroho, Y.; Prasetyo, A. Assessing information systems success: A respecification of the DeLone and McLean model to integrating the perceived quality. *Probl. Perspect. Manag.* **2018**, *16*, 348–360. [CrossRef]
49. Petter, S.; DeLone, W.; McLean, E. Measuring information systems success: Models, dimensions, measures, and interrelationships. *Eur. J. Inf. Syst.* **2008**, *17*, 236–263. [CrossRef]
50. Bakhit Jaafreh, A. Evaluation Information System Success: Applied DeLone and McLean Information System Success Model in Context Banking System in KSA. *Int. Rev. Manag. Bus. Res.* **2017**, *6*, 829–845.
51. Zhu, K.; Kraemer, K.L.; Xu, S. The process of innovation assimilation by firms in different countries: A technology diffusion perspective on e-business. *Manag. Sci.* **2006**, *52*, 1557–1576. [CrossRef]

52. Adeyemi, I.O.; Issa, A.O. Integrating Information System Success Model (ISSM) And Technology Acceptance Model (TAM): Proposing Students' Satisfaction with University Web Portal Model. *Rec. Libr. J.* **2020**, *6*, 69–79. [CrossRef]
53. Rahimi, B.; Nadri, H.; Afshar, H.L.; Timpka, T. A systematic review of the technology acceptance model in health informatics. *Appl. Clin. Inform.* **2018**, *9*, 604–634. [CrossRef] [PubMed]
54. Fathali, S.; Okada, T. Technology acceptance model in technology-enhanced OCLL contexts: A self-determination theory approach. *Australas. J. Educ. Technol.* **2018**, *34*, 138–154. [CrossRef]
55. Malik, B.H.; Shuqin, C.; Qamar, S.; Mattiullah, B. Examining Success of Land Record Information Systems (LRMIS) in Pakistan: Validating an incorporated IS success model. *Eur. Sci. J. ESJ* **2016**, *12*, 258. [CrossRef]
56. Malik, B.H.; Shuqin, C.; Mastoi, A.G.; Ahmed Ghais, A.H.A. Citizen's Adoption Of Mobile Land Record Information Systems (mLRMIS): A Case of Pakistan. *Eur. Sci. J. ESJ* **2016**, *12*, 393. [CrossRef]
57. Ramírez-Correa, P.; Alfaro-Peréz, J.; Cancino-Flores, L. Meta Analysis of the DeLone and MacLean IS sucess model at individual level: An examination of the heterogeinity of the studies. *Espacious* **2015**, *36*, 11.
58. McLean, E.; Sedera, D.; Tan, F. Reconceptualizing System Use For Contemporary Information Systems. In Proceedings of the 15th Pacific Asia Conference on Information Systems (PACIS), Brisbane, Australia, 7–11 July 2011.
59. Yu, P.; Qian, S. Developing a theoretical model and questionnaire survey instrument to measure the success of electronic health records in residential aged care. *PLoS ONE* **2018**, *13*, e0190749. [CrossRef]
60. DeLone, W.H.; McLean, E.R. The DeLone and McLean Model of Information Systems Success: A Ten-Year Update. *J. Manag. Inf. Syst.* **2003**, *19*, 9–30.
61. Contractor, N.S.; DeChurch, L.A. Integrating social networks and human social motives to achieve social influence at scale. *Proc. Natl. Acad. Sci. USA* **2014**, *111*, 13650–13657. [CrossRef] [PubMed]
62. Barry, M.; Muhsen, A.; Molero, R.; Muhsen, A.-R. Evolutionary Land Tenure Information System Development: The Talking Titler Methodology Evolutionary Land Tenure. In Proceedings of the 8th FIG Regional Conference 2012 Surveying towards Sustainable Development, Montevideo, Uruguay, 26–29 November 2012.
63. Enemark, S.; McLaren, R.; Lemmen, C. Fit-for-Purpose Land Administration—Providing Secure Land Rights at Scale. *Land* **2021**, *10*, 972. [CrossRef]
64. Ijadi Maghsoodi, A.; Saghaei, A.; Hafezalkotob, A. Service quality measurement model integrating an extended SERVQUAL model and a hybrid decision support system. *Eur. Res. Manag. Bus. Econ.* **2019**, *25*, 151–164. [CrossRef]
65. Lepmets, M.; Mesquida, A.L.; Cater-Steel, A.; Mas, A.; Ras, E. The evaluation of the IT service quality measurement framework in industry. *Glob. J. Flex. Syst. Manag.* **2014**, *15*, 39–57. [CrossRef]
66. Peluso, N.L.; Kelly, A.B.; Woods, K. Context in Land Matters: Access Effects and History in Land Formalization. In Proceedings of the Fourteenth Biennial Conference of the International Association for the Study of the Commons, Mt. Fuji, Japan, 3–7 June 2013.
67. Bollen, K.A. *Structural Equations with Latent Variables*; Wiley: Hoboken, NJ, USA, 2014. [CrossRef]
68. Population Projection. Available online: https://www.statsethiopia.gov.et/population-projection/ (accessed on 3 December 2021).
69. World Bank. *Ethiopia Urbanization Review*; World Bank: Washington, DC, USA, 2020.
70. Gavonel, M.F. *Patterns and Drivers of Internal Migration Among Youth in Ethiopia, India, Peru and Vietnam*; Young Lives: Oxford, UK, 2017.
71. Gashu, A.; Bahir, A. Urbanization and the Struggle for Land in the Peri-Urban Areas of Ethiopia. Available online: http://cega.berkeley.edu/assets/miscellaneous_files/22_-ABCA_Urbanization-research_paper-ABCA.pdf (accessed on 23 September 2021).
72. Knippenberg, E.; Jolliffe, D.; Hoddinott, J. Land Fragmentation and Food Insecurity in Ethiopia. *Am. J. Agric. Econ.* **2019**, *102*, 1557–1577. [CrossRef]
73. Soper, D.S. A-priori Sample Size Calculator for Structural Equation Models [Software]. 2019. Available online: http//www.danielsoper.com/statcalc (accessed on 10 June 2021).
74. Ministry of Agriculture and Natural Resources. Technical Specifications-NRLAIS PILOT Note: These Technical Requirements Are Presented in a Format Based on the World Bank International Competitive Bidding template for "Supply and Installation of Information systems". Section VI. Technical Requirements, Responsible and Innovative Land Administration in Ethiopia, Addis Ababa, Ethiopia.
75. Schuberth, F. Confirmatory composite analysis using partial least squares: Setting the record straight. *Rev. Manag. Sci.* **2021**, *15*, 1311–1345. [CrossRef]
76. Rose, N.; Wagner, W.; Mayer, A.; Nagengast, B. Model-based manifest and latent composite scores in structural equation models. *Collabra Psychol.* **2019**, *5*, 9. [CrossRef]
77. Voorhees, C.M.; Brady, M.K.; Calantone, R.; Ramirez, E. Discriminant validity testing in marketing: An analysis, causes for concern, and proposed remedies. *J. Acad. Mark. Sci.* **2016**, *44*, 119–134. [CrossRef]
78. Bentler, P.M.; Bonett, D.G. Significance tests and goodness of fit in the analysis of covariance structures. *Psychol. Bull.* **1980**, *88*, 588–606. [CrossRef]
79. Hamdan, M.N.M.; Al-Hajri, N.J. The effect of information systems success factors on user satisfaction in accounting information systems. *Manag. Sci. Lett.* **2021**, *11*, 2045–2052. [CrossRef]
80. Subaeki, B.; Rahman, A.A.; Putra, S.J.; Alam, C.N. Success model for measuring information system implementation: Literature review. *J. Phys. Conf. Ser.* **2019**, *1402*, 077015. [CrossRef]

81. Al-Fraihat, D.; Joy, M.; Masa'deh, R.; Sinclair, J. Evaluating E-learning systems success: An empirical study. *Comput. Hum. Behav.* **2020**, *102*, 67–86. [CrossRef]
82. Kosicka, E.; Gola, A. The Use of QFD for the Design of a Maintenance Service Support System. *MATEC Web Conf.* **2019**, *252*, 06012. [CrossRef]
83. Floropoulos, J.; Spathis, C.; Halvatzis, D.; Tsipouridou, M. Measuring the success of the Greek Taxation Information System. *Int. J. Inf. Manage.* **2010**, *30*, 47–56. [CrossRef]
84. Machdar, N.M. The effect of information quality on perceived usefulness and perceived ease of use. *Bus. Entrep. Rev.* **2019**, *15*, 131–146. [CrossRef]
85. Chaula, J.A.; Institutionen för data-och systemvetenskap (Stockholm). *A Socio-Technical Analysis of Information Systems Security Assurance: A Case Study for Effective Assurance*; Department of Computer and Systems Sciences, Stockholm University/KTH DSV: Stockholm, Sweden, 2006; ISBN 9171553398.
86. Hsieh, J.J.P.A.; Wang, W. Explaining employees' extended use of complex information systems. *Eur. J. Inf. Syst.* **2007**, *16*, 216–227. [CrossRef]
87. Kulkarni, U.R.; Ravindran, S.; Freeze, R. A knowledge management success model: Theoretical development and empirical validation. *J. Manag. Inf. Syst.* **2006**, *23*, 309–347. [CrossRef]
88. Chaudhry, B.; Wang, J.; Wu, S.; Maglione, M.; Mojica, W.; Roth, E.; Morton, S.C.; Shekelle, P.G. Systematic review: Impact of health information technology on quality, efficiency, and costs of medical care. *Ann. Intern. Med.* **2006**, *144*, 742–752. [CrossRef] [PubMed]

land

Article

Resource Opportunity in China's Market Transition and Governance: Time Factor in Urban Housing Inequality

Jiawen Zhou [1] and Jing Xiong [2,*]

1 School of International Economics and Trade, Shanghai Lixin University of Accounting and Finance, 995 Shangchuan Road, Fifth Teaching Building 411, Shanghai 201209, China; 20180083@lixin.edu.cn
2 China Institute for Urban Governance, School of International and Public Affairs, Shanghai Jiao Tong University, 1954 Huashan Road, Xinjian Building 329, Shanghai 200030, China
* Correspondence: bearnear@sjtu.edu.cn; Tel.: +86-156-0195-9586

Abstract: Since China's reform and opening up, the country's rapid marketization process has been accompanied by the rapid growth of inequality, which has been significant for all classes of society. In terms of its impact, housing inequality is particularly noticeable. In this paper, we discuss the influence of real-estate purchase time, organization, human capital, and political capital on the value of real estate and the appreciation of real estate in China by using a conditional mean model and a quantile regression model. The differences in the degree of influence of these factors on different quantile levels are also investigated. We found that, after adding the time factor, the prior possession of resources in the early stage of market transformation will benefit the long-term marketization process. Organizations that can penetrate "market-redistribution" and professions that directly participate in the distribution of real-estate resources also have significant advantages in this regard.

Keywords: market transition; urban housing; time factor; resource allocation

Citation: Zhou, J.; Xiong, J. Resource Opportunity in China's Market Transition and Governance: Time Factor in Urban Housing Inequality. *Land* **2021**, *10*, 1331. https://doi.org/10.3390/land10121331

Academic Editor: Shiliang Su

Received: 26 October 2021
Accepted: 1 December 2021
Published: 3 December 2021

1. Introduction

Since China's reform and opening up in 1978, the country's high-speed process of marketization has resulted in unprecedented advancements. Additionally, concerns, such as imbalanced and insufficient developments, ensued, with social inequality becoming the most serious issue. In the early 1980s, China and other Eastern European socialist countries were the most equal countries in the world. Then, the Gini coefficient in China was less than 0.3, which was lower than that of developed countries and other developing countries [1]. Despite official statistics indicating a drop in the Gini coefficient in China during these years—0.49 in 2008 to 0.46 in 2015 [2]—the nation has undeniably transformed from one of the most equal countries in the world into one of the world's most unequal countries over the last 40 years. This transition has exerted a profound influence on every stratum of society, and housing inequality is the most remarkable problem.

Several issues account for this phenomenon of inequality, one of which is the commodification of housing. The homeownership rate of Chinese urban residents increased from 15% in 1995 to 80% in 2002, and the latter number was even higher than the figure for the United States in 2003, which was 68% [3,4]. Thus, a real-estate market emerged from the old system of socialist redistribution.

Residents have gone from a period marked by "houses allocated by work units" to a period characterized by "buying houses on their own" within a generation. The old institutional arrangements and the force of marketization are intertwined, forming a medley of houses with different historical origins on the market. Additionally, the housing allocation circumstances of society and the complex arrangements in economics and society are intertwined, resulting in housing inequality.

Notably, housing prices are rapidly increasing, and two aspects account for this trend. First, many people in China leave their hometowns to seek employment in big cities. As a result, there is considerable demand for housing in these cities. Second, under circumstances marked by inflation and a sluggish manufacturing industry, housing hash has become the fastest-growing value-added asset of Chinese families. The average profitability rates of first-, second-, and third-owned urban residences are 340.31%, 143.25%, and 96.70%, respectively [5]. Therefore, housing inequality is more than a matter of social resource distribution, such as income inequality. The accompanying high profitability is more likely to become an institutionalized source of rent-seeking. Individuals who invest early tend to receive a substantial amount of added value through real-estate appreciation and residential rental properties, creating even wider inequality. This will probably lead to a more rigorous problem—the Matthew Effect. The Matthew Effect means that any individual, group, or region that achieves success and progress in a certain aspect (such as money, reputation, status, etc.) will produce an accumulated advantage and have more opportunities to achieve greater success and progress [6]. What are the characteristics of housing inequality? Which stratum of society owns high-value properties? Who can benefit from the property market? All of these questions must be studied and solved in greater detail.

2. Theory and Hypothesis

Since the 1980s, the relation between China's market transition and social stratification has been a main focus in academic research, and scholars have performed productive research in this respect. Although judgments vary regarding social inequality in early socialist China, many scholars have emphasized that the redistribution of collective goods is likely to form greater inequality than once thought against the background of "egalitarianism" in a redistribution structure bonded by a unit system [7–10]. Correspondingly, such a habit of redistribution of collective goods is likely to continue even after market transition due to institutional inertia or path dependence [11,12].

Housing is a clear example of this. In the early days of the reform, many urban residents could purchase houses at a price lower than the market price through their work unit ties or social relations [10,13,14]. In recent years, many empirical studies have indicated a gradual widening of housing inequality in urban residents from every stratum of society [15,16]. From the perspective of market transition theory [17,18], housing inequality among different strata is inevitable and will decline because the market transition will form a new means of resource allocation under a redistribution system. In this emerging market, the influence of old political capital declines, and market mechanisms focused more on individual capabilities and fair competition assume greater importance. Because of the Matthew Effect from competition, housing inequality will widen in the short-term. In the process, the return on human capital, such as education, will increase, and the return on political capital, such as party membership and cadre position, will decrease. Compared with old powerful elites, current market elites have gained new access to real-estate profits. Market transition will eventually create a fairer and more equal housing supply mechanism. With the further deepening of the market transition, ever more people with substantial human capital are able to profit from it, and inequality will eventually decline due to the more thorough social security system established by the government [19].

Another theory is the Power Continuance Theory or the Power Transition Theory [20], which underscores that the state's power creates an innate advantage and a stronger ability to expand in the process of market transition. Additionally, compared with other governments, local governments in China tend to participate in economic construction more actively under the pressure of political achievement in terms of economic development [21]. These factors justify the further expansion of bureaucracy in the market, and government functions also become stronger. Under such circumstances, factors such as political capital may assume greater importance in the process of market transition. Empirical studies have also demonstrated the positive effect of the work unit system and public power on property area and purchasing opportunity [22]. Additionally, income studies have shown that China ranks relatively low among developing countries regarding the monetary return on one extra year of schooling. Additionally, the income advantage of Chinese Communist Party membership is remarkable [23]. Chinese Communist Party membership is often regarded as a kind of political capital [24], similar to a kind of social capital (relationship), which can obtain many resources (information, influence, or operation power) and opportunities [25]. In other words, party members have the opportunity to contact people who are highly beneficial to their future careers [26]. The aforementioned statistics support the argument of power continuance.

The major difference between market transition theory and power continuance theory is that the resources brought by market reform are acquired by new market elites or old powerful elites. The common assumption of the two theories is as follows: China's marketization reform cannot be achieved instantly; by contrast, it continues gradually. The marketization level of China seems to increase continuously and is now in its preliminary stage. Marketization is bound to offer a dynamic aspect to economic development. The advantage of such an assumption is that the stimulating effect of the market on economic development has been clearly expounded in economics [27]. Needless to say, the market improves the efficiency of the economy in a broader sense. Although market transition advances step by step, means of production, such as commodity, workforce, and property, slowly form their respective markets simultaneously. However, these markets almost complete the commodification process overnight, which is characterized by "punctuated" institutional changes (Punctuated-Equilibrium Theory, for reference). The time factor has tended to be ignored in the literature. Based on CHIP long-term statistics, Li concluded that the pronounced wealth inequality of Chinese residents began between 1995 and 2002 [28].

Additionally, based on mean regression studies [29,30], one problem that can easily be ignored in many cases is that unequal models vary among different groups of people. The progressive reform functioning as "China's experience" is the fundamental logic of Chinese urban residence institutional reform and avoids overall turbulence in the reform process. Notably, something else ensues; because the basic attribute of urban residence and the concept of equality of urban residents have been changed, the difference in housing benefits acquired by individuals from every stratum of society has accumulated and strengthened, and this allows the market capability of residents to play a bigger role in social stratification [31]. Due to the ongoing process of marketization and social stratification, housing will continue to strengthen the benefits of individuals who already own a house. Additionally, people in a relatively low-income class will find profiting from the property market more difficult. Another technical problem is that plenty of papers on housing inequality have been published based on survey data from around 2005 [32,33]. Additionally, the independent variable has tended to be living space; thus, property value has not been fully studied. As such, there is a gap in the literature that is worthwhile filling.

To sum up, some classic hypotheses and propositions in previous theories can be borrowed to illustrate housing inequality:

Hypothesis 1a. *The higher the educational level, the higher the housing value.*

Hypothesis 1b. *The higher the income, the higher the housing value.*

Hypothesis 2a. *The cadre position has a greater advantage than other occupations in terms of housing value.*

Hypothesis 2b. *People with party membership have a greater advantage than non-party people in terms of housing allocation and acquirement.*

Hypothesis 3a. *It is easier for non-market organizations, such as state-owned enterprises and the party and government organs, to acquire high-value housing than market units in a manner similar to private enterprises and self-employed individuals.*

Hypothesis 3b. *Non-market organizations, such as state-owned enterprises and the party and government organs, profit more from the property market than market units, such as private enterprises and self-employed individuals.*

This paper proposes the following research hypotheses:

Hypothesis 4a. *Time factor. Against the backdrop of progressive reform, individuals who enter the property market late receive relatively high returns due to fully developed marketization.*

Hypothesis 4b. *Time factor. Against the backdrop of progressive reform, individuals who enter the real-estate market late receive relatively low returns due to a limited holding period for housing.*

Above all, a close and hierarchical analysis is conducted for people receiving different levels of housing profits.

3. Research Methods

3.1. Methodology

Regarding research methodology, this paper adopts a quantile regression model and the common ordinary least squares (OLS) linear regression model found in the literature. Compared with an OLS linear regression model based on mean, a quantile regression model has two advantages. First, the quantile is less likely to be affected by extreme values than the mean. Although such a problem can be alleviated by taking logarithms of variables in the mean regression, quantile regression is better in terms of managing skewed variables, such as income and housing value. Second, regression equations can be established according to different quantile levels of dependent variables. Therefore, all factors can be closely studied under the influence of all sorts of housing profitability levels. Furthermore, the changing patterns of these factors in different quantile levels can be analyzed, which is better than the over-general conclusion reached under the guidance of the OLS model.

This paper adopts 19 quantiles from 5 to 95 at an interval of 5; additionally, it selects a multivariable linear regression model, in which every dependent variable has 20 regression equations. Thus, an analysis based on regression excels can be conducted as performed in classic quantitative research. Additionally, regression coefficient tables in different quantiles can be created to inspect the changing patterns of independent variables in different quantile levels.

3.2. Data Collection

This paper adopts data from the Social Development and Social Construction national survey conducted by Shanghai University in 2012. The survey applies simple random sampling in three stages, and the data are from various geographical locations in China, namely Henan Province, Jilin Province, the municipality of Shanghai, Guangdong Province, Yunnan Province, and Gansu Province. The total number of samples is 5745. Due to the substantial differences between a rural homestead self-built house system and a commoditized urban property market, 2482 rural samples were excluded from this research. Among the remaining samples, 2041 people either purchased or built houses on their own. These people in question claimed the right to their property, know the price of their property, and are the main research objects in this paper.

3.3. Measures

Dependent variable 1 is the total value of the main residence. Because the property market in China is relatively mature, the housing price of the respondents includes information about housing quality and housing location, which is a suitable variable to represent the distribution of housing resources. Dependent variable 2 is the added value of the house after purchase or construction. Because respondents are asked on the questionnaires to provide the date when they purchased or built their house and the cost of doing so, the added value of the house in question can be calculated from the date of the contract to the year 2012, when the survey was conducted. This variable indicates the profit of respondents from the property market and allows us to more accurately operationalize incremental resources in the process of market transition. Logarithms have been taken of the two variables in the OLS model because the two variables are skewed to the right. Although it is necessary to take logarithms in the quantile model, natural logarithms of dependent variables have been used improve the comparison of the results of the OLS model. The main independent variables are as follows:

- Education level: The education of the respondents is re-classified as elementary school and below, secondary school, high school, and university and above.
- Party membership: Membership includes Communist Party membership and membership in democratic parties. In China, democratic parties have their own political implications.
- Work unit type: Due to the joint-stock system reform in recent years, some state-owned enterprises and collective enterprises have begun to include private capital, and some public institutions are owned and run by individuals. Therefore, first, the work unit types offered in questionnaires are roughly classified. Next, the work unit types marked as enterprise are further classified as state-owned, collective, or privately owned in terms of ownership.
- Occupation and administration position: Occupation is a major factor affecting housing distribution. To better compare the variable of administration position to other occupations, respondents with an administration position are encoded into two professions: senior cadres with a title of section chief and above, and junior cadres with a title of section chief and below. These two professions are added to the variable of occupation. Additionally, samples that have both administration positions and other occupations are ruled out.
- Control variable: According to the literature, demographic variables are included as control variables to increase the accuracy of the model. These variables are age and the square of age, marital status, number of family members, province, and natural logarithm, taken as the total income of the previous year.

Another independent variable warranting attention in this paper is the time when the house was purchased or built. Although this independent variable is a continuous variable on the questionnaire, this paper de-dimensionalizes it into four nominal variables: before 1998, 1999–2003, 2004–2008, and 2009–2012. The years 1998, 2003, and 2008 are selected as key division points due to theoretical and historical reasons. According to the research conducted by Wu Xiaobo [34], the then incumbent Zhu Rongji administration suspended the policy of selling state capital to private businesses, which targeted the state-owned small and medium enterprises with poor performance. After 1998, state-owned capital withdrew from competitive industries, such as textiles, home appliances, and food, while playing a dominant and monopolistic role in strategic industries, such as resources, energy, and heavy chemicals. State-owned businesses started to retreat to the upper-stream industries, forming an advantage of an oligarchy or multi-oligarchy operation.

After 1998, the housing of urban residents transformed from work–unit-distributed houses into commodity houses for transactional purposes. At the end of 2003, the State Council released *Provisions on Curbing Blind Investment in Steel, Electrolytic Aluminum, Cement and other industries*, to manage the over-popular investments in energy industries. Additionally, in 2003, the State-Owned Assets Supervision and Administration Commission of the State Council was founded. In the following three years, the main business income of enterprises directly under central authorities increased by 78.8%, the profit increased by 140%, the tax revenue increased by 96.5%, and the hedge ratio of state-owned assets increased to 144.4%. Feng Lun, then chairman of Wangtong Group, said, "Faced with state-owned capital, private capital has to stick to the principle of cooperation over competition, supplement over substitution, affiliation over dominance. Only by accomplishing this can the private capital advance continuously and fare well".

In 2008, because exports were hindered by the financial crisis, the central government implemented a proactive fiscal policy—a USD 4 trillion economy stimulus package. As a result, state-owned enterprises gained 90% of the new loans. Above all, 1998 was the starting point of the commodification of housing, and 2003 and 2008 were significant years because the administration increased the investment in state-owned capital and private capital withdrew from the production field during the process of market transition. Furthermore, the private capital withdrawn in 2003 and 2008 was mostly transferred to the financial market, in which the real-estate market, as the fastest-growing value-added investment, received substantial attention. In the process of property investment, local governments also profited from land finance, namely collecting land transaction fees by transferring land use rights [35]. Additionally, state-owned enterprises can profit from the upstream industries resulting from property construction. A prohibitive housing price scenario is the result of the common interests of private capital, local governments, and state-owned capital. Therefore, the years 1998, 2003, and 2008 are particularly important because they have different implications for housing prices. The time when the property market was entered is divided into four stages.

The descriptive statistics of the aforementioned variables are in Table 1.

Table 1. Descriptive statistics of variables.

Continuous Variable						
Variables	**Sample Numbers**	**Mean**	**Gini Coefficient**	**Standard Deviation**	**Minimum Value**	**Maximum Value**
The market value of property [1] (CNY 10,000)	2791	80.67	0.59332	128.46	0.01	3000
The added value of property (CNY 10,000)	1931	55.88	0.57414	70.977	−49	990
Income (CNY)	3165	44,700	0.58174	534,295	300	3×10^7
Number of properties	2860	1.203	——	0.5986	1	15
Year of purchase or construction	1998	2000	——	9.6015	1812	2013
Age	3263	42.54	——	14.224	17	70
Discrete variable [2]						
Variables	Sample numbers	Frequency	Percentage (%)	Standard deviation	Minimum value	Maximum value
Time of Purchase or Construction						
Before 1998	1998	659	32.983	0.4703	0	1
1999–2003	1998	555	27.7778	0.448	0	1
2004–2008	1998	479	23.974	0.427	0	1
After 2009	1998	305	15.2653	0.3597	0	1
Work Unit Types						
Party and government organs	2891	94	3.25147	0.1774	0	1
Public institutions	2891	426	14.7354	0.3545	0	1
State-owned enterprises	2891	669	23.1408	0.4218	0	1
Collective enterprises	2891	142	4.9118	0.2162	0	1
Private enterprises	2891	969	33.51781	0.4721	0	1
Self-employed business	2891	591	20.44275	0.4034	0	1
Education Level						
Elementary school	3261	477	14.62741	0.3534	0	1
Secondary school	3261	869	26.64827	0.4422	0	1
High school	3261	908	27.84422	0.4483	0	1
University and above	3261	1007	30.8801	0.4621	0	1
Occupation						
Senior cadre	2904	145	4.99311	0.2178	0	1
Junior cadre	2904	130	4.47658	0.2068	0	1
Senior management	2904	49	1.68733	0.1288	0	1
Junior management	2904	117	4.02893	0.1967	0	1
Intermediate and senior technician	2904	125	4.30441	0.203	0	1
Ordinary technician	2904	327	11.26033	0.3162	0	1
Organization clerk	2904	215	7.40358	0.2619	0	1
Salesman of enterprises and institutions	2904	147	5.06198	0.2193	0	1
Business service personnel	2904	466	16.04683	0.3671	0	1
Skilled worker	2904	154	5.30303	0.2241	0	1
Ordinary worker	2904	630	21.69421	0.4122	0	1

Table 1. *Cont.*

Continuous Variable						
Variables	**Sample Numbers**	**Mean**	**Gini Coefficient**	**Standard Deviation**	**Minimum Value**	**Maximum Value**
Self-employed entrepreneur	2904	363	12.5	0.3308	0	1
Worker after retirement [3]	2904	36	1.23967	0.1107	0	1
Others	2904	4	0.00138	0.03712	0	1
Party member (Yes = 1, No = 0)	3263	510	15.62979	0.3632	0	1
Marriage status (Yes = 1, No = 0)	3258	2536	77.83917	0.4154	0	1
Family member number	3260	7564	232.0245	1.2309	0	9
Gender (Male = 1, Female = 0)	3263	1602	49.09592	0.5	0	1
Province						
Shanghai Municipality	3263	837	25.6512	0.43677	0	1
Yunnan Province	3263	368	11.278	0.31637	0	1
Jilin Province	3263	575	17.6218	0.38106	0	1
Guangdong Province	3263	735	22.5253	0.41781	0	1
Henan Province	3263	392	12.0135	0.32517	0	1
Gansu Province	3263	356	10.9102	0.31181	0	1

[1] Values less than 0 were taken of the three variables—housing value, housing added value, and income—and the natural logarithms were taken from the model. [2] All discrete variables were changed into dummy variables and then added into the following model. [3] The group working after retirement is not included in the regression model.

4. Analysis and Results

Regarding the two variables—housing value and housing added value—two of the same sets of independent variables were put into two respective models: an OLS model based on mean estimation and a QR model based on 19 quantiles.

The parameter estimation of quantile regression parameters adopts the bootstrap method. Every quantile model conducts 500 samplings with replacements of initial samples. Clustered standard error is used with different cities as different clusters to estimate standard deviation.

Compared with the general robust standard error, a clustered standard error presumes that the random error terms in the regression equations are uncorrelated among cities but correlated within cities, and this helps to accurately estimate the vast difference among Chinese regions and strengthen the model; however, it has one disadvantage. The *p*-value of a clustered standard error is higher than that of a robust standard error and is less significant.

4.1. Housing Value Model

Because independent variables are mostly dummy variables, the basic condition of the model and the result of control variables are displayed first in Table 2.

In general, because of the proper selection of the covariate, R^2 (coefficient of determination) in the OLS model is over 0.5, which indicates that the model fits the data well. Because the values of different variables in the questionnaires are not exactly the same, 1648 cases were eventually selected for the model. In contrast to the life course theory, age and housing value do not increase first and then decrease. Age basically exerts zero influence on the housing price, which probably occurs because marriage status and family member numbers are controlled. In terms of gender, the income returns on housing for males compared with females is lower. A possible reason for this result is that the gender advantages of males are demonstrated in other variables, which—in turn—serve as the control variables of gender. Regarding the difference among provinces, the municipalities of Shanghai and Guangdong Province have a greater advantage on housing price than other provinces, and Jilin Province ranks lowest. The model results are listed, respectively, according to independent variables in Table 3.

Table 2. Basic condition of the housing price model and control variables.

Model Type/Quantiles	OLS	0.1	0.25	0.5	0.75	0.9
Age	−0.0183	−0.0388 !	−0.0253	−0.0117	−0.0104	−0.00454
Square of age	0.000303 *	0.000538 *	0.000384 *	0.000215	0.000182	0.000135
Marriage status	−0.0088	0.00175	0.00708	0.000983	0.0423	−0.00369
Number of family members	0.0535 ***	0.0253	0.0532 **	0.0537 ***	0.0526 **	0.0493 *
Gender	−0.120 ***	−0.125 !	−0.139 **	−0.125 ***	−0.0472	−0.0695
Province						
Yunnan Province	−1.126 ***	−1.306 ***	−1.244 ***	−1.110 ***	−1.076 ***	−1.070 ***
Jilin Province	−1.618 ***	−1.924 ***	−1.615 ***	−1.449 ***	−1.463 ***	−1.467 ***
Guangdong Province	−0.479 *	−0.801 **	−0.637 *	−0.498	−0.328	−0.0866
Henan Province	−1.057 ***	−0.988 ***	−1.031 ***	−1.046 ***	−1.147 ***	−1.223 ***
Gansu Province	−1.391 ***	−1.473 ***	−1.416 ***	−1.321 ***	−1.379 ***	−1.376 ***
Constant term	1.923 ***	0.6	1.469 !	1.580 **	2.792 ***	3.493 ***
Number of cases	1648	1648	1648	1648	1648	1648
R^2	0.577	0.3503	0.3589	0.3807	0.3959	0.3866

*** $p < 0.001$, ** $p < 0.01$, * $p < 0.05$, ! $p < 0.1$ Double underline is $p < 0.05$. Single underline is $p < 0.1$.

Table 3. Time factor of housing price model.

Model Type/Quantiles	OLS	0.1	0.25	0.5	0.75	0.9
Time of Purchase or Construction Reference Group: Before 1998						
1999–2003	0.329 ***	0.485 ***	0.327 ***	0.288 ***	0.211 ***	0.197 *
2004–2008	0.341 ***	0.526 ***	0.373 ***	0.252 ***	0.195 **	0.133
After 2009	0.398 **	0.556 **	0.383 ***	0.276 **	0.224 *	0.255 *

*** $p < 0.001$, ** $p < 0.01$, * $p < 0.05$.

As can be generally observed from the OLS model (Table 3), the time of purchase or construction continues to have a significantly positive effect on housing value after controlling other variables. Compared with individuals who entered the property market before 1998, individuals who entered the property market from 1999 to 2003, 2004 to 2008, and after 2009 have an increased housing value of 39.0%, 40.6%, and 48.9%, respectively. In other words, the later an individual acquires housing, the higher the housing price. This phenomenon can be attributed to the ever-increasing housing price. Due to the ongoing process of China's market transition, the more fully developed the marketization, the higher the housing value. Hypothesis 4a is thus proven true.

Additionally, such a positive effect varies at different quantile levels. To study the difference and changing trend of the effect in question, a quantile regression coefficient line graph is used. In Figure 1, 19 quantiles are at an interval of 0.05 on the x-axis. There are quantile regression coefficients on the y-axis. Only those regression coefficients with a p-value less than 0.1 are marked.

At the quantile levels from 0.1 to 0.25, which are mainly low-price houses, individuals who enter the real-estate market later obtain higher returns. One possible explanation for this result is that low-price houses are mostly government-subsidized housing units. Due to the commodification reform, houses acquired after 1999 can better satisfy the low-end living needs than before. With the increase of quantile, the value of low-price houses decreases gradually. At the quantile levels from 0.4 to 0.5, the housing value of individuals entering the property market from 2004 to 2008 is slightly lower than that from 1999 to 2003. This result shows that, in the early days of market transition, before a substantial amount of private capital flowed into the property market, the acquisition of middle-end housing could result in some advantages. At the quantile levels from 0.85 to 0.9, a number of high-price houses emerge after 2009 compared with 1998. Such a phenomenon is rare before 2009.

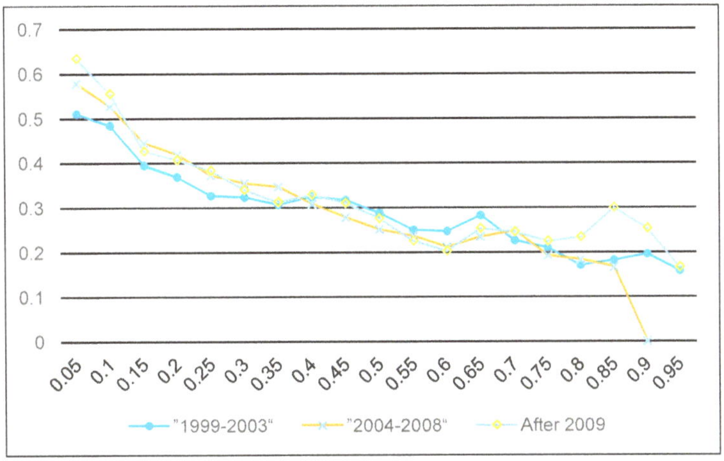

Figure 1. Quantile regression coefficient line: time factor of housing price model.

In terms of organizational factor, in Table 4 and Figure 2, the OLS model shows that different work units continue to exert influence on the distribution mechanism of housing, if not a great influence in the redistribution era. Unexpectedly, the housing value of collective enterprises ranks the highest, and the regression coefficients of other enterprises are negative. This is because of the many personal factors under control, and it is different from the results of early empirical studies. Based on the statistics from 1999 [26,36], the income of those from collective enterprises was substantially lower than from private enterprises, the party, and administration organs because collective enterprises were stuck between redistribution and the market. However, in housing distribution in recent years, the identity of collective enterprise has become an advantage rather than a disadvantage. Because party and administration organs do not show significance, and the difference between state-owned enterprises, public institutions, and private enterprises is small, Hypothesis 3a is not directly proven.

Table 4. Organizational factor of housing price model.

Model Type/Quantile	OLS	0.1	0.25	0.5	0.75	0.9
Work Unit Type Reference Group: Collective Enterprises						
Party and government organs	−0.196	−0.128	−0.206	−0.276 !	−0.169	−0.045
Public institutions, social organizations	−0.291 **	−0.261	−0.335 *	−0.289 **	−0.220 *	−0.181
State-owned enterprises	−0.330 **	−0.427!	−0.414 **	−0.312 **	−0.235 *	−0.15
Private enterprises	−0.281 **	−0.304	−0.332 **	−0.269 **	−0.224 *	−0.194
Self-employed individuals	−0.275	−0.0655	−0.304	−0.299 !	−0.229	−0.299

*** $p < 0.001$, ** $p < 0.01$, * $p < 0.05$, ! $p < 0.1$.

Although the difference in the regression coefficients among work organizations is small in the OLS model and at quantile levels from 0.4 to 0.55, the regression coefficient of state-owned enterprises is remarkably lower than other types of organization at the quantile levels from 0.1 to 0.3. In general, the regression coefficient of state-owned enterprises is lower than that of private enterprises. Additionally, the regression coefficient of public institutions is higher than that of state-owned enterprises and lower than that of private enterprises. The p-values of party and government organs, as well as self-employed individuals, are relatively low; thus, an overall pattern is difficult to observe. In general, Hypothesis 3b is proven partly false even though the housing prices of party and government organs and public institutions are higher than those of other organizations at high quantile levels from 0.65 to 0.75; this is illustrative of institutional inertia. However, the regression coefficient of state-owned enterprises ranks the lowest among low-quantile housing, which reflects the brunt of market transition toward old institutional arrangements. As state-owned enterprises have borne most of the brunt, party and government organs and public institutions remain advantageous in middle- to high-price housing.

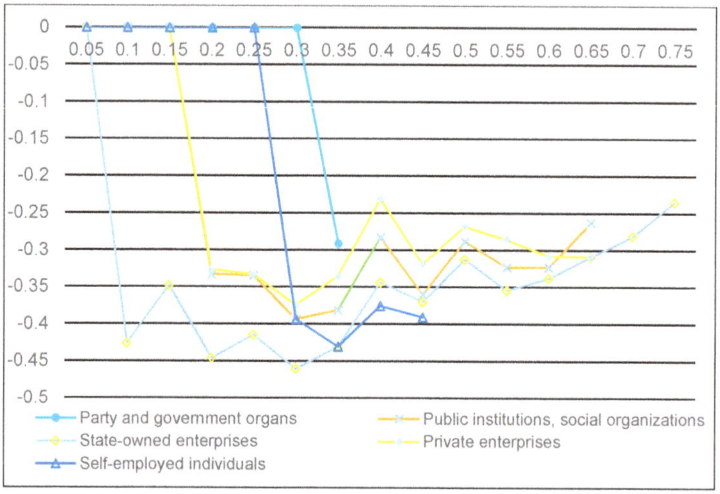

Figure 2. Quantile regression coefficient line: time factor of housing price model.

In terms of human capital factors, Table 5 and Figure 3 show that, the higher the education level, the higher the housing value. There is a remarkable advantage in college education compared with other levels of education, which corresponds to the conclusion of the OLS model. Therefore, Hypotheses 1a and 1b are proven true. With the increase in quantile, the two variables—income and education level—tend to decrease. This result demonstrates two things. First, as a type of investment, houses, especially high-price houses with a quantile over 0.7, are not directly related to income but related to the investment ability of investors. Second, regarding the acquisition of low-price and middle-price houses with a quantile below 0.65, high-school education has a greater advantage than secondary school education. However, such an advantage becomes less significant when the quantile is over 0.7, and sometimes secondary school education tends to be more rewarding. In terms of human capital, the returns on education mainly differ between individuals with or without a college education. The returns on income are higher in low-value and middle-value houses.

Table 5. Human capital factor of housing price model.

Model Type/Quantile	OLS	0.1	0.25	0.5	0.75	0.9
Education Level Reference Group: Elementary School and Below						
Secondary school	0.325 **	0.283	0.311 *	0.266 **	0.233 **	0.296 **
High school	0.412 ***	0.524 *	0.440 **	0.345 **	0.263 **	0.261 *
University and above	0.584 ***	0.564 *	0.633 ***	0.507 ***	0.471 ***	0.493 ***
Logarithms taken of income	0.233 ***	0.316 ***	0.268 ***	0.270 ***	0.184 ***	0.131 **

*** $p < 0.001$, ** $p < 0.01$, * $p < 0.05$.

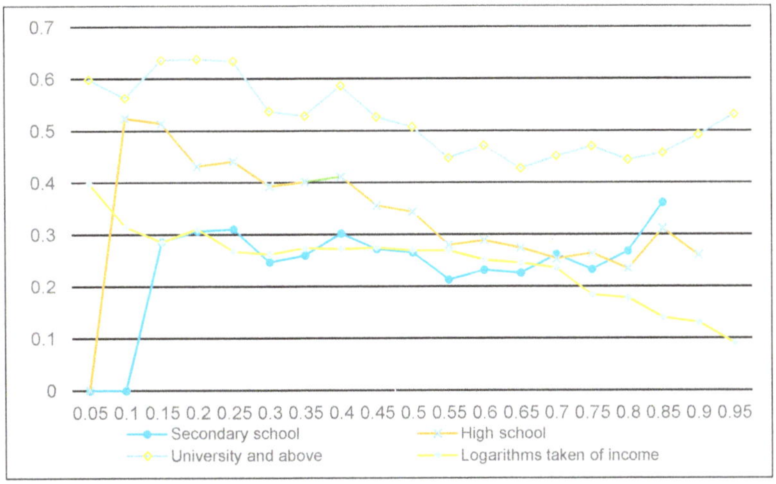

Figure 3. Quantile regression coefficient line: human capital factor of housing price model.

In terms of political capital in Table 6, the housing value of individuals with a title of section chief or above is 32.0% higher than that of ordinary workers, 15.5% higher than that of senior technicians, and 27.8% higher than that of junior management. Hypothesis 2a is thus proven true. Because the p-value of party membership is mostly low and even negative in some quantiles, the influence of party membership on housing value is exerted through other factors, such as occupation, work unit, and education level. When the aforementioned factors are under control, party membership is not as significant as in models with fewer variables. Sometimes, the p-value of party membership in the quantile regression is negative.

Table 6. Political capital factor of housing price model.

Model Type/Quantile	OLS	0.1	0.25	0.5	0.75	0.9
Occupation Reference Group: Ordinary Workers						
Senior cadre	0.278 *	0.494 *	0.203	0.200 !	0.236 **	0.158
Junior cadre	0.231 !	0.41	0.134	0.182 !	0.176 !	0.250 **
Senior management	0.226 !	0.155	−0.158	0.161	0.223	0.537 *
Junior management	0.245 *	0.236	0.231	0.219 *	0.179 !	0.247
Intermediate and senior technician	0.144 !	0.273	0.0567	0.139 !	0.0942	0.0203
Ordinary technician	0.0382	−0.0529	−0.0257	0.0389	0.0635	0.0896
Organization clerk	0.365 **	0.455 *	0.301 *	0.301 **	0.267 *	0.356 **
Salesman of enterprises and institutions	0.0868	0.161	−0.0476	0.0701	0.0754	0.241
Business service personnel	0.118 !	0.362 *	0.0763	0.0648	0.0316	0.0173
Skilled worker	0.147 !	0.299	−0.0214	−0.0241	0.222 *	0.214 !
Self-employed entrepreneur	0.168	0.121	0.0945	0.135	0.184	0.261
Others	−0.126	−0.212	−0.393	0.323	0.068	−0.205
Party membership	−0.0928	−0.0499	−0.126 *	−0.092	−0.0467	−0.0422

*** $p < 0.001$, ** $p < 0.01$, * $p < 0.05$, ! $p < 0.1$.

The returns of housing value for senior cadre rank are higher than those for junior cadre rank, but the difference is not substantial. The influence and changing trend of political capital in different quantiles can be clearly seen in Figure 4. The occupation variable levels off in different quantiles. However, at the quantile levels from 0.75 to 0.9 and from 0.2 to 0.55, organization clerks have a higher return than other occupations, which has rarely been observed in the literature.

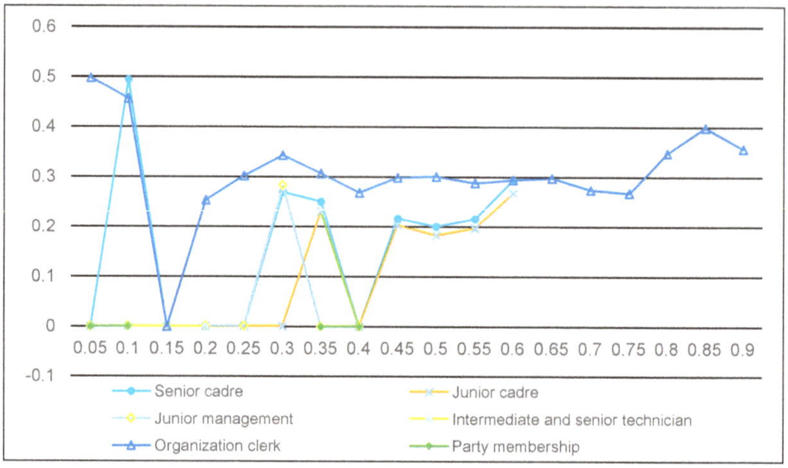

Figure 4. Quantile regression coefficient line: human capital factor of housing price model. The criterion of career choice in this study is to choose career variables with significance that can form the changing trends of different quantiles.

4.2. Housing Added-Value Model

The housing added-value model adopts the same independent variables and methodology as the housing value model. The basic condition of the model and control variables is presented in Table 7.

Table 7. Basic condition of housing added-value model and control variables.

Model Type/Quantile	OLS	0.1	0.25	0.5	0.75	0.9
Age	−0.00913	−0.013	−0.0041	−0.018	0.00834	−0.00282
Square of age	0.000226	0.000309	0.000216	0.000305 !	0.000017	9.27×10^{-5}
Political status	−0.0973	−0.0939	−0.155 *	−0.0935	−0.0736	0.0206
Marriage status	−0.0108	−0.00724	−0.0281	−0.0275	0.0245	0.0134
Family member number	0.0551 **	0.0415	0.0640 **	0.0467 *	0.0536 **	0.0615 *
Gender	−0.0787 *	−0.195 *	−0.112 !	−0.0898 *	−0.0268	0.0177
Province Reference Group: Shanghai Municipality						
Yunnan Province	−1.371 ***	−1.970 ***	−1.526 **	−1.273 ***	−1.212 ***	−1.088 **
Jilin Province	−1.751 ***	−1.854 ***	−1.777 ***	−1.693 ***	−1.697 ***	−1.629 ***
Guangdong Province	−0.648 *	−0.945 **	−0.906 **	−0.766 !	−0.461	−0.28
Henan Province	−1.279 ***	−1.203 ***	−1.241 ***	−1.315 ***	−1.235 ***	−1.357 ***
Gansu Province	−1.568 ***	−1.665 ***	−1.674 ***	−1.492 ***	−1.514 ***	−1.572 ***
Logarithms taken of income	0.229 ***	0.307 ***	0.260 ***	0.264 ***	0.239 ***	0.128 **
Constant term	1.713 **	−0.0158	0.822	1.848 **	1.851 **	3.495 ***
Number of cases	1.560	1.560	1.560	1.560	1.560	1.560
R^2	0.533	0.3002	0.3242	0.3623	0.3859	0.3728

*** $p < 0.001$, ** $p < 0.01$, * $p < 0.05$, ! $p < 0.1$.

The coefficient of determination in the OLS model is 0.533, which indicates that the model fits the data well. The model of control variables corresponds to the housing value model.

Regarding time factor, the housing added-value model is different compared with the housing value model. The housing condition from 2004 to 2008 is not significant in Table 8.

As houses before 1998 were relatively cheap and held for a longer period, new houses after 2009 had a lower added value than those before 1998. However, houses between 1999 and 2003 had a higher added value than houses before 1998. Therefore, Hypothesis 4b has been proven to be partly false.

Table 8. Time factor of housing added-value model.

Model Type/Quantile	OLS	0.1	0.25	0.5	0.75	0.9
The Time of Purchase or Construction Reference Group: Before 1998						
1999–2003	0.195 **	0.320 *	0.218 **	0.165 *	0.102 !	0.123 !
2004–2008	−0.00078	0.155	−0.0199	−0.0915	−0.082	−0.0658
After 2009	−0.478 **	−0.772 **	−0.711 ***	−0.476 ***	−0.376 **	−0.424 ***

*** $p < 0.001$, ** $p < 0.01$, * $p < 0.05$, ! $p < 0.1$.

From the distribution of quantile regression coefficients in Figure 5, small differences are observed in high-return housing after 2009 compared with high-return housing of 1998. This result indicates that low-price housing after 2009 has not fully appreciated and lags behind housing held for a longer time. Another discovery is that housing between 1999 and 2003 had a higher added value than housing before 1998. Taking the historical background into account, 1999 was the starting point of the commodification reform of housing, and premium housing resources that had been prevented from transaction were released into the redistribution system. Therefore, the purchase or construction of property at this time would result in a first-mover advantage. These have become the most value-added properties, even if they are not those selling at the highest price. Such an advantage is most remarkable at quantile levels from 0.05 to 0.15, indicating that, in low-return housing, the added value of purchase or construction from 1999 to 2003 is 23.7% to 61.4% higher than that of purchase or construction before 1998. Due to the commodification of housing, a number of people with low added-value housing have acquired the trading right of housing without much cost, allowing them to gain substantial added-value advantages. Additionally, such an advantage results in 20% more added value in other quantiles and rarely decreases as quantiles increase.

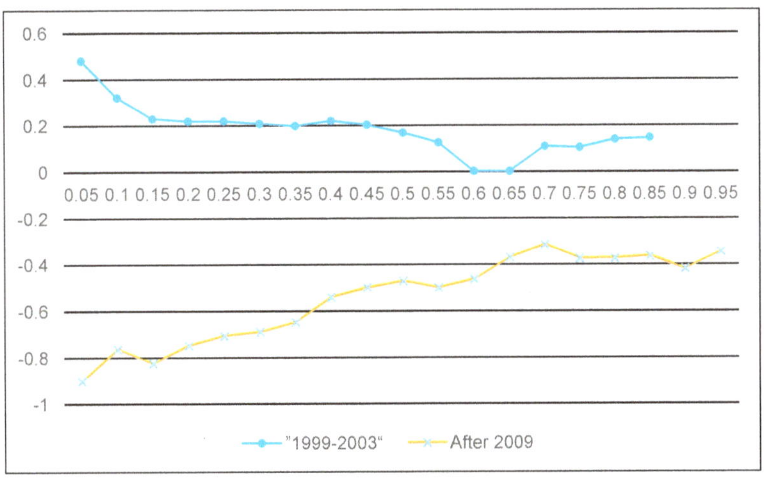

Figure 5. Quantile regression coefficient line: time factor of housing added-value model.

In terms of work unit type, from the overall OLS model in Table 9, collective enterprises have the highest value-added; this is similar to the total housing value model, which followed party and government organs and public institutions. However, the p-value of party and government organs is between 0.05 and 0.1, which is not steady. Regarding the level of added value, the lower the marketization level, the higher the added value. This phenomenon applies to all types of work units except for collective enterprises. In other words, the work units that do not have a fully developed marketization similar to party and government organs and public institutions tend to have higher added value. After considering the exception of collective enterprises, Hypothesis 3b is proven to be partly true.

Table 9. Organizational factor of housing added-value model.

Model Type/Quantile	OLS	0.1	0.25	0.5	0.75	0.9
Work Unit Type Reference Group: Collective Enterprises						
Party and government organs	−0.275 !	−0.0186	−0.204	−0.251	−0.255 *	−0.293 *
Public institutions	−0.324 *	−0.266	−0.329 !	−0.307 *	−0.244 !	−0.212
State-owned enterprises	−0.393 **	−0.521 *	−0.382 *	−0.358 **	−0.303 **	−0.257 !
Private enterprises	−0.409 **	−0.576 *	−0.434 **	−0.365 **	−0.325 **	−0.269 !
Self-employed individuals	−0.446 *	−0.372	−0.433 !	−0.526 *	−0.470 *	−0.423

*** $p < 0.001$, ** $p < 0.01$, * $p < 0.05$, ! $p < 0.1$.

From the changing pattern of quantile regression coefficients in Figure 6, a similar pattern can be observed. The group of self-employed individuals is the work unit type with the highest level of marketization and the lowest added value. In quantiles after 0.45, the middle- and high-return housing with high added value do not fit into the group of self-employed individuals. Although there is a breakpoint in coefficients of party and government organs, these factors remain advantageous in low- and middle-return housing on quantile levels from 0.35 to 0.45. The income of non-corporate organizations, such as party and government agencies and public institutions from real-estate appreciation, is between 0.25 and 0.85 points, which is greater than private and state-owned enterprises. In general, in housing quantiles with higher returns, the difference between non-enterprise organizations and collective enterprises tends to be smaller.

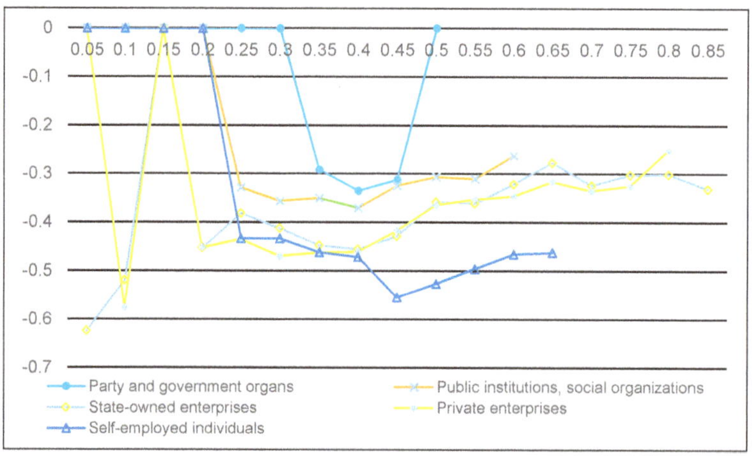

Figure 6. Quantile regression coefficient line: organizational factor of housing added-value model.

Regarding the human capital factor, from the OLS model in Table 10, the higher the education level, the higher the income and the higher the housing added value. However, the advantage of a college education is less remarkable than it is in the housing value model. The housing added value of high-school education and college education is 68.3% and 58.4% higher, respectively, than elementary school education and below.

Table 10. Human capital factor of housing added-value model.

Model Type/Quantile	OLS	0.1	0.25	0.5	0.75	0.9
Education Level Reference Group: Elementary School and Below						
Secondary school	0.291 *	0.215	0.270 !	0.189	0.327 **	0.370 **
High school	0.460 ***	0.596 **	0.562 **	0.302 *	0.298 **	0.342 **
University and above	0.521 ***	0.470 *	0.604 **	0.436 **	0.441 ***	0.493 ***
Senior cadre	0.301 *	0.482 !	0.303 *	0.243 *	0.158 !	0.117
Junior cadre	0.148	0.248	0.194	0.109	0.048	−0.0077
Logarithms taken of income	0.229 ***	0.307 ***	0.260 ***	0.264 ***	0.239 ***	0.128 **

*** $p < 0.001$, ** $p < 0.01$, * $p < 0.05$, ! $p < 0.1$.

In the quantile regression model in Figure 7, the added value of college education is higher than that of other education levels at high quantile levels from 0.6 to 0.95. At low-quantile levels from 0.05 to 0.35, the difference between college education and high-school education is not substantial. The added value of high-school education is even higher than that of college education, which probably reflects that the human capital factor plays an insignificant role in low-return housing. Therefore, high-school education can also result in high returns on low-return housing. However, college education is a must to gain returns on high-return housing. With the increase in quantiles, the influence of income on housing added value decreases, which is the same as in the housing price model.

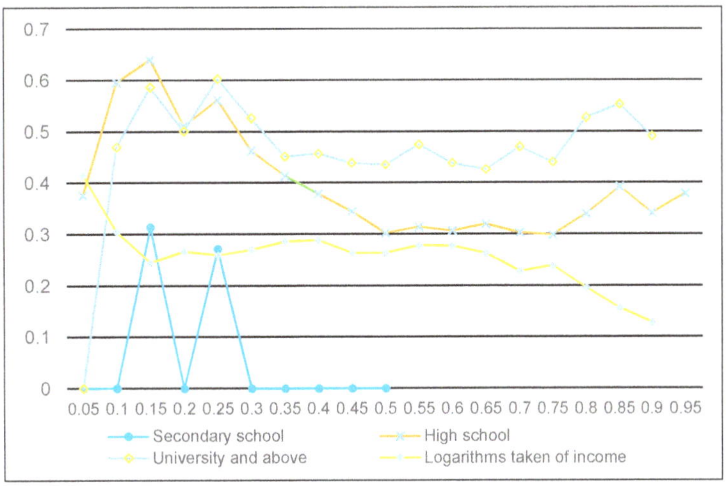

Figure 7. Quantile regression coefficient line: human capital factor of housing added-value model.

In Table 11, any variables in occupation and political capital factor are not significant, but the returns of organization clerks still rank the highest in terms of housing added value. The difference between the occupation of section chief and above and the occupation of organization clerk is smaller than that in the housing price model. The housing added value of section chief and above and organization clerk is, respectively, 35.1% and 44.9% higher than that of an ordinary worker.

Table 11. Political capital factor of housing added-value model.

Model Type/Quantile	OLS	0.1	0.25	0.5	0.75	0.9
Occupation Reference Group: Ordinary Worker						
Senior management	0.255	0.339	0.269	0.101	0.198	0.33
Junior management	0.214	0.412	0.270 !	0.11	0.0392	0.0763
Intermediate and senior technician	0.176 !	0.452 *	0.0507	0.178	0.103	−0.0691
Ordinary technician	0.0294	0.119	−0.0687	0.083	0.0638	−0.0609
Organization clerk	0.372 **	0.604 **	0.368 **	0.265 *	0.278 *	0.256 *
Salesman of enterprises and institutions	0.138	0.28	0.0814	0.133	0.0808	−0.0676
Business service personnel	0.157 **	0.363 !	0.224 *	0.118 !	0.0847	−0.0145
Skilled worker	0.126	0.351	−0.162	0.0261	0.0939	0.0748
Self-employed entrepreneur	0.214	0.341	0.321 *	0.237	0.119	0.122
Others	0.195	1.059 **	0.48	0.329	0.152	−0.118
Party membership	−0.0973	−0.0939	−0.155 *	−0.0935	−0.0736	0.0206

*** $p < 0.001$, ** $p < 0.01$, * $p < 0.05$, ! $p < 0.1$.

In the quantile model in Figure 8, the *p*-value of party membership is partly negative at the low and middle quantile levels, which is similar to the housing value model. Therefore, Hypothesis 2b is proven to be partly false. The difference in the returns among various occupations is not large and decreases with the increase in quantiles. However, in the housing price model, the regression coefficients of various occupations in different housing prices are more or less steady. This result indicates that a negative correlation between the returns from the high-return property market and occupation, which is especially remarkable at quantile levels from 0.1 to 0.25.

Figure 8. Quantile regression coefficient line: political capital factor of housing added-value model.

5. Conclusions and Discussion

5.1. Conclusions

The research results show that human capital exerts significant influence. Individuals with a high education level are more likely to acquire high-value housing, especially individuals with a college education. Additionally, individuals with a high education level are more likely to profit from the added value of housing. However, a small difference was observed between individuals with a college education and individuals with a high-school education regarding low added-value housing. Individuals with low-quality and low-return housing profit more from factors such as education and income. Therefore, improving education level remains a reliable means to reduce housing inequality, especially for individuals comprising the lower classes. We found that political capital exerts less influence. When a cadre position is considered as a type of occupation, it does not result in the greatest returns, and the returns of an organization clerk outweigh those of cadres with a title of section chief or above. Party membership may exert influence through other variables. When other variables are under control, party membership even plays a negative role under some circumstances. The influence of political capital on the distribution of housing is not as strong as it was before. In the property market, the quantiles of state-owned enterprises are not as significant as those of private enterprises, and this indicates that traditional work organizations have borne the brunt of market transition. Although collective enterprises rank the highest in returns on the property market, non-enterprise organizations are better than enterprises in terms of housing added value. This result shows that the redistribution system of the work unit era has a substantial effect on current housing inequalities, and patience is required as the marketization process plays out. The later an individual acquires housing, the higher the housing value; however, this phenomenon does not mean that the longer an individual holds the property the higher the added value. The housing added-value returns of individuals who entered the property market from 1999 to 2003 are not only greater than individuals entering after 2009, but also greater than individuals owning a house before 1998. This is why first-mover advantage is critical in the early days of market transition.

5.2. Discussion

From its genesis, China's market reform has been guided and intervened in by the government. In contrast to the "shock therapy" of the drastic changes in Eastern Europe, China's gradual reforms are carried out under the action of improving the socialist system and exerting the institutionalization of socialism. This is the continuity of the political system and reality [37] (Liu 2003). The current social and economic state of China is not a completely self-disciplined market economy system, and all irrational interventions, including interventions of power and privileges, have not all been driven out of the market system [38,39].

On one hand, in the field of market economy, emerging economic elites have gradually emerged whose privileges are based on asset ownership. On the other hand, cadres, or at least some of them, have also learned how to use the market and successfully transfer their own bureaucrats. Privileges are therefore "commercialized" [40] (Szelenyi, 1987). This dual stratification system is particularly obvious in the housing field. On the one hand, according to the logic of market operation, housing resources are differentiated based on an individual's financial ability, and the market provides more incentives for direct producers. Market incentive mechanisms will reflect the return of human capital indicators, such as education, skill level, and housing resources. On the other hand, the powerful elite in the original redistribution system not only use the "privatization of public housing" reform process to share advantages under the planned economic system, as housing resources are legally "privatized" and "commercialized", but they also seize favorable political opportunities in the housing reform process or seek housing benefits through administrative power to enjoy market privileges.

While market transition theory and power continuance theory have a common assumption that China is undergoing a long process of marketization, the common concern of the two theories is how new resources resulting from the market transition are used and divided. Market elites gain resources fairly with better human capital. Powerful elites gain resources by transforming previous power advantages into market ability and status. The possible problem of such an assumption is that the speed of resources emerging from the market transition is misinterpreted into a steady speed, and a fight over resources between market elites and powerful elites is based on an unchanging number of resources.

However, when the time factor is included in this research, the speed of resources emerging from the market transition is unsteady. In the early days of market transition, such resources tend to emerge in a substantial number. Housing acquired between 1999 and 2003 has the highest added value in the housing market. This result means that the question of which party occupies the new resources first is not as critical as the question of which party first gains access to the new resources. An example of this is housing inequality. Once an individual acquires premium housing early, she or he has the capability to increase added value in the long term and profits from renting the house, deepening the stratification and forming a greater Matthew Effect. The answer to this can be found in the fundamental debate of market transition: which party can gain access to the new resources the earliest?

Based on the advantage of collective enterprises and organization clerks on housing value and housing added value, this paper proposes a third approach to answering the above question. Greater human capital and political capital do not necessarily result in more market resources. However, the work units between the market and redistribution system, such as collective enterprises, and occupations, such as organization clerk, which can directly participate in the housing redistribution process deliver the highest housing returns. Although collective enterprise did not have substantial advantages in the early days of market transition, collective enterprise is likely to profit from its vague identity between the market and redistribution system in the long process of reform. Organization clerks can profit because of their early and direct contact with the distribution of housing. These two show the competitive advantage outside the binary contrast between the market and redistribution. Regardless of which main resource distribution system is chosen, it is possible to bypass or avoid these two and then gain specific advantages. A prediction for this process is that, if such a possibility were not managed properly, a troubling group that could neither be affected nor directly influenced by marketization might emerge, further compounding the issue of housing inequality in the long term.

Author Contributions: Formal analysis, J.Z.; investigation, J.Z.; data curation, J.Z. and J.X.; writing—original draft preparation, J.Z.; writing—review and editing, J.Z. and J.X. All authors have read and agreed to the published version of the manuscript.

Funding: This research has received funding from National Social Science Fund of China (Grant No. 18BGL257), the Shanghai Philosophy Society Planning Project (Grant No. 2018EJB011) and the Chenguang Scholar Program in Shanghai (Grant No. 19CG70).

Conflicts of Interest: The authors declare no conflict of interest.

References

1. Wang, F.; Ma, L. *Boundaries and Categories: Rising Inequality in Post-Socialist Chin*; Zhejiang People's Publishing House: Hangzhou, China, 2013.
2. Reuters Beijing. The First Time the Bureau of Statistics Released Gini Coefficiet in China. Available online: http://cn.reuters.com/article/cn-gini-population-reform-idCNCNE90H04A20130118 (accessed on 25 October 2021).
3. Davis, D.S. Urban Chinese homeowners as citizen-consumers. In *The Ambivalent Consumer edited by Sheldon Garon and Patricia Maclachlan*; Cornell University Press: Ithaca, NY, USA, 2006; pp. 281–299.
4. Davis, D. Urban consumer culture. *China Q.* **2005**, *183*, 692–709. [CrossRef]
5. Gan, L. *China Household Finance Survey Report*; Southwestern University of Finance and Economics Press: Chengdu, China, 2012.
6. Merton, R.K. The matthew effect in science, ii: Cumulative advantage and the symbolism of intellectual property. *Isis* **1988**, *79*, 606–623. [CrossRef]
7. Walder, A.G. Property rights and stratification in socialist redistributive economies. *Am. Sociol. Rev.* **1992**, *57*, 524–539. [CrossRef]
8. Parish, W.L.; Ethan, M. Politics and markets: Dual transformations. *Am. J. Sociol.* **1996**, *101*, 1042–1059. [CrossRef]
9. Gerber, T.P.; Hout, M. More shock than therapy: Market transition, employment, and income in russia, 1991–1995. *Am. J. Sociol.* **1998**, *104*, 1–50. [CrossRef]
10. Bian, Y.; Logan, J.R. Market transition and the persistence of power: The changing stratification system in urban china. *Am. Sociol. Rev.* **1996**, *61*, 739–758. [CrossRef]
11. Zhou, X.G. Economic transformation and income inequality in urban china: Evidence from panel data. *Am. J. Sociol.* **2000**, *105*, 1135–1174. [CrossRef]
12. Bian, Y.J.; Zhang, Z.X. Marketization and income distribution in urban china, 1988 and 1995. *Res. Soc. Strat. Mobil.* **2002**, *19*, 377–415. [CrossRef]

13. Walder, A.G. *Communist Neo-Traditionalism*; University of California Press: Auckland, CA, USA, 1988.
14. Bian, Y.; Liu, Y. Social stratification, home ownership, and quality of living: Evidence from china's fifth census. *Sociol. Res.* **2005**, *3*, 82–98.
15. Liu, X. Market transition and social stratification: Theoretial debates and further research issues. *Soc. Sci. China* **2003**, *5*, 102–110.
16. Liu, Z.; Hu, R. Urban housing stratification: An analysis based on cgss2006 data. *Chin. J. Sociol.* **2010**, *30*, 164–192.
17. Nee, V. Social inequalities in reforming state socialism: Between redistribution and markets in china. *Am. Sociol. Rev.* **1991**, *56*, 267–282. [CrossRef]
18. Nee, V. The emergence of a market society: Changing mechanisms of stratification in china. *Am. J. Sociol.* **1996**, *101*, 908–949. [CrossRef]
19. Szelenyi, I. Social inequalities in state socialist redistributive economies. *Int. J. Comp. Sociol.* **1978**, *19*, 63–87. [CrossRef]
20. Chen, Z. Market transition or power conversion: The wealth distribution and the regional disparity of urban residents in mainland china. In *Market, Classes and Politics: Chinese Society during Transition*; Liu, Z., Ed.; Hong Kong Institute of Asia-Pacific Studies, The Chinese University of Hong Kong: Hongkong, China, 2000.
21. Walder, A.G. Local governments as industrial firms: An organizational analysis of china's transitional economy. *Am. J. Sociol.* **1995**, *101*, 263–301. [CrossRef]
22. Liu, X. Housing inequality in urban china. *Sociol. Forum Fudan Univ.* **2005**, *1*, 149–171.
23. Xie, Y.; Hannum, E. Regional variation in earnings inequality in reform-era urban china. *Am. J. Sociol.* **1996**, *101*, 950–992. [CrossRef]
24. Liu, Z.Q. The economic impact and determinants of investment in human and political capital in china. *Econ. Dev. Cult. Change* **2003**, *51*, 823–849. [CrossRef]
25. Song, S.G. The delayed effect of power conversion-an explanation of the regeneration and circulation of elites in the process of transforming socialist countries to markets. *Sociol. Res.* **1998**, 3, 26–36.
26. Li, B.; Walder, A.G. Career advancement as party patronage: Sponsored mobility into the chinese administrative elite, 1949–1996. *Am. J. Sociol.* **2001**, *106*, 1371–1408. [CrossRef]
27. Fan, G.; Wang, X. Marketization index for China's provinces. *Econ. Res.* **2003**, 3, 9–18.
28. Li, S.; Wei, Z.; Ding, S. Empirical analysis on the inequality and the reason for china's residents' property distribution. *Econ. Res. J.* **2005**, *6*, 4–15.
29. Zhang, J. *Study on the Housing Inequality in China*; Dongbei University of Finance and Economics: Dalian, China, 2013.
30. Hu, R. Housing inequality during market transition: Analysis of housing value and housing quantity. *J. Gansu Admin. Ins.* **2014**, *6*, 108–116.
31. Li, B. *Housing Policies in Differentiatio*; Social Sciences Academic Press: Beijing, China, 2009.
32. Luo, C.L. Housing area inequality in urban residents: An analysis based on 2000 and 2005 census in chin. *Acad. Bras.* **2014**, *1*, 80–90.
33. Hu, R. Housing inequality during the market transition: Evidence from the data of cgss2006. *Soc. Chin. J. Sociol. Shehui* **2012**, *32*, 126–151.
34. Wu, X.B. *Gains and Losses of Economic Reform in the Course of History*; Zhejiang University Press: Hangzhou, China, 2013.
35. Zhou, F.Z. *Fiscal Relations and Local Governance*; Shanghai SDX Joint Publishing Company: Shanghai, China, 2012.
36. Xie, Y.; Wu, X. Danwei profitability and earnings inequality in urban china. *China Q.* **2008**, *195*, 558–581. [CrossRef]
37. Liu, X. Market transformation and social stratification: The focus of theoretical debate and problems to be studied. *Chin. Soc. Sci.* **2003**, 102–110.
38. Giddens, A. *The Class Structure of the Advanced Society*; Hutchinson: London, UK, 1973.
39. Goldthorpe, J. *Social Mobility and Class Structure in Modern Britain*; Clarendon: Oxford Shire, UK, 1987.
40. Szelnyi, I.; Manchin, R. Social policy under state socialism: Market, redistribution, and social inequalities in East European socialist societies. In *Stagnation and Renewal in Social Policy*; Rein, M., Esping-Anderson, G., Rainwater, L., Eds.; M.E. Sharpe: White Plains, NY, USA, 1986; pp. 102–139.

land

MDPI

Article

Land Tenure Disputes and Resolution Mechanisms: Evidence from Peri-Urban and Nearby Rural Kebeles of Debre Markos Town, Ethiopia

Sayeh Kassaw Agegnehu [1,*], Tilahun Dires [1], Worku Nega [1] and Reinfried Mansberger [2]

[1] Institute of Land Administration, Debre Markos University, Debre Markos 269, Ethiopia; tilahun_dires@dmu.edu.et (T.D.); worku_nega@dmu.edu.et (W.N.)

[2] Institute of Geomatics, University of Natural Resources and Life Sciences, A-1190 Vienna, Austria; mansberger@boku.ac.at

* Correspondence: sayeh_kassaw@dmu.edu.et

Abstract: In Ethiopia, like in other developing countries, land disputes are critical problems both in peri-urban and rural areas. Handling such disputes requires scientific and evidence-based interventions. This study analyzes the nature, types, and causes of land tenure disputes and the resolution mechanisms thereof in peri-urban and nearby rural kebeles of Debre Markos town. Interviews for the investigation were conducted with sample landholders and concerned legal experts in Debre Markos town's peri-urban area and Gozamin Wereda of Amhara National Regional State in Ethiopia. Compared to rural areas, the incidence of land tenure disputes is high in peri-urban areas. The land tenure disputes identified in the study areas are boundary trespassing disputes, landholding disputes, land rental disagreements, divorce-related land disputes, bequeath disputes, parcel exchange disputes, and land use-related disputes. The land tenure disputes are resolved mainly by formal means such as court litigations and administrative decisions, or by informal means known as alternative dispute resolution mechanisms (ADRMs). In both study areas, negotiation, mediation/conciliation, and arbitration are the most frequently employed ADRMs. In particular, mediation plays a significant role in resolving symmetrical land tenure disputes both in peri-urban and rural areas.

Keywords: dispute resolution; land dispute; land tenure; peri-urban

Citation: Agegnehu, S.K.; Dires, T.; Nega, W.; Mansberger, R. Land Tenure Disputes and Resolution Mechanisms: Evidence from Peri-Urban and Nearby Rural Kebeles of Debre Markos Town, Ethiopia. *Land* **2021**, *10*, 1071. https://doi.org/10.3390/land10101071

Academic Editors: Uchendu Eugene Chigbu, Ruishan Chen and Chao Ye

Received: 3 September 2021
Accepted: 8 October 2021
Published: 11 October 2021

Publisher's Note: MDPI stays neutral with regard to jurisdictional claims in published maps and institutional affiliations.

1. Introduction

Dispute, as defined by sociologists, is a social fact, which involves at least two parties with differences either in interests or in social position [1,2]. Disputes can arise either from actual or perceived competition of interests for resources such as land [3] and can be symmetric or asymmetric according to the power balances between the conflicting parties. Symmetric conflicts are conflicts between relatively similar parties with respect to power, whereas asymmetric conflicts are conflicts between dissimilar parties [4]. Land dispute can be defined as a social fact involving at least two parties in an actual or perceived competition of interests over the property rights to land [5–7]. Conflicts of interests may be "the right to use the land, to manage the land, to generate an income from the land, to exclude others from the land, to transfer the land and the right to get compensation from it" [6,8–10]. Therefore, a land dispute can be understood as misuse, restriction, or dispute over property rights to land [10,11].

Land tenure disputes may occur in rural, peri-urban, and urban areas [12]. Peri-urban areas are those surrounding cities and towns with high rates of land tenure transformation, and often with multiple agents exhibiting various disputing interests [11–13]. Peri-urban areas are focal regions for municipalities, since a considerable percentage of land needed for urban expansion comes from the rural–urban interface [7,14–16]. As such, peri-urban areas have received attention from governmental institutions and industries due to their proximity to the urban centers [15–17]. Nevertheless, peri-urban areas are also primary settings for small-scale agriculturalists who depend on these areas for their livelihood, often on a small-scale or subsistence level [15,18,19]. Given these circumstances, land disputes are typical in peri-urban areas, especially in countries with high rates of urban expansion [10].

Ethiopia is a country exhibiting a high rate of horizontal urban expansion, with 4.1% per annum on average [14,20,21]. Thus, the competition for land between agricultural and non-agricultural sectors is becoming very intense [14]. In Ethiopia, land is owned by the people and the government. Farmers have a constitutional right to use the land for undetermined periods, and they have right of protection against eviction from their landholdings [22,23]. Land required for urban expansion is expropriated by the government upon the payment of compensation to the affected farmers and is transferred to developers. Thus, the government is the sole provider of land for public purposes. Therefore, in peri-urban areas of Ethiopia, at least three parties are competing for the land. These are the government, which is the provider of the land, the private individual/company who needs the land, and the peri-urban landholders who are losing their land rights due to expropriation as a consequence of urban expansion [14]. Land dispute may occur between these three parties or within the same party [14,24,25]. For example, there are land disputes between the rural land administration organization and urban land administration organization, as the two institutions act on behalf of two different land legislations (rural land proclamation and urban land proclamation). Various studies show that land dispute is prevalent in both rural and peri-urban areas of Ethiopia even though there are some differences in the extent and cause [11,26–28]. It is challenging to effectively manage peri-urban zones since different stakeholders are involved with competing interests [29]. However, there are different resolving mechanisms in different jurisdictions which are intended to mitigate such disputes. All the mechanisms can be categorized into formal and informal ways of settling land disputes [1,2,30,31]. Formal dispute resolution mechanisms are state-based dispute resolution mechanisms which are often referred to as judicial dispute resolution mechanisms [21,32–35]. In contrast, there are the informal ways of resolving land tenure disputes, also known as alternative dispute resolution (ADR) mechanisms, which propose approaches to resolve disputes without ordinary court proceedings [36]; these include negotiation, facilitation, mediation, conciliation, arbitration, community conferencing, fact-finding, and so on [24].

Although many scientific articles examine the impact of title on land tenure security, there is a lack of independent research on land tenure disputes and on the role of informal and formal conflict resolution mechanisms for land tenure disputes in peri-urban and rural areas. Therefore, the assessment of land tenure disputes and dispute resolution systems in peri-urban areas vis-à-vis rural areas in Ethiopia is an increasingly important issue that needs to be addressed when designing appropriate policy interventions to enable the sustainable development of urban areas while considering the property rights of peri-urban subsistence farmers. The purpose of this research paper is to assess the nature, types, and causes of land tenure disputes as well as dispute resolution mechanisms in the peri-urban area of Debre Markos town and the neighboring rural kebeles[1] in Gozamin Wereda[2] of Amhara National Regional State in Ethiopia. In this paper, Section 2 provides a brief overview of the literature related to the topic. Section 3 documents the study areas and the research methodology applied. In Section 4, the results of the study are presented and then are discussed under Section 5. Finally, Section 6 contains the conclusion and recommendations.

2. Land Disputes and Mechanisms to Resolve Land Disputes

Land is the basis of all forms of human activity. From it we find everything important for life, i.e., the food we eat, the shelter we need, the space to work, and the room to relax [10,37–41]. The accessibility to land is vital for human life. The need for thoughtful and careful stewardship of the land together with the sustainable use of its resources is crucial both for the present and future generations. As a result, disputes over land-related issues are prevalent worldwide [2,10]. Land disputes have been a major source of disturbance and civil wars in many parts of the world. Anseeuw et al. [42] documented 71 civil wars and insurgencies in agrarian world states, from which more than 84% were caused by land-related issues. Therefore, land disputes have always been an integral part of all human societies. The instigation of land disputes is closely linked with the competition amongst people for controlling scarce land resources available for consumption [43,44].

Land disputes can take different forms [38,45,46]. In some land disputes, there are only two parties and hence they are relatively easy to resolve. Inheritance disputes between siblings over a particular piece of land and boundary trespassing disputes are the most common two-party dispute types [10]. Land disputes become more complex and difficult to resolve where more parties are involved. Group invasions or evictions of entire settlements are common examples of these types of land disputes [10,47]. Land disputes in rural areas are often found between different interest groups, e.g., between farmers and investors and/or the state, as well as between farmers themselves [48].

In peri-urban areas, there are numerous simple to complex construction works starting from heavy industries to small legal and illegal residential houses [7,20]. In all such situations, the rural land has been transformed to urban land use types [49]. Changes in these areas, caused mainly by urban expansion, make land one of the most controversial issues and the main source of disputes. One of the most important issues is the competition for land for various purposes of urban development, which may lead to changes in land tenure and use [10]. When land is converted from rural to urban use without designing alternative business strategies for the peri-urban subsistence farmers [50], it leads to disputes, contestations, and in some cases to violence. The peri-urban environment may include parts of urban areas or the edges of urban areas, and areas far from the city. The region may also include urban and rural land which is occupied formally or informally [51].

Land tenure disputes in peri-urban areas can be clustered into three dimensions: "dispute of interest", "dispute of power", and "legal and normative dispute" related to the interests of land tenure, power, and domination of beneficiaries under contradictory norms and rules [6]. The fundamental dimension affecting land tenure disputes in peri-urban areas is the dispute of interest, which includes revealed and hidden disputes between individuals, groups, and institutions related to the benefits resulting from land rights, people's relationships with the land, and the mechanisms of the institutions that affect it. It involves a wide range of types of disagreements and inconsistencies that can lead to disputes, violence, or antagonism.

The interests and motives associated with land are so widespread that sometimes disputants are confronted with one another, especially when the two sides are in the same position with respect to different interests [10,33]. These interests can result from human aspirations and motivations. Ignoring the needs of individuals due to limited resources, social status, and power or value systems leads to the hostile and conflicting behaviors of the involved actors. Land disputes have negative effects on individual households as well as on the nation's economy. They increase costs, slow down investments, and may result in the loss of the property of the disputants. Land disputes also increase social and political instability. People lose confidence in the state and distrust each other.

Although land issues are amongst the most prominent causes and driving factors for the outbreak of armed conflicts, there is a lack of adequate attention to address those issues with appropriate approaches and strategies in a timely manner [13,30,48]. Presently, due to growing competition over diminishing land resources, many developing states have found the resolution and management of land disputes to be some of their most critical challenges. This situation is "being further aggravated by environmental degradation, population growth and climate change" [52]. According to Wehrmann [10], the most important prerequisites for resolving land disputes are a comprehensive understanding of the causes of disputes as well as the positions, needs, and interests of the disputing parties. In addition, other factors also play pivotal roles in the successful resolution of land disputes, such as understanding the types of land disputes, the identity of the parties involved in a particular land dispute and their perceptions on how to resolve it, the complexity of the causes of the land dispute, and the driving factors that escalate the dispute [10].

The experiences of some countries such as South Africa, Zimbabwe, and many Latin American countries suggest that the leadership, good land policies, and the quality of land institutions and land governance are important factors to prevent violent disputes or to amicably resolve disputes. As an example from another region, in Norway, there are a large number of land boundary disputes compared to other Nordic countries [53]. However, Norwegian farmers use mediation as the first and best mechanism to resolve these land disputes. If unable to resolve the dispute via mediation, the next step is to handle the case by means of the "land consolidation court". Thus, there are bundle of formal and informal mechanisms for resolving land-related disputes. The formal mechanisms of solving land-related disputes follow official procedures guided by government rules, regulations, and laws. They can be administrative and judicial. The administrative methods are applied by semi-judicial organizations, such as government resource offices, police, and local government organs. Judicial mechanisms to solve land disputes are carried out by courts.

Informal procedures comprise the so-called alternative dispute resolution (ADR) mechanisms. ADRs refer to the procedure of setting land disputes by means other than litigation. ADR mechanisms normally accelerate the solving of the dispute and prevent future disputes. Therefore, informal mechanisms help to reduce the costs of dispute processing. However, to amicably resolve disputes by using ADR mechanisms, disputants must be willing to participate and they must all feel that resolving disputes by means of ADRM is more valuable than by court proceedings [33]. In addition, for the successful resolution of land disputes, different stakeholders who have concerns with the land have to collaborate rather than focusing solely on their positions [54].

The development of dispute resolution mechanisms is characterized as pragmatic and political rather than theoretical and scientific [34]. In the United States of America, in the mid-twentieth century, the legal and academic communities began to be seriously concerned about the pitfalls of increasing litigation because, although the laws of the day granted a wide range of rights and personal protections, seeking remedies for these rights while they were being violated by the legal system became a complex exercise [55].

There are many potential ways to resolve a dispute, ranging from the formality of legal proceedings to physical violence [34,56]. The law deals with all these means, but not all means of dispute resolution are "legal" in form or acceptability. Litigation as used in many areas of traditional law is too costly, creates divisions, is inaccessible or inefficient, and requires long hours spent in court [57]. As a result, alternatives to litigation are often so regulated or perverted by litigation-oriented lawyers, courts, and lawmakers that they become alternative methods of litigation rather than alternatives to litigation. Instead of the alleged gulf between 'legal' and 'non-legal' dispute resolution methods, a unified dispute settlement theory that establishes the types of disputes in which each alternative dispute resolution method is most effective is essential [58]. One of the tasks of a unified dispute settlement theory is to integrate the various alternative dispute resolution methods into a coherent framework. Sometimes physical violence is an appropriate way to resolve a dispute, other times it is not; furthermore, sometimes a dispute can be resolved by negotiation between the involved parties only, other times it cannot. Sometimes a dispute can be resolved with what Felsteiner calls avoidance, which means the parties sever the relationship. Sometimes one party forms an alliance with a third party. Therefore, this paper is framed by the above concepts of dispute settlement mechanisms [59].

Land disputes arise in all property regimes, though the extent differs. There are four types of property regimes: private property, state property, communal property, and open access property [60]. Ostrom has described what each one means. Private property is a privately owned property regime which relies on the availability and enforcement of rules describing the control, use, and exclusion rights of the landowner, whereas communal property is a property regime owned by groups and its use and appropriation depend on the rules invented by communal property users. On the other hand, a state property regime is a property type owned and controlled by the state mostly in the form of national reserves and parks. Open access is a property regime type when the property is open to all, and no one has the exclusive right to forbid others. In an open access property regime, users try to maximize their private interest at the expense of others.

3. Research Methodology

3.1. Description of the Study Area

The research was conducted in the peri-urban area of Debre Markos town, the capital of East Gojjam. East Gojjam is one of the 11 administrative zones in Amhara National Regional State (ANRS) of Ethiopia (see Figure 1). Debre Markos currently demonstrates a high rate of urban expansion [49].

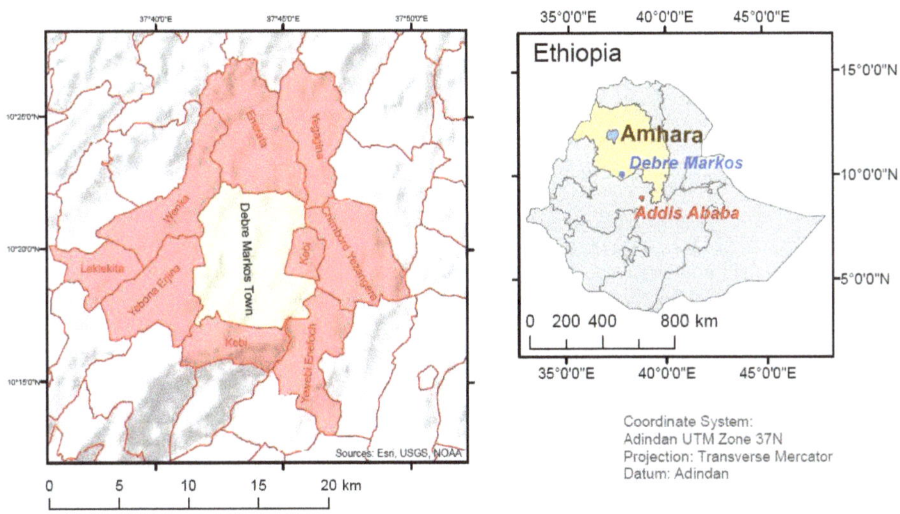

Figure 1. Map of the study areas.

The study areas comprise the peri-urban kebeles of Debre Markos town within the range of 5 km. The land use pattern of the study areas is dominated by a heterogeneous mosaic of agricultural, forest, meadow, and residential land. Due to urban expansion, the municipality has begun to expropriate and transform agricultural land in the peri-urban areas at a rapid pace. The main reasons for selecting this area for the investigations were the rapid urbanization and the current high rate of land tenure transformation.

In addition, three rural kebeles, Chimbord Yezangera, Leklekita, and Kebi, were selected for the investigation in order to compare land tenure disputes in both peri-urban and rural scenarios. The kebeles are close to the peri-urban fringe with an expected changing socio-demographic situation. Initially, the intention was to investigate land tenure disputes during the previous 10 years, but it was difficult for some respondents to remember details of disputes from a decade earlier. Thus, the potential cases were limited to those within the past five years.

3.2. Methods of Data Collection

In this study, both quantitative and qualitative data were collected from primary and secondary sources. The quantitative data were obtained from respondents whereas qualitative data were collected from focus group discussion, legal documents, and key informant interviews.

A preliminary study was carried out in the sample area with the objective of obtaining a general overview of the peri-urban areas and input data for the selection of respondents. Discussions with relevant offices were conducted to obtain basic information and retrieve secondary data. The lists of all landholders in the study's peri-urban areas were recorded, and the lists of landholders in the selected rural kebeles were obtained from the Gozamin Environmental Protection, Land Administration, and Land Use Office. From these lists, 200 respondents from the peri-urban areas and 200 respondents from the rural kebeles were selected randomly. There was an equal probability of selecting respondents who had/had not been involved in land tenure disputes. In addition, during the selection of households, the age and gender distributions in the area were taken into account.

The techniques of data collection in survey research can be face-to-face interviews, telephone interviews, mail questionnaire, self-administered questionnaire, and web surveys [61]. This research involved face-to-face-interviews to gather the necessary data. Face-to-face interviews are preferable as the interviewer communicates personally with each respondent, asks questions, and records their responses [61]. The face-to-face interaction in survey research has many advantages. It stimulates a higher response rate and helps to properly manage longer and more complex questionnaires. In addition, the interviewer has control over the process, and enables respondents to better understand the questions. The interviews were based on a prepared questionnaire. Pre-testing of the questionnaire (see the Appendix A) was conducted before beginning the primary data collection process.

Two focus group discussions (FGDs) were conducted with six members each from the peri-urban areas and the rural kebeles. The participants were experts of the rural and urban wereda courts as well as from justice offices. The experts provided general information about land dispute cases in peri-urban areas, as the legal bodies are particularly responsible for land disputes that are heard in court. A checklist was prepared to collect qualitative data during the FGD. This facilitated the assessment of detailed information and the triangulation of data from the respondents' surveys. Secondary data were collected by reviewing published and unpublished documents from the relevant offices.

Descriptive statistics were employed to analyze quantitative as well as qualitative data by using the Statistical Package for Social Sciences (SPSS Version 20). Mean values, standard deviations, and percentages of the collected data were calculated for the final investigations. In the analysis of data, every questionnaire was coded and checked.

4. Results

4.1. Socio-Economic Characteristics of Respondents

As can be seen in Table 1, the majority of the respondents (76%) were from 30–60 years of age. Due to the small number of female-headed households, there were fewer female interviewees than male, and most of the female interviewees living in peri-urban areas were engaged in low income earning activities besides agriculture. While most of the interviewees were married in both study locations, more divorced and widowed interviewees lived in the peri-urban areas (19%) compared to the rural ones (14%). This is because peri-urban areas have locational advantages for widowed or divorced females, especially for poor rural women, who can find non-agricultural income generating activities, such as daily labor work and retail. Women's labor force participation is higher in urban areas [18].

There are high land tenure transformation and agricultural land losses in peri-urban areas due to urban expansion [15]. In the rural areas, the landholdings of almost all the study participants were registered. In the peri-urban areas, about 13% of the interviewees had unregistered landholdings and 23% were without any landholding certificate. Unregistered landholdings usually have a higher potential for land tenure disputes.

Table 1. Respondents' characteristics and landholding.

	Peri-Urban	Rural
Gender		
Total number of respondents	200	200
Percent of female respondents	19.5	18.5
Age Groups (%)		
Respondents with age below 30 years	4.5	5.5
Respondents with age 30–40 years	24.5	29.0
Respondents with age 40–50 years	27.5	29.5
Respondents with age 50–60 years	24.0	19.0
Respondents with age above 60 years	19.5	17.0
Marital Status (%)		
Not married	3.0	2.5
Married	78.5	83.5
Divorced and widowed	18.5	14.0
Land registration and certification (%)		
Respondents whose land is not registered	13.2	0.0
Respondents without landholding certificates	23.0	1.0

4.2. *The Extent of Land Tenure Disputes*

As documented in Table 2, a higher frequency of land tenure disputes was investigated in the peri-urban areas (65%) compared to the rural areas (47%). The test of the hypothesis, analyzed by cross tabs, also indicates the existence of a significant difference in land disputes in these two locations (p-value < 0.05). Participants in the focus group discussions also reported that the land tenure disputes from peri-urban areas are higher in number compared to rural ones, which reinforces the findings analyzed in the interview.

Table 2. Extent of land tenure disputes.

Location	Count			% within Location			Pearson Chi-Square		
	Yes	No	Total	Yes	No	Total	Value	df	Sig (2-Tailed)
Peri-urban	130	70	200	65.0	35	100.0			
Rural	94	106	200	47.0	53	100.0	13.149 [a]	1	0.000
Total	224	176	400	56.0	44	100.0			

Note: The bases are those farmers interviewed for land tenure disputes: peri-urban, $N = 200$; rural, $N = 200$. Statistical significance: Fischer's exact test $p < 0.05$. ($\chi^2 = 13.149$, df = 1, $p < 0.001$).

4.3. *Types of Land Disputes*

The study respondents revealed that there are disputes in relation to boundary trespassing, landholding, land rental, divorce, land inheritance, parcel exchange, and land use (Table 3). Generally, these can be classified as interpersonal level (micro-social dimension) or societal level (meso/macro-social dimension) disputes [47]. The data show that most of the conflicts identified are at interpersonal levels, involving one-time incidents, and most of these are resolved.

From the total number of respondents involved in different types of land tenure disputes, boundary disputes were the most common type of dispute (about 49% in both study areas). In this study, landholding disputes, also known as land rival disputes, describe a situation "when several people claim the same parcel" [10]. In situations where property rights are ill-defined or where land adjudication is not carried out, the claims of different parties on the same parcel of land increases. Landholding disputes have decreased substantially (by 15%) after land registration in both locations, though peri-urban areas accounted for a high share (21%) next to boundary disputes. In rural areas, the landholding disputes have been reduced after land registration. Instead, in these areas, inheritance disputes constitute the next most frequent type of dispute. Land inheritance disputes, also known as land bequeath disputes, involve intra-family disputes relating to the inheritance of land by children when their parents die [10].

Table 3. Types of land tenure disputes (multiple responses).

Typologies	Peri-Urban		Rural		Total	
	Count	% within	Count	% within	Count	% Total
Boundary disputes	60	46.2	49	52.1	109	48.7
Landholding disputes	28	21.5	6	6.4	34	15.2
Land rental disputes	7	5.4	5	5.3	12	5.4
Divorce-related land tenure disputes	7	5.4	6	6.4	13	5.8
Land bequeath disputes	19	14.6	20	21.3	39	17.4
Parcel exchange disputes	10	7.7	8	8.5	18	8.0
Land use-related disputes	16	12.3	8	8.5	24	10.7

Note: The bases are those farmers involved in land tenure disputes: peri-urban, $N = 130$; rural, $N = 94$.

About 11% of the respondents described land use-related disputes as the main challenges in their locality. Though land use-related disputes are complex [45], three specific topics were identified by the respondents as the primary sources for land use-related disputes in the peri-urban and rural study areas: inappropriate waste disposal, misuse of land, and communal land encroachments. Communal land encroachment is a problem both in peri-urban and rural areas, whereas inappropriate waste disposal and misuse of land are challenges peculiar to peri-urban areas.

4.4. Land Tenure Disputes on Property Regimes

Table 4 documents the ranked assessment of the respondents' perceptions of the frequency of land tenure disputes related to the four property regimes (state, common, private, and communal). Drawing from the data, the perceptions of the respondents about the land tenure dispute situations on these four property regimes were assessed and ranked in order of frequency. Disputes associated with communal land property regimes were enormous (76%) in both rural and peri-urban areas. The results show that communal land is facing severe encroachment and misuse problems. The data show that the perceived high rank of disputes on common land (about 10%) is actually small and similar for both rural and peri-urban respondents. For state lands in both study areas, about 85% of respondents reported that landholding disputes are relatively few, and only about 6–7% of the respondents in both areas reported a high frequency of land disputes. Thus, there are low frequencies of disputes related to state-owned lands because these are often found far from residential sites in lowland areas.

Table 4. Land tenure disputes on different property regime types (multiple responses).

Property Regimes	Respondents' Judgements (%)					
	Peri-Urban ($N = 200$)			Rural ($N = 200$)		
	High	Medium	Low	High	Medium	Low
Private	62.5	23.1	14.4	38.5	32.1	29.4
Common	10.4	42.9	46.7	10.1	11.8	78.1
Communal	76.2	18.2	5.6	76.2	16.9	6.9
State	6.1	8.2	85.7	6.8	8.0	85.2

4.5. Resolving Land Tenure Disputes

For both peri-urban and rural areas, dispute resolution mechanisms were assessed and analyzed in terms of their success and applications. The Pearson chi-square test results in Table 5 show that the differences between the rural and the peri-urban areas are significant. Though most of the disputes have been resolved in both areas, the percentage of solved cases in rural areas is higher (about 86%) compared to that of peri-urban areas (about 71%).

Table 5. Locational scenarios of land tenure dispute resolution.

Location	Count			Frequency (%)			Pearson Chi-Square		
	Yes	No	Total	Yes	No	Total	Value	df	Sig (2-Tailed)
Peri-urban	92	38	130	70.8	29.2	100			
Rural	81	13	94	86.2	13.8	100	7.359 [a]	1	0.007
Total	173	51	224	77.2	22.8	100			

Note: Statistical significance: Fischer's exact test $p < 0.05$. The test results have demonstrated marked statistical significance ([a] $\chi^2 = 7.359$, df = 1, $p < 0.007$).

Table 6 shows that from the available systems of conflict resolution, communities typically use alternative dispute resolutions mechanisms as these are seen to be the most effective methods of land tenure dispute resolution in both study sites (69%). Though ADR mechanisms play a pivotal role in resolving land tenure disputes in both study areas, analysis shows that higher frequencies of disputes are being resolved by ADR mechanisms in rural areas (76%) compared to peri-urban areas (63%). This is likely because of the parties involved in disputes in these two locations. When there are power inequalities, the application of negotiation, negotiation-assisted processes, and arbitration is not sufficient, as outcomes usually will benefit the more powerful party. Legislative processes play pivotal roles in such circumstances [34]. That is why, in peri-urban areas, the frequency of litigation in land dispute resolution is higher (21%) than in rural areas (16%).

The most commonly employed ADR mechanism to resolve land tenure disputes in the study areas is mediation (Figure 2) with a higher rate in rural areas. The data show that compared to peri-urban areas, more land tenure disputes are resolved by mediation in the rural areas. The second most employed ADR mechanism for land tenure dispute resolution in both locations is arbitration, followed by negotiation.

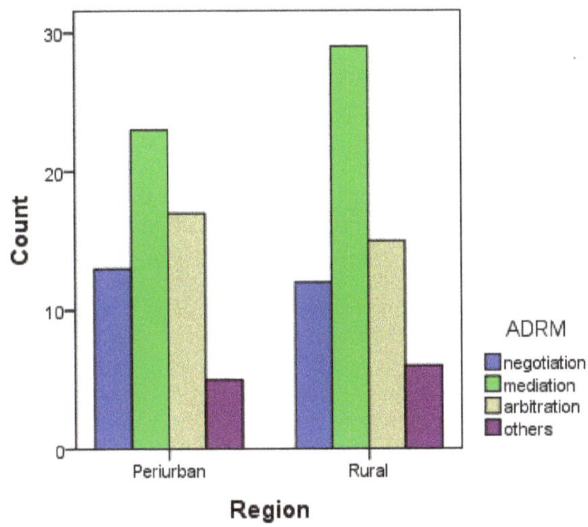

Figure 2. Employed ADR mechanisms.

Table 6. Dispute resolution systems.

System of Dispute Resolution	Peri-Urban		Rural		Total	
	Count	% within	Count	% within	Count	% Total
ADR mechanisms	58	63.0	62	75.6	120	69.0
Administrative decisions	15	16.3	7	8.5	22	12.6
Litigation	19	20.7	13	15.9	32	18.4
Total	92	100	82	100	174	100

5. Discussion

5.1. Land Tenure Disputes

As noted in the results part of this study, high frequencies of respondents in peri-urban areas are victims of land tenure disputes compared to those from rural areas. This finding has also been documented in other studies. For instance, Mbiba and Huchzermeyer [62] noted that peri-urban areas in Africa are dominated by land-related disputes. Though there are locational variations, land can be seen as the direct or indirect source of most disputes in agrarian societies [52,63,64]. In most developing countries, even if horizontal urban expansion is the major cause of peri-urban land disputes [38], the problem is exacerbated in situations where there is a lack of sound spatial planning and where there are institutional problems, inadequate compensation payments, and the absence of effective land tenure security for all landholders [6,49,65]. In addition, in a study conducted in China, it was reported that the lack of institutional competence, i.e., the exercise of power by lower governments without getting into agreements with farmers and inadequate compensation payments, are the main causes of land expropriation disputes [66]. Different types of land disputes were identified in the study. These are boundary disputes, landholding disputes, land rental disputes, divorce-related land tenure disputes, land bequeath disputes, parcel exchange disputes. and land use-related disputes. Boundary disputes are disputes which arise due to trespassing of the parcel boundary between bordering parcel owners/holders and can arise between individuals, between groups, between public agencies, and between individuals and public institutions [10]. From the identified plethora of land disputes, boundary trespassing disputes are the most dominant. The lack of surveying to fix cadastral boundaries is the main challenge even if land has already been registered. Researchers in other African countries also confirm that boundary disputes constitute a significantly high share of land dispute types [12,46,47]. As confirmed in this study, the asymmetric boundary conflicts between municipalities and individuals are a complex challenge for peri-urban landholders. The boundary between urban and peri-urban territory is obscure. Urban municipalities usually want to have control over the surrounding land, assuming that it is municipal land, while suburban landholders claim rights to challenge the supremacy intentions of municipalities. As confirmed by the participants of the focus group discussions, this has increased the incidence of landholding disputes in peri-urban areas and leads to greater tenure insecurity for peri-urban subsistence farmers.

Since 2007, the territory of Debre Markos town has expanded by about 5 km in all directions, based on the municipal master plan. As a result, former rural areas have been designated as urban areas and landholders whose houses are within the expansion radius are assigned as residents of the urban areas. These are the areas which face institutional problems of land administration. For example, as reported by a farmer in the interview, some of his parcels are now dedicated to urban territory, while other parcels are still registered as rural areas. One parcel even is bisected into urban and rural territory. As the rural land administration institution is responsible for the portion of the parcel in the rural territory and the urban land administration institution is responsible for the portion designated to the urban territory, the landholder sometimes faces contradictory rules. For instance, the rural land administration legislation allows the landholder to use the land for houses, agriculture, animal husbandry, forest management, or other activities, if no written land use plan has been prepared by a competent body (often land use plans are absent in peri-urban areas). In contradiction, the urban land administration rule prohibits constructing or renovating buildings or fences without building permits. Constructions of any sort may lead to sanctions and even demolition. Thus, peri-urban landholders are sometimes subject to two different sets of regulations from two different authorities, which jeopardize the security of their land tenure and leads to disputes. Deininger et al. [40] and Nega et al. [67] have also noted that the unclear assignment of institutional responsibilities causes land tenure conflicts. Peri-urban expansion by a municipality is often not based on a sound spatial plan but on the spontaneous desires of a municipality [49]. Property rights in peri-urban areas are often uncertain due to a lack of institutional arrangements to manage peri-urban land [67]. To manage these areas by urban administration is challenging since the focus of the municipalities is to expand urban territory by expropriating peri-urban land. Most of these land disputes are interest-based disputes since they are competing for the scarce resource of land.

As noted in the result part of the study, the three predominant types of peri-urban land use disputes are inappropriate waste disposal, misuse of land, and communal land encroachment. The municipality of Debre Markos town disposes waste materials in the surrounding peri-urban areas haphazardly without taking into account the consequences of their actions for the well-being of peri-urban dwellers. Both liquid and solid waste materials are disposed of at open-air waste sites, which are expanding over time by consuming productive agricultural land. Even the land near the disposed waste material cannot be properly cultivated because of the noxious smell of the waste. Some wastes, like plastic cases, are also blown by wind and pollute other properties. Face-to-face interviews carried out in the study gave evidence that peri-urban landholders are concerned by this abhorrent a situation, as they are affected in multiple ways. They are victims of expropriation, and their landholdings are being polluted and spoiled by waste. Consequently, landholders are suffering from the unpleasant smells due to environmental pollution. These findings are in line with those of other studies. Allen et al. [51] also noted that peri-urban areas are often used as waste disposal sites, while Alemu et al. [68] noted that many peri-urban dwellers spend their whole lives in polluted environments. The municipality, responsible for waste disposal, has never consulted the peri-urban farmers about the suitability of sites for waste disposal. Even though the active involvement of the affected farmers and discussion with the relevant stakeholders are essential to reduce land disputes, the lack of participatory land use planning is one of the basic challenges of peri-urban land management in most African countries [69,70].

The current economic growth in Ethiopia is exhibiting promising results. The government desires to transform the country from an agrarian economy to industrialization based on agricultural development. Thus, the policies promote small enterprises as well as medium- and large-scale industries. Most of the private and public enterprises find it the most feasible to establish their firms close to towns. Thus, peri-urban areas become primary choices for investment by many governmental and non-governmental organizations, and the municipalities incorporate industrial zones in their master plans in the peri-urban areas. Accordingly, Debre Markos has delineated industrial zones along the urban fringe in its master plan. For the establishment of this industrial zone, the authorities expropriated many parcels and transferred individual land to municipal ownership. After the transfer, the municipality usually makes this land available to those coming with development projects and with the intention to establish firms. However, some of those firms are abusing land. For instance, enterprises carry out earthworks to construct company buildings and infrastructure and indiscriminately dump the excavated material, polluting peri-urban grazing land. It would be better to transport the excavated soils to the hinterland and use them for filling eroded gullies and ravines, even though this would be more expensive.

In Amhara National Regional State of Ethiopia, there are four types of landholdings: private, common, communal, and state property [71]. Private, communal, and state property regimes are similar in definition to those terms in different literature [72,73], though landholders lack exclusive rights in the case of Ethiopia. The use of the term common property regime differs in Ethiopia from other countries: this refers to a property type commonly held by small groups of individuals. In order to minimize land fragmentation, the minimum parcel size (0.25 hectares) is specified in the rural land administration Regulation No. 159/2018 of Amhara National Regional State [74]. Thus, if any property formation measure results in an area of less than 0.25 hectares, two or more individuals will own the parcel commonly as common property [23,71,74]. As noted in the results of this study, disputes on this type of property regime are few in number. The main reason for this is that the common landholding regime is found in small amounts in peri-urban areas and the nearby rural kebeles. Disputes on this property regime are expected to increase in the future, however, because land fragmentation is legally prohibited and there will be many common landholders.

As with common property regimes, disputes on state property regimes are also few in number. However, in some areas, state property regimes in lowland areas are becoming sources of dispute between the local community and investors engaged in agricultural production [16,75–77]. This is because some farmers have possessed land in these lowland areas informally, and because of increasing immigration to these formerly remote areas.

Although the rural land administration proclamations grant full rights to communal landholders [23,71], these rights are not properly practiced. The absence of clearly defined property rights to communal land [23] and a lack of enforcement contribute to increased competition and encroachment on the communal land. The registration for these property regimes is not well managed and even the book of possession has not yet been issued, and as a consequence, compensation is not often paid for the expropriated communal lands. When farmers raise the issue to the municipality, they are told that it is the right of the municipality to use communal land for urban expansion without paying compensation. Communal property regimes in peri-urban areas are targets of expropriation by municipalities. This finding is confirmed by Puppim de Oliveira [78], who also observed that municipalities focus on expropriating communal property regimes without titles.

The encroachment of communal land generally increases problems in peri-urban areas. This is because farmers, who have parcels adjacent to the boundary of the communal land, assume that converting pastures into arable land is a way to get at least some compensation in the case of expropriation. For this reason, farmers whose plots border communal land are encroaching on it and converting communal grazing land into arable land. During the past few decades, communal forests have been encroached upon to such an extent that there are only sparse areas left. Currently, communal grazing lands are becoming new areas of interest for encroachment, even though this has a detrimental effect on the livestock sector.

5.2. *Land Tenure Dispute Resolution Mechanisms*

Land tenure disputes in the study areas have been resolved by employing various processes and mechanisms. These processes include both alternative dispute resolution mechanisms (ADRMs), which are informal in nature, and administrative decision and litigation, which are formal in nature.

Though, as noted in the results, most of the disputes have been resolved in both areas, a higher percentage of disputes are resolved in rural areas compared to peri-urban areas. The difference was justified by the experts in the focal group interviews, who said that the rural disputes are usually symmetrical (between farmers) and can be easily resolved by employing ADR methods. As identified by Moore and Jayasundere [5], mediation is most effective when the parties have symmetrical power relationships. In peri-urban areas, most of the land tenure disputes are between farmers and municipalities or other institutions. These asymmetrical disputes are usually more difficult to resolve.

Communities tend to employ alternative dispute resolution mechanisms, as these are seen to be the most effective methods of land tenure dispute resolution. This finding is confirmed by results of other studies documenting that land disputes can be solved within a short period using local dispute resolution mechanisms rather than being handled by courts [8,28,29,56]. Additionally, less formal techniques, especially mediation and to some extent negotiation, are encouraging mechanisms to resolve disputes, in contrast to adversary litigation [79–81].

This research identified that the majority of the landholders in both peri-urban and rural areas used the ADR mechanisms more than the formal means to resolve land tenure disputes (Table 6). State missionaries, such as court and government administrative organs, are normally involved in the formal ways of land tenure dispute resolution mechanisms [21,32]. These land tenure dispute resolution mechanisms are often noted as JDR (juridical dispute resolution), court litigations, or administrative means of land dispute resolution mechanisms. Court litigations are applied to resolve symmetric land tenure disputes in situations when the resolution by the other methods is not feasible. As reported by the discussants and informants, court litigation is only applied when the parties are confident enough to win the dispute. At the same time, parties who are not confident in winning the dispute are likely to send older neighbors to the opponent to negotiate with them. The confident opponent feels honored by the involvement of the older neighbors and agrees to settle the dispute out of court. However, this does mean that the numbers of respondents going to court is insignificant. The discussions with civil judges revealed that the majority of the civil cases lodged to the civil benches have been directly or indirectly related to land issues. In the group discussions, judges described that most of the instigated caseloads in each year are land-related disputes. Administrative authorities may also resolve issues which are not very controversial. For example, at the kebele level, land administration committees play an important role in resolving land disputes. However, formal means of land tenure dispute resolution mechanisms are recommended as the last resort of dispute resolution mechanisms in Ethiopia in general and in Amhara National Regional State in particular [23,71,74]. The discussants and the interviewees indicated that the formal procedures for resolving land tenure disputes are mainly applied when one of the parties in the dispute is a government organ or a judicial person[3].

The Federal Democratic Republic of Ethiopia's (FDRE) constitution gives regional states the option to incorporate the informal land tenure dispute resolution mechanisms in their land administration and use legislations [22]. Moreover, the Federal Land Administration and Use Proclamation No. 456/2005, used as a framework for regional states to enact their own laws, explicitly recognizes negotiation, conciliation, and arbitration as appropriate means of setting rural land disputes [23]. Almost all the rural land administration and use legislations emphasize informal dispute settling institutions in addition to administrative tribunals, regular courts, or special courts. As the Amhara National Regional State (ANRS) recognized the informal resolution mechanisms for land tenure disputes in the ANRS rural land administration and use proclamation No. 252/2017 and the consequent Regulation No.159/2018 and prioritizes informal land tenure dispute resolution mechanisms [71,74], negotiation, mediation, conciliation, and arbitration are practised in the study area. In ANRS, informal land tenure dispute resolution mechanisms have equal (sometimes favored) status with the formal means of land tenure dispute resolution mechanisms.

As noted in the results of the study, ADRMs are most widely used to resolve land tenure disputes, though the extent widely differs from place to place. Negotiation, mediation/conciliation, and arbitration are the most commonly used ADR mechanisms to solve land tenure disputes in the study area. This finding is according to Article 52(1) of the Amhara National Regional State's revised rural land administration and use proclamation No. 252/2017. Gowok [24] also reported that Ethiopia has been using alternative dispute resolution methods for centuries, and that negotiation, mediation, conciliation, and arbitration are the key dispute resolution mechanisms in Ethiopia.

Arbitration, also known as 'shimaglle'[4], is one of the oldest forms of dispute resolution practiced in Ethiopia. It relies on solving disputes by appointing arbitrators who are persons with particular knowledge on custom or have experience in shimglina [24].

Farmers and experts interviewed in the study confirmed that parties usually try to solve land tenure disputes by negotiation. Sometimes the parties solve the land tenure disputes by negotiation, but there are many cases where this is not possible. For example, disputants find it difficult to solve problems in negotiation when they are entangled in competing positions or when there are interpersonal problems between the parties [80]. Due to these problems, negotiation assistance processes (i.e., mediation/conciliation) were applied to effectively resolve land tenure disputes. Mediation is seen as an integral model to dispute resolution [5,35,81]. This is because mediation has the advantage of empowering parties and improving communication [82]. Though the practice of mediation has often been used to settle labour disputes or family and divorce disputes, the technique is now widely in use in interpersonal disputes [5]. As an example, mediation has been espoused as a central land dispute resolution model in Ghana [37]. Norwegians also widely use mediation to resolve boundary disputes [53]. Though mediation is not a panacea for all land tenure disputes, it plays a significant role in managing problems between disputants.

6. Conclusions

This paper addressed the extent of land tenure disputes and the most commonly used land tenure dispute resolution mechanisms based on investigations in peri-urban areas and in nearby rural areas of Debre Markos town in Ethiopia. Though the rural–urban interface is a conflict prone area, land conflicts in such areas are exacerbated in situations where there is a lack of timely assurances of peri-urban land rights and when urban areas are spatially expanding and consuming agricultural land [49,50]. To avoid disputes, governments must take into consideration the land property rights of peri-urban subsistence farmers. In this way, the sustainable development of urban areas can be attained in the peri-urban and surrounding rural areas.

For the effective management of peri-urban land and to support the urban sprawl of a town, it is essential to set up sound spatial planning techniques considering various spatial planning support systems and to integrate affected persons and different stakeholders. Moreover, spatial planning should restrict the outward growth of urban areas by designing strategies for the development of the inholdings of vacant land in the town and by implementing proper land use planning strategies.

This study has identified different types of land tenure disputes. Boundary trespassing disputes are the most frequent type both in rural and peri-urban areas. This is because of a lack of progress in defining and demarcating the boundary points. Therefore, conducting cadastral survey and mapping boundaries are of paramount importance to reduce these disputes.

Land tenure disputes are high in communal land and private land property regime types. Communal land is experiencing severe levels of encroachment. This suggests that land property rights must be secured for communal landholders. In addition, boundaries must be clearly defined and demarcated in both peri-urban and rural areas. However, peri-urban areas should be given special attention since they are the primary focus of municipalities for expropriation, which also aggravates the encroachment problem.

In the study area, the situation of land administration is not well defined. Landholders may be liable to rules from two different land administration authorities. This creates confusion and stress in the lives of peri-urban subsistence farmers. The remedy for this problem is to organize peri-urban land administration institutions. Alternatively, it would be possible to manage these areas by rural land administration and land use rules, since the land predominantly exhibits a rural character, and the majority of the landholders are subsistence farmers. In addition, the merging of the rural and urban land administration authority would be an option. Research on this issue must be carried out.

Alternative dispute resolution mechanisms are the most preferred systems to resolve land tenure dispute, especially in symmetric dispute situations. This is because they are less expensive, faster, and more efficient than other approaches to dispute resolution. Many ADR mechanisms are available to resolve land tenure disputes, irrespective of the location, and of those, negotiation, mediation/conciliation, and arbitration are the most widely used methods of dispute resolution as an alternative to litigation. Strengthening alternative dispute resolution mechanisms is of paramount important for countries like Ethiopia, where most farmers rely on subsistence living.

This study attempted to identify the land tenure disputes and resolution mechanisms in Debre Markos peri-urban study areas and nearby rural kebeles. The studied areas may be representative for the peri-urban land dispute cases and resolution mechanisms in other areas, since peri-urban areas have more or less similar scenarios with respect to land dispute as a consequence of fast rate urban sprawl. However, further studies in other regions of the country must be conducted to be used as additional inputs for policy makers who have stakes in the amicable resolution of land tenure disputes.

Author Contributions: Conceptualization, S.K.A., T.D., W.N. and R.M.; formal analysis, T.D., S.K.A., W.N. and R.M.; funding acquisition, S.K.A. and R.M.; methodology, S.K.A., T.D., W.N. and R.M.; project administration, S.K.A. and R.M.; supervision, R.M.; validation, S.K.A., T.D., W.N. and R.M.; visualization, S.K.A., T.D., W.N. and R.M.; writing—original draft, S.K.A.; writing—review and editing, S.K.A., T.D., W.N. and R.M. All authors have read and agreed to the published version of the manuscript.

Funding: This study is funded by EduLAND2 project fund.

Institutional Review Board Statement: All respondents of questionnaire and all participants of focus group discussions and key experts' interviews are not mentioned by name. Results are aggregated and cannot be traced back to individual persons.

Informed Consent Statement: All persons involved in the study participated voluntarily and agreed the study results derived from their responses.

Data Availability Statement: Not applicable.

Conflicts of Interest: We confirm that this work is original and has not been published elsewhere nor is it currently under consideration for publication elsewhere and the authors declare that there is no conflict of interest.

Appendix A

Land disputes are one of the main problems which hinder the socio-economic development of a country. Even though these disputes are common phenomenon globally, the extent is severe particularly in developing countries where land is the basic asset of economic development and property rights are ill defined. Conducting research on these land dispute situations is paramount important for socio-economic development of a nation by designing appropriate context oriented dispute resolution mechanisms. The objectives of this research paper is to identify the extent and causes of land conflicts, the effect of land registration in reducing these conflicts and to design low costly dispute resolution mechanisms. For the sake of addressing these objectives, data can be collected from different stakeholders and this questionnaire is for data collection from respondents. I hope that completing this survey may not take much of your time and you will deliver necessary information within the survey period of time. I want also to thank in advance for your cooperation.

Part I: Please give your opinion on the extent of land tenure related disputes

1. Did you come across with any land tenure related conflicts during the past 5 years?
   1.1 Yes
   1.2 No
2. If yes, when is it?
   2.1 Before land registration and certification
   2.2 During land registration and certification
   2.3 After land registration and certification
3. What type of land conflict is it?
   3.1 Boundary dispute
   3.2 Inheritance dispute
   3.3 Rival dispute
   3.4 Divorce dispute
   3.5 Land rent contract dispute
   3.6 Land use related dispute
   3.7 Others, please specify, ——————————
4. With whom was your land tenure dispute?
   4.1 With another individual farmer
   4.2 With governmental bodies such as municipalities
   4.3 With private developers
   4.4 Others, please specify ——————————
5. Why you became to such land dispute ——————————————————————————
6. In your kebele, which segments of the society are more victim to land conflicts
   6.1 Females
   6.2 Poor farmers
   6.3 Others
   6.4 No difference
7. On which property regime types land disputes are more frequent
   7.1 Private property regimes
   7.2 Common property regimes
   7.3 Communal property regimes
   7.4 State property regimes
8. What do you think the reason might be for such property regime dispute to be more frequent?
   ——

Part II: About land tenure conflict resolution mechanisms and transparency

9. How are the land tenure related conflict cases resolved?
   9.1 By formal (legal) procedures
   9.3 Through administrative decisions
   9.4 By alternative dispute resolution mechanisms (ADRM),
   9.5 Others, specify ——————————
10. If your answer to the above question is ADRM, what type of alternative dispute resolution mechanism/s is used to solve your problem?
    10.1 Negotiation
    10.2 Mediation
    10.3 Arbitration
    10.4 Others, specify ————————

11. Why this method is mostly preferred? ——

12. If via administrative bodies, which one of those resolved your disputes?
    12.1 Kebele administrative bodies,
    12.2 Kebele land administration committees
    12.3 Woreda administrative office
    12.4 Woereda land administration office,
    12.5 Others, specify ——————————
13. How do you evaluate the decision made by the administrative bodies?
    13.1 Very good
    13.2 Good
    13.3 Poor
    13.4 Very poor
    13.5 I don't know
14. If your answer to the above question is poor and below, what do you think the reason might be?
    14.1 Lack of accountability
    14.2 Lack of equity
    14.3 Inefficiency
    14.4 Others, please specify ——————————
15. When you compare the decision by previous social courts and shimaglewoch shengo, which one you prefer and why?
    ——

16. How do you evaluate the conflict resolution ability of shemaglewoch shengo?
    16.1 Very good
    16.2 Good
    16.3 Poor
    16.4 Very poor
    16.5 I don't know
17. If your answer to the above question is poor and below, what do you think the reason might be?
    17.1 ——————————————————
    17.2 ——————————————————
    17.3 ——————————————————
18. If your answer to question number 9 is litigation, which court gives you decision
    18.1 Social courts
    18.2 Wereda courts,
    18.3 Others, specify ——————————
19. What is your satisfaction by the decision of courts?
    19.1 Very satisfied
    19.2 Satisfied
    19.3 Fair
    19.4 Unsatisfied
    19.5 Very unsatisfied
20. If your response to question number 19 is fair and below, what do you think the reason may be for this low service delivery? ——

21. How do you rate the cost of litigation

21.1 Very expensive
21.2 Expensive
21.3 Fair
21.4 Cheap
21.5 Very cheap

22. How do you rate the time of decision making for litigation

22.1 Very high
22.2 High
22.3 Fair
22.4 Low
22.5 Very low

23. What about the appeal cases for unsatisfied party in decisions?

23.1 Transparent and responsive
23.2 Time consuming and expensive
23.3 No well-established appeal system
23.4 Others, specify ————————

24. In general assessment, are you satisfied with courts' decision

24.1 Yes
24.2 No

Part III: Please give your opinion about land registration

25. Are all of your parcels registered

25.1 Yes
25.2 No

26. Have you received certificate of holding for all of your parcels

26.1 Yes
26.2 No

27. Were the land registration and your certificate of holding valuable in resolving the conflict?

27.1 Yes
27.2 No

28. If no, why?

28.1 Undefined boundaries
28.2 Not well adjudicated
28.3 Not being used by decision makers
28.4 Others, specify ————————

29. Are you clear about the rights and responsibilities of land registration?

29.1 Yes
29.2 No

30. If no why? ————————
31. During adjudication, have you participated in public hearings?

31.1 Yes
31.2 No

32. If yes, did you notice land tenure double claims identified during public hearing?

32.1 Yes
32.2 No

33. How do you evaluate these public hearings in reducing land disputes

33.1 Very good
33.2 Good
33.3 Fair
33.4 Poor

33.5 Very poor

34. If your response to the above question is fair and below, what do you think the reason is? ——————————

35. When you compare land disputes before the land registration and after registration, how do you evaluate the land dispute situation?
    35.1 Strongly improved
    35.2 Improved
    35.3 Moderately improved
    35.4 Not improved
    35.5 Worsened
36. If your response for the above question is not improved and below, what do you think the reason may be for this?
    36.1 Multiple claims
    36.2 Undefined boundaries
    36.3 Reducing the role of conflict resolving local institutions
    36.4 Others, specify ——————
37. What is your opinion in order to reduce these land related conflicts easily? ——————————

IV. Personal data questions

38. General description of the respondent?

38.1	sex	Male	☐	Female	☐				
38.2	Age	30–40	☐	40–50	☐	50–60	☐	>60	☐
38.3	Marital status	Marrried	☐	Single	☐	Divorced	☐		
38.4	Landholding in Ha	————							

Thank you very much for your contribution and patience!

Notes

1 Kebele is the lowest administrative organization in Ethiopia, consisting of on average 5000 inhabitants.

2 Wereda is the lower administrative organization in Ethiopia, consisting of many kebeles.

3 Judicial persons are a creation of the human mind and have legal personalities to exercise judicial acts.

4 There are debates about whether shimaglle is the Amharic translation of arbitration or not, since shimaglle covers issues wider than arbitration. However, in this research shimaglle is used synonymously with arbitration.

References

1. Harbom, L.; Högbladh, S.; Wallensteen, P. Armed conflict and peace agreements. *J. Peace Res.* **2006**, *43*, 617–631. [CrossRef]
2. Peter, W. *Understanding Conflict Resolution: War, Peace and the Global System*; SAGE Publications: Thousand Oaks, CA, USA, 2002.
3. Moore, S.A. Defining "successful" environmental dispute resolution: Case studies from public land planning in the United States and Australia, Environ. *Impact Assess. Rev.* **1996**, *16*, 151–169. [CrossRef]
4. Miall, H.; Ramsbotham, O.; Woodhouse, T. *Contemporary Conflict Resolution: The Prevention, Management and Transformation of Deadly Conflicts*; Polity: Cambridge, UK, 1999.
5. Moore, C.W.; Jayasundere, R.; Thirunavukarasu, M. *The Mediation Process: Trainees' Manual Community Mediation Program*; Sri Lanka Ministry of Justice: Colombo, Sri Lanka, 2009; Volume 1.
6. Dadashpoor, H.; Somayeh, A. Land tenure-related conflicts in peri-urban areas: A review. *Land Use Policy* **2019**, *85*, 218–229. [CrossRef]
7. Adam, A.G. Informal settlements in the peri-urban areas of Bahir Dar, Ethiopia: An institutional analysis. *Habitat Int.* **2014**, *43*, 90–97. [CrossRef]
8. Lombard, M. Land conflict in peri-urban areas: Exploring the effects of land reform on informal settlement in Mexico. *Urban Stud.* **2016**, *53*, 2700–2720. [CrossRef]
9. Ansah, B.O.; Chigbu, U.E. The Nexus between Peri-Urban Transformation and Customary Land Rights Disputes: Effects on Peri-Urban Development in Trede, Ghana. *Land* **2020**, *9*, 187. [CrossRef]
10. Wehrman, B. *GTZ Land Conflicts: A Practical Guide to Dealing with Land Disputes*; Deutsche Gesellschaft für Technische Zusammenarbeit (GTZ) GmbH: Eschborn, Germany, 2008.

11. Wubie, A.M.; de Vries, W.T.; Alemie, B.K. Synthesizing the dilemmas and prospects for a peri-urban land use management framework: Evidence from Ethiopia. *Land Use Policy* **2021**, *100*, 105122. [CrossRef]
12. Barry, M.; Danso, E.K. Tenure security, land registration and customary tenure in a peri-urban Accra community. *Land Use Policy* **2014**, *39*, 358–365. [CrossRef]
13. Hotte, L. Conflicts over property rights and natural-resource exploitation at the frontier. *J. Dev. Econ.* **2001**, *66*, 1–21. [CrossRef]
14. Adam, A.G. Peri-Urban Land Tenure in Ethiopia. Ph.D. Thesis, Royal Institute of Technology (KTH), Stockholm, Sweden, 2014.
15. Haregeweyn, N.; Fikadu, G.; Tsunekawa, A.; Tsubo, M.; Meshesha, D.T. The dynamics of urban expansion and its impacts on land use/land cover change and small-scale farmers living near the urban fringe: A case study of Bahir Dar, Ethiopia. *Landsc. Urban Plan.* **2012**, *106*, 149–157. [CrossRef]
16. Cochrane, L.; Legault, D.D. The rush for land and agricultural investment in Ethiopia: What we know and what we are missing. *Land* **2020**, *9*, 167. [CrossRef]
17. Simon, D.; McGregor, D.; Thompson, D. Contemporary Perspectives on the Peri-Urban Zones of Cities in Developing Areas. In *The Peri-Urban Interface: Approaches to Sustainable Natural and Human Resource Use*; Earthscan: London, UK, 2012. [CrossRef]
18. Cohen, B. Urbanization in developing countries: Current trends, future projections, and key challenges for sustainability. *Technol. Soc.* **2006**, *28*, 63–80. [CrossRef]
19. Simmons, C.S. The political economy of land conflict in the eastern Brazilian Amazon. *Ann. Assoc. Am. Geogr.* **2004**, *94*, 183–206. [CrossRef]
20. Wubie, A.M.; de Vries, W.T.; Alemie, B.K. A socio-spatial analysis of land use dynamics and process of land intervention in the peri-urban areas of bahir dar city. *Land* **2020**, *9*, 445. [CrossRef]
21. Ayano, M.F. Law and land conflict in emerging market economies: Ethiopia, 2014–2018. *Int. J. Const. Law* **2020**, *18*, 988–1012. [CrossRef]
22. Federal Democratic Republic of Ethiopia. *Constitution of the Fedral Democratic Republic of Ethiopia*; Negarit Gazata; Federal Democratic Republic of Ethiopia: Addis Ababa, Ethiopia, 1995.
23. Federal Democratic Republic of Ethiopia. *Federal Democratic Republic of Ethiopia Rural Land Administration and Land Use Proclamation No 456/2005*; Negarit Gazata, Ethiopia; Federal Democratic Republic of Ethiopia: Addis Ababa, Ethiopia, 2005.
24. Gowak, S.M. Alternative Dispute Resolution in Ethiopia—A Legal Framework. *Afr. Res. Rev.* **2008**, *2*, 265–285. [CrossRef]
25. Bennett, R.M.; Alemie, B.K. Fit-for-purpose land administration: Lessons from urban and rural Ethiopia. *Surv. Rev.* **2016**, *48*, 11–20. [CrossRef]
26. Bogale, A.; Taeb, M.; Endo, M. Land ownership and conflicts over the use of resources: Implication for household vulnerability in eastern Ethiopia. *Ecol. Econ.* **2006**, *58*, 134–145. [CrossRef]
27. Holden, S.T.; Deininger, K.; Ghebru, H. Impact of Land Registration and Certification on Land Border Conflicts in Ethiopia. In Proceedings of the Annual Bank Conference on Land Policy and Administration, Washington, DC, USA, 26–27 April 2010.
28. Lavers, T. Responding to land-based conflict in Ethiopia: The land rights of Ethnic minorities under federalism. *Afr. Aff.* **2018**, *117*, 462–484. [CrossRef]
29. Barry, M. Periurban Tenure Management in South Africa. In Proceedings of the Second FIG Regional Conference, Marrakech, Morocco, 2–5 December 2003; pp. 1–14.
30. Adenew, B.; Abdi, F. *Land Registration in Amhara Region, Ethiopia*; Securing Land Rights in Africa; International Institute for Environment and Development: London, UK, 2005; pp. 1–34.
31. Deininger, K.; Castagnini, R. Incidence and impact of land conflict in Uganda. *J. Econ. Behav. Organ.* **2006**, *60*, 321–345. [CrossRef]
32. Tesfay, H. Rural Land Dispute Settlement Mechanisms in Tigray: The case of Humera. Master's Thesis, Addis Ababa University, Addis Ababa, Ethiopia, 2011.
33. Rose, M.; Suffling, R. Alternative dispute resolution and the protection of natural areas in Ontario, Canada. *Landsc. Urban Plan.* **2001**, *56*, 1–9. [CrossRef]
34. Effron, J. Alternatives To Litigation: Factors in Choosing. *Mod. Law Rev.* **1989**, *52*, 480–497. [CrossRef]
35. Radulescu, D.M. Mediation-An Alternative way to Solve Conflicts in the International Business Environment. *Procedia Soc. Behav. Sci.* **2012**, *62*, 290–293. [CrossRef]
36. Cappelletti, M. Alternative Dispute Resolution Processes within the Framework of the World-Wide Access-to-Justice Movement. *Mod. Law Rev.* **1993**, *56*, 282–296. [CrossRef]
37. Fred-Mensah, B.K. Capturing ambiguities: Communal conflict management alternative in Ghana. *World Dev.* **1999**, *27*, 951–965. [CrossRef]
38. Hui, E.C.M.; Bao, H. The logic behind conflicts in land acquisitions in contemporary China: A framework based upon game theory. *Land Use Policy* **2013**, *30*, 373–380. [CrossRef]
39. Bittner, C.; Sofer, M. Land use changes in the rural-urban fringe: An Israeli case study. *Land Use Policy* **2013**, *33*, 11–19. [CrossRef]
40. Deininger, K.; Hilhorst, T.; Songwe, V. Identifying and addressing land governance constraints to support intensification and land market operation: Evidence from 10 African countries. *Food Policy* **2014**, *48*, 76–87. [CrossRef]
41. Alemu, B.Y. Expropriation, Valuation and Compensation Practice in Amhara National Regional State (ANRS)—The Case of Two Cities (Bahir-Dar and Gonder). *Nord. J. Surv. Real Estate Res.* **2012**, *9*, 30–58.
42. Anseeuw, W.; Wily, L.A.; Cotula, L.; Taylor, M. *Land Rights and the Rush for Land*; International Land Coalition: Rome, Italy, 2011.

43. Qin, M.; Lin, W.; Li, J.; Yu, Z.; Wachenheim, C. Impact of land registration and certification on land rental by Chinese farmers. *Land Use Policy* **2020**, *99*, 104875. [CrossRef]
44. Deininger, K.; Ali, D.A.; Holden, S.; Zevenbergen, J. Rural Land Certification in Ethiopia: Process, Initial Impact, and Implications for Other African Countries. *World Dev.* **2008**, *36*, 1786–1812. [CrossRef]
45. von der Dunk, A.; Grêt-Regamey, A.; Dalang, T.; Hersperger, A.M. Defining a typology of peri-urban land-use conflicts—A case study from Switzerland. *Landsc. Urban Plan.* **2011**, *101*, 149–156. [CrossRef]
46. Colin, J.P. Securing rural land transactions in Africa. An Ivorian perspective. *Land Use Policy* **2013**, *31*, 430–440. [CrossRef]
47. Wehrmann, B. *Cadastre in Itself Won't Solve the Problem: The Role of Institutional Change and Psychological Motivations in Land Conflicts—Cases from Africa*; International Federation of Surveyors: Copenhagen, Denmark, 2006.
48. Dufwenberg, M.; Köhlin, G.; Martinsson, P.; Medhin, H. Thanks but no thanks: A new policy to reduce land conflict. *J. Environ. Econ. Manag.* **2016**, *77*, 31–50. [CrossRef]
49. Agegnehu, S.K.; Fuchs, H.; Navratil, G.; Stokowski, P.; Vuolo, F.; Mansberger, R. Spatial Urban Expansion and Land Tenure Security in Ethiopia: Case Studies from Bahir Dar and Debre Markos Peri-Urban Areas. *Soc. Nat. Resour.* **2016**, *29*, 311–328. [CrossRef]
50. Agegnehu, S.K.; Mansberger, R. Community involvement and compensation money utilization in ethiopia: Case studies from bahir dar and debre markos peri-urban areas. *Sustainability* **2020**, *12*, 4794. [CrossRef]
51. Allen, A.; a da Silva, N.L.; Corubolo, E. *Environmental Problems and Opportunities of the Peri-Urban Interface and Their Impact Upon the Poor*; The Development Planning Unit: London, UK, 1999; 46p.
52. Augustinus, C.; Lewis, D.; Leckie, S. *A Post-Conflict Land Administration and Peace-Building Handbook*; Series 1; UN-Habitat: Nairobi, Kenya, 2007.
53. Goodale, M.R.G.; Sky, P.K. A comparative study of land tenure, property boundaries, and dispute resolution: Case studies from Bolivia and Norway. *J. Rural Stud.* **2001**, *17*, 183–200. [CrossRef]
54. Saarikoski, H.; Raitio, K.; Barry, J. Understanding "successful" conflict resolution: Policy regime changes and new interactive arenas in the Great Bear Rainforest. *Land Use Policy* **2013**, *32*, 271–280. [CrossRef]
55. Twining, W. Alternative to What? Theories of Litigation, Procedure and Dispute Settlement in Anglo-American Jurisprudence: Some Neglected Classics. *Mod. Law Rev.* **1993**, *56*, 380–392. [CrossRef]
56. Mamo, A.B. Three ways of looking at dispute resolution. *Wake For. L. Rev.* **2019**, *54*, 1399.
57. Kalande, W. Kenyan Land Disputes in the Context of Social Conflict Theories. In Proceedings of the FIG Commission 7th Annual Meeting and Open Symposium on Environment and Land Administration 'Big Works for Defence of The Territory', Verona, Italy, 11–15 September 2008; pp. 1–16.
58. Reuben, R.C. Constitutional gravity: A unitary theory of alternative dispute resolution and public civil justice. *UCLA Law Rev.* **2000**, *47*, 949.
59. Trubek, D.M.; Kritzer, H.; Holst, K.; Felsteiner, W. *Costs, Processes, and Outcomes: Lawyers' Attitudes to Courts and Other Dispute Processing Options*; U.S. Department of Justice: Washington, DC, USA, 1984.
60. Ostrom, E.; Hess, C. Private and Common Property Rights. In *Property Law and Economics*; Edward Elgar Publishing: Cheltenham, UK, 2010; pp. 53–106. [CrossRef]
61. Neuman, W.L. *Basics of Social Research Methods: Qualitative and Quantitative Approaches*, 2nd ed.; Pearson Education: Boston, MA, USA, 2007.
62. Mbiba, B.; Huchzermeyer, M. Contentious development: Peri-urban studies in sub-Saharan Africa. *Prog. Dev. Stud.* **2002**, *2*, 113–131. [CrossRef]
63. Cotula, L.; Toulmin, C.; Hesse, C. *Land Tenure and Administration in Africa: Lessons of Experience and Emerging Issues*; International Institute for Environment and Development: London, UK, 2004; 50p.
64. Singletary, L.; Smutko, L.S.; Hill, G.C.; Smith, M.; Daniels, S.E.; Ayres, J.S.; Haaland, K. Skills needed to help communities manage natural resource conflicts. *Confl. Resolut. Q.* **2008**, *25*, 303–320. [CrossRef]
65. Payne, B.G.; Durand-lasserve, A.; Payne, G. *Holding On: Security of Tenure—Types, Policies, Practices and Challenges*; Office of the United Nations High Commissioner for Human Rights: Geneva, Switzerland, 2012; pp. 1–78. Available online: http://www.ohchr.org/Documents/Issues/Housing/SecurityTenure/Payne-Durand-Lasserve-BackgroundPaper-JAN2013.pdf (accessed on 2 September 2021).
66. Lin, Q.; Tan, S.; Zhang, L.; Wang, S.; Wei, C.; Li, Y. Land Use Policy Conflicts of land expropriation in China during 2006–2016: An overview and its spatio-temporal characteristics. *Land Use Policy* **2018**, *76*, 246–251. [CrossRef]
67. Nega, W.; Tenaw, M.; Hunie, Y.; Agegnehu, S.K. Evaluating Institutional Dichotomy between Urban and Rural Land Administration in Amhara Region, Ethiopia. *Sustainability* **2021**, *13*, 9431. [CrossRef]
68. Alemu, G.T.; Ayele, Z.B.; Berhanu, A.A. Effects of Land Fragmentation on Productivity in Northwestern Ethiopia. *Adv. Agric.* **2017**, *2017*, 1–9. [CrossRef]
69. Magigi, W.; Drescher, A.W. The dynamics of land use change and tenure systems in Sub-Saharan Africa cities; learning from Himo community protest, conflict and interest in urban planning practice in Tanzania. *Habitat Int.* **2010**, *34*, 154–164. [CrossRef]
70. Morran, R.M.; Scott, A.J.; Price, M.F. Reconstructing sustainability; participant experiences of community land tenure in North West Scotland. *J. Rural Stud.* **2014**, *33*, 20–31. [CrossRef]

71. Amhara National Regional State. *The Revised Rural Land Administration and Use Determination Proclamation of the Amhara National Regional State, Proclamation No. 252/2017*; Zikre Hig; Amhara National Regional State: Bahir Dar, Ethiopia, 2017.
72. Feder, G.; Feeny, D. Land Tenure and Property Rights: Theory and Implications for Development Policy. In *The World Bank Economic Review*; Oxford University Press: Oxford, UK; Volume 5, pp. 135–153.
73. Bromley, D.W. *Environment and Economy: Property Rights and Public Policy*; Basil Blackwell Ltd.: Oxford, UK, 1991.
74. Amhara National Regional State. *The Revised Amhara National Regional State Rural Land Administration and Use System Implementation, Council of Regional Government Regulation No. 159/2018*; Zikre Hig; Amhara National Regional State: Bahir Dar, Ethiopia, 2018.
75. Riedel, K. Land Grabbing in Ethiopia—Welfare or Farewell. 2011. Available online: https://farmlandgrab.org/uploads/attachment/Land%20grabbing%20in%20Ethiopia%20-%20Riedel,%20Sommerstein.pdf (accessed on 2 September 2021).
76. Stebek, E. Between 'Land Grabs' and Agricultural Investment: Land Rent Contracts with Foreign Investors and Ethiopia's Normative Setting in Focus. *Mizan Law Rev.* **2012**, *5*, 175–214. [CrossRef]
77. Lavers, T. "Land grab" as development strategy? The political economy of agricultural investment in Ethiopia. *J. Peasant. Stud.* **2012**, *39*, 105–132. [CrossRef]
78. de Oliveira, J.A.P. Property rights, land conflicts and deforestation in the Eastern Amazon. *For. Policy Econ.* **2008**, *10*, 303–315. [CrossRef]
79. van Veen, D.H.; Kreutzwiser, R.D.; de Loë, R.C. Selecting appropriate dispute resolution techniques: A rural water managment example. *Appl. Geogr.* **2003**, *23*, 89–113. [CrossRef]
80. Fisher, R.; Ury, W. *Getting to Yes: Negotiating Agreement Without Giving In*; Patton, B., Ed.; Penguin Publishing Group: London, UK, 1991.
81. Andrew, J.S. Potential Application of Mediation to Land Use Conflicts in Small-Scale Mining. *J. Clean. Prod.* **2003**, *11*, 117–130. [CrossRef]
82. Menkel-Meadow, C. Lawyer Negotiations: Theories and Realities—What We Learn from Mediation. *Mod. Law Rev.* **1993**, *56*, 361–379. [CrossRef]

land

Article

Agricultural Land Transition in the "Groundnut Basin" of Senegal: 2009 to 2018

Bonoua Faye [1] and Guoming Du [2,*]

[1] School of Economics and Management, Northeast Agricultural University, Harbin 150030, China; bonoua.faye2021@neau.edu.cn
[2] School of Public Administration and Law, Northeast Agricultural University, Harbin 150030, China
* Correspondence: duguoming@neau.edu.cn; Tel.: +86-13384657203

Abstract: The study aims to reveal the transition features of agricultural land use in the Groundnut Basin of Senegal from 2009 to 2018, especially the impact of urbanization on agricultural land and the viewpoint of farmland spatiotemporal evolution. Integrated data of time series MCD12Q1 land-use images of 2009, 2012, 2015, and 2018 were used to provide a land transition in agricultural and urban areas through the synergistic methodology. Socio-economic data was also used to serve as a basis for the argument. The results highlight that: (1) Agricultural land increased by 14.53%, with a dynamic index of 1.45 from 2009–2018. (2) Over the same period, urbanization increased by 2.80%, with a dynamic index of 0.28. (3) In different regions, the transition of agricultural land in Kaffrine is most intense (expansion rate: 22.80%). The same situation of urbanization happened in Thiès Region with a value of 7.94%. Except for Thiès, agricultural land in other regions has not yet been subject to major pressure due to urbanization. Overall, the farming system in Groundnut Basin is an extensive model, the recommendations from the point of view of land-use planning and land law are necessary to ensure efficient agricultural land management in the area.

Keywords: agricultural land transition; urbanization; land management; groundnut basin; Senegal

Citation: Faye, B.; Du, G. Agricultural Land Transition in the "Groundnut Basin" of Senegal: 2009 to 2018. *Land* **2021**, *10*, 996. https://doi.org/10.3390/land10100996

Academic Editors: Uchendu Eugene Chigbu, Ruishan Chen and Chao Ye

Received: 15 August 2021
Accepted: 16 September 2021
Published: 22 September 2021

1. Introduction

Land-use changes are consistently the principal driver of habitat change on approximately half of the earth's terrestrial surface [1]. From 2000 to 2015, the global land cover trend indicated a net loss of natural/semi-natural lands. These losses of land resulted from many processes. Among the processes, we can identify deforestation, unsustainable agricultural practices, urbanization, land tenure, and poverty [2]. This situation is more visible in industrialized countries, in Europe or Asia. Therefore, the land-use study is of fundamental significance, as the land resources play a strategic role in the determining man's economic, social and cultural progress [3].

Indeed, the global land system faces unprecedented pressures from growing human populations and climatic change [4]. Urban sprawl on agricultural land has become a global phenomenon plaguing all countries of the world, rich or poor [5]. This phenomenon has attracted the attention of social researchers since the mid-20th century [6]. Thus, many researchers are interested in the causes and consequences of urbanization. Some authors such as G. Duranton and D. Puga highlight that economic growth drives urban expansion in constructing businesses, dwellings, roads, leisure centers, transportation, etc. [7]. Therefore, the metropolitan regions face the growing problems of urban sprawl, including a decline in natural vegetation, wildlife habitats, and agricultural land [3]. More than 8.8% of European Union (EU) land is used for residential, commercial, and industrial purposes [8]. Other very important factors are the quality of the environmental conditions of agricultural production, population density, and net migration [9], which also affect agricultural land. Therefore, it is important to note that the life of more than 7.7 billion human beings today [10], 9.73 billion by 2050, and 11.2 by 2100 will depend on the soil availability for food production [11]. As long as urban populations continue to grow, the challenge of maintaining food security is increasing [12]. In Eastern and Central European countries, namely in Hungary, Slovak Republic, Poland, Estonia, Lithuania, and Latvia, agricultural land prices were gradually increasing in these countries during the past decade [13], and this price depends on the market forces of the supply and the demand [14]. In Romania, the political and socio-economic changes began in the early 1990s had a strong impact on land, especially on the quantity and quality of agricultural terrains [15]. Furthermore, we found that in Bulgaria arable land abandoned without cultivation, represents a very large area because of a shortage of financial resources [16]. In addition, land privatization and farm restructuring are inseparable issues for agriculture and the rural sector in the transformation to a market economy [17]. The case of Slovakia illustrates this. Indeed, the pressure from investors has increased in creating new residential, commercial services and shopping centers, logistic and industrial parks, which is consequently reflected in land-use changes [18]. In summary, in Europe the areas with the most visible impacts of urban sprawl are in countries or regions with high population density and economic activity like Belgium, northern Italy, the Paris and Madrid regions, and Germany [19].

The south Asian region as a whole is experiencing expansion and intensification of cropland and urbanization, shrinking of forests and grassland [20]. On the other hand, in China, for example, the problems in urban-rural spatial structure and food security have been the hot spots of land use research [21]. It is because China has urbanized rapidly over the past three decades, underpinned by rapid economic growth. It has many rapidly growing cities and some that have had declining populations [22]. As such, the per capita area of farmland fell from 0.106 ha in 1996 to 0.092 ha in 2008, raising concerns about food security [23]. Therefore, migration, rural economic development, and urbanization are the primary forces driving the conversion from farmland to non-agricultural uses in China [24]. Other researchers attest that immigration plays an important role in land-use transitions because it leads to urban expansion and farmland occupation [25].

The review of the literature also conduced us to examine land tenure. Indeed, land tenure security refers to the right of individuals and groups of people to effective protection by their government against forcible evictions [26]. Therefore, in the agricultural sector, securing land tenure has regularly been prioritized by policy-makers to ensure and develop more productive agriculture [27]. In that sense, secure tenure is widely recognized as an essential foundation for achieving a range of rural economic development goals [28]. However, land tenure security is central to agricultural production and sustainable use of natural resources [29], and the links between tenure security and agricultural productivity are of primary interest [30]. This may imply that land tenure systems can affect agricultural productivity by influencing the efficient use of inputs and the adoption of modern technology [31]. On another note, secure access to productive land reduces the vulnerability to hunger and poverty to the millions of poor people living in rural areas and depending on agriculture. It influences their capacity to invest in their productive activities and the sustainable management of their resources [32]. From then on, land tenure security is important not only for agricultural production, but it also allows people to diversify their livelihoods by using their land as collateral, renting it out, or selling it [33]. Finally, strengthening land tenure security is key to achieving efficient land allocation among farmers in both in land abundant and land-constrained areas as it facilitates land markets [34].

In the above background, it is important to analyze the evolution of the population to in order to understand rapid urbanization. Globally, more people live in urban areas than in rural areas [35]. We found that 54 percent of the world's population live in urban areas. In view of this, we noted that the urban population of the world has grown rapidly since 1950, from 746 million to 3.9 billion in 2014 [10]. Along with this, the medium-variant projection indicates that the global population could grow to around 8.5 billion in 2030 [36]. Hence, prior work has highlighted that urbanization is a problem because urban expansion inevitably covers some agricultural land.

In contrast, changes in land values and land markets around cities often result in vacant land as the owners anticipate their gains from selling it or using it for non-agricultural uses [22]. Therefore, the changes in land use arise from competing for economic, political, social, and environmental goals [37], and urban sprawl dynamics will also play an important role in the future land-use change in some countries like India [38], and Africa countries. In summary, land-use change is characterized by a high diversity of change trajectories depending on the local conditions, regional context, and external influences [39]. Therefore, it is important to understand the underlying technological, institutional, and economic drivers of land-use change and how they play out in different environmental, socio-economic, and cultural contexts [40].

According to the literature review, the drivers of the agricultural land transition are mainly non-agricultural activities—for instance, residential construction. The more human activities intensify, the more agricultural land is threatened. Therefore, as Senegal is a developing country, this study is important from the point of view of understanding the mechanisms and the main factors influencing the transition of agricultural land. Given this worrying situation, the objective of this work is to (i) understand the degree and trend of transition and evolution of agricultural land space and time. Then, our interest is to (ii) identify the keys drivers of this transition and evolution of agricultural land in the Groundnut Basin. In this study, we will focus on these important issues.

2. Study Area

Senegal is located in the West African continent, with a land area of 196,722 km^2. It is located between the latitudes 12°20′ and 16°20′ N and the longitudes 11°20′ and 17°30′ W [41]. Ecologically and agriculturally, the country is subdivided into six geographical eco-zones [42]. Our research concerned the Groundnut Basin. It covers the administrative regions of Diourbel, Thiès, Kaolack, Fatick, Kaffrine, part of the Tambacounda region (departments of Koumpentoum and part of the department of Tambacounda) and the department of Kébémer [43]. Therefore, our study focuses only on the five administrative regions (Figure 1). They are considered most important in this area because they occupy almost all the arable land in the Groundnut Basin, and constitute an area of very high agricultural production in Senegal [44]. It has a population of 6,436,912 people according to the population censuses in 2013 [45], and covers a total area of 34,964.36 km^2, with a density of 184.10 people per km^2. The distribution of arable land by agro-ecological zone shows that the Groundnut Basin represents 70% of arable land [46]. Throughout the Groundnut Basin, the cropping systems are mainly cereal-leguminous rotations [47], and it is dominated by subsistence production of millet, maize, groundnuts, cowpeas, and bissap (hibiscus) [48]. Groundnut Basin is characterized by degraded and patchy open forests dominated by *Bombax costatum*, *Lannea acida*, *Pterocmpuserinaceus*, *Sterculia setigera*, *Khaya senegalensis*, *Daniellia oliveri*, *Detarium senegalensis* [49].

With a poverty rate of 47% in 2011 [50], agriculture in Senegal has always been seen as the foundation of the country's socio-economic development [51]. In terms of economic activities, this sector is dominated by agriculture and occupies 74% of the population in the area [52]. Its average in terms of gross domestic product (GDP) ranged from 21.24% (1960–1989) to 15.26% (1990–2011) [53], and further to 16.1% in 2017 [54]. According to data from the National Agency for Statistics and Demography (NASD) site, in 2020, the areas planted (in hectares) in peanuts represent, respectively, 55.16% (2017) and 63.15% (2018) of the national areas (in hectares) planted. In 2013, the population censuses showed that 70% of farms were small family farms with an area of fewer than five hectares [45]. In the study area, the median value of annual rain-fed crop sales per household in 2018 is around $246.61. According to the NASD site, the percentage of farm household members with agricultural education vary from 5.34% in 2017 to 0.67% in 2018.

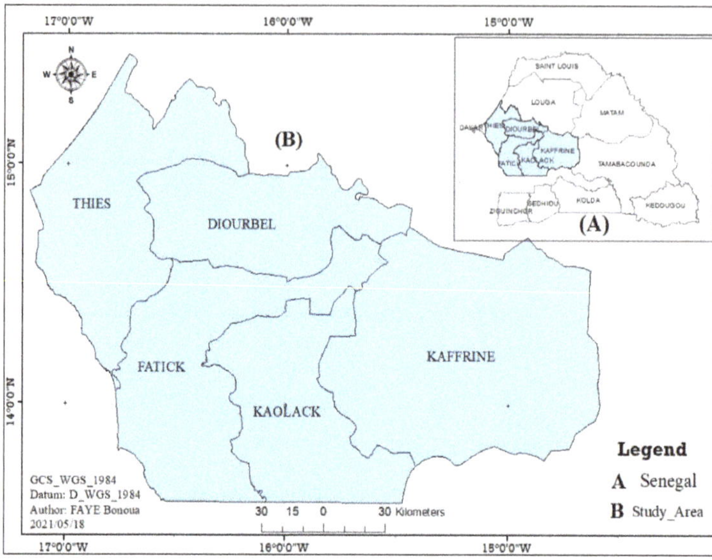

Figure 1. Geographic information: (**A**) represents the blue color represents the localization of the study area within Senegal; (**B**) represents the Groundnut Basin with five regions.

In fact, in Africa, particularly in the Sahel, after the rainy periods of the 1960s, many researchers noted anomalies of rain in the early 1970s [55]. The consequences of this rainfall deterioration are reflected in Senegal by the degradation of the natural environment, with drought leading to the degradation of the plant cover, the soils being subjected to erosion and runoff, and the accentuation acidification and salinization [56]. In addition, the factors of low annual rainfall, frequent dry spells, and the rainy season shortening affect the vegetative cycle of crops [57]. Severe droughts, especially in northern regions, appear as the biggest risk in estimated aggregate losses to crop and livestock [58]. In Senegal, agriculture is mainly rain-fed and depends heavily on seasonal rainfall amounts, and distribution [59]. Therefore, the combined effect of rainfall, land surface temperature, and solar radiation explain approximately 40% of the variation in cropland productivity over West Africa at the 95% significance level [60]. This situation underlines the fact that the evolution of agricultural land, and climate are closely linked. Drought is a recurring phenomenon in the coastal zone of Senegal and its hazards affect the economies of the predominantly agricultural population [61].

The main factors affecting land-use in Senegal are manifold. Firstly, lots of land areas were changed by the urban setting and associated rapid growth and transformation of human societies in Senegal [42]. In 2015, Senegal's population was estimated at 14,356,575 people with an average annual growth rate of 2.7% [45]. The growth of the rural population has brought greater pressure on land and natural resources and contributed to land fragmentation, particularly in densely populated and high-potential areas with easy access to markets [62]. As a result, in Sangalkam, around Dakar (capital of Senegal), some 7.64 km^2 of agricultural land use was officially developed in the area between 2003 and 2009 [63]. The law n° 64–46 of 17 June 1964, governs land management in Senegal, which stipulates, land does not belong to the State, territorial communities, or users, but the "Nation" [63]. Local governments are responsible for the allocation/dedication of land in the national domain for rural activities [64]. Secondly, the chronological summary of agricultural policy in Senegal has gone through several phases. The literature shows the Agricultural Program (1960–1980), the New Agricultural Policy (1985–1994) and the Agricultural Development Policy Programs, letters and Declaration (1995–2003). Since the 2000s, we have seen a reconfiguration of the situation regarding agriculture. For example, we have the Law of

Agro-Sylvo-Pastoral (LOASP) in 2004. Since 2013, Senegal has defined a new agricultural policy called the Program to Accelerate the Cadency of Senegalese Agriculture (PRACAS). Despite all these agricultural policies, Senegalese agriculture still faces difficulties [65]. Third, soils in Africa affected by water erosion ranging from medium to high effect cover an area of more than 12 million hectares, or 18.5% of the total national territory [66]. The Groundnut Basin is today confronted with chemical and physical-biological degradation which has become more intense. Thus, the soils are impoverished, restructured, chemically exhausted by wind and water erosion, recurrent droughts [67].

3. Materials and Methods

3.1. Data Sources

The data presented in this study are derived from different databases (please see Table 1). The analysis model aims at analyzing the agricultural land evolution and transition in terms of area and soil occupation change over ten years chronological series between 2009 to 2018.

Table 1. Data sources information.

Types		Information
Spatial data	Source	The spatial data used for this study are obtained from USGS (United States Geological Survey) and NASA (National Aeronautics and Space Administration), Modis land cover-Modis MCD12Q1V6 (acquisition data: 26 March 2021) https://earthexplorer.usgs.gov/
	Types	cropland; grassland; urban and built-up; permanent wetland (four images in 2009; 2012; 2015, and 2018 between June and October)
Socio-economic data	Source	National Agency for Statistics and Demography (NASD) (acquisition data: 30 March 2021) https://satisfaction.ansd.sn/
	Types	Evolution of the population; characteristics of farm household members; farm household members with agricultural education.
Agricultural data	Sources	World Bank (W.B) and Food and Agriculture Organization (FAO) (acquisition data: 30 March 2021) http://www.fao.org/faostat/fr/#data https://donnees.banquemondiale.org/indicator
	Types	Fertilizer consumption; permanent cropland (% of land area); agriculture value added (% of gross domestic product)
Climatic data	Sources Types	Data Portal: Center for Hydrometeorology and Remote sensing—available online: https://chrsdata.eng.uci.edu/ (acquisition data: 28 August 2021) Rainfall data (between June and October)

3.2. Methodology

The spatial data used in this study are obtained from NASA LPDAAC Collections-Modis Land Cover—MCD12Q1V6 (Earth Explorer USGS). However, due to their characteristics, pre-processing is necessary to have more clarity. Therefore, several steps have been taken. First, to optimize the quality of the images, the layers were re-projected according to the reference projection system of the study area, which is World Geodetic System (WGS)_1984_Complex_UTM_Zone_28N (EPSG:31028). In addition, the strips were cut according to the size of the study area. To make the classified land cover images comparable, we resampled the images to 50 m, which is the common resolution for all images [68]. This resampling allows us to obtain common results between the processed images. Second, after this geometric correction, we used supervised classification to categorize the land cover components used in our study. The land-use types are mainly classified into four categories: cropland, grassland, urban and built-up, and permanent wetland. Then, concerning the spatial analysis, four temporal remotely sensed images were selected for evolution and transition land use detection, namely, Modis Land Cover of USGS in 2009, 2012, 2015, and 2018. All four images, respectively, were used to examine the area's evolution and transition land-use dynamics. The area information was used as a basis for analyzing the quantitative change in land use. After the conversion of raster data to vector data, we had to use ArcGIS 10.6 platform to analyze the change pattern of cropland, grassland, urban and built-Up, and permanent wetland.

3.2.1. Analysis of Land Use Dynamics

Dynamics of land use are expressed as an increase or decrease in area. Therefore, this method positively affects the analysis of the evolutionary pattern of land use in the study area. In addition, it gives an idea of the future dynamics of transition and evolution of land use over time.

$$KT = \frac{Ub - Ua}{Ua} \times \frac{1}{T} \tag{1}$$

where *KT* is the dynamic attitude of the pattern using during the study period, *Ua* is the area of the pattern at the beginning of the study period, *Ub* is the area of the pattern at the end of the study period *T* is the time interval of the study years [69].

3.2.2. Calculation of the Land Evolution

The degree of evolution in each tenure category will be assessed by calculating the rate of evolution *E* (*i*, *k*) in the area of land use as follows [70].

$$E\ (i,k) = \frac{Sk - Si}{Si} \times 100 \tag{2}$$

(*Si*) the area of a land-use category of the year *i* and (*Sk*) the area of a land-use category of year *k*, with *k* > *i*. *E* (*i*, *k*) will be equal to: If *E* (*i*, *k*) = 0, it is concluded that this land use category is stable, if *E* (*i*, *k*) < 0, it is concluded that there is a regression of this category, and if *E* (*i*, *k*) > 0, there is an extension or evolution of this category.

3.2.3. Analysis of Agricultural Land Transition

To visualize the land cover transition, we merged the layers using the intersection tool in the ArcGIS platform. Then, we used the equation below to show the different transitions between the spatial data and used the pivot table function in Excel to produce statistics according to a couple of classes. Initially, the results of the four components studied showed fourteen different land-use change classes. Since not all land-use change classes have the same degree of importance, those representing the main changes were kept and the others were unified. For example, when we considered the relationship between grassland and permanent wetland, we obtained two different couple of classes change: grassland to permanent wetland and permanent wetland to grassland. Accordingly, the difference in change between these two couple of classes is maintained. In addition, we opted for this methodology to highlight the most important or the dominate transition between couple classes. This operation allowed us to keep the four most representatives, land-use change classes.

$$\mathrm{T} = \mathrm{Layer\ A} + (-) + \mathrm{layer\ B} \tag{3}$$

Layer A = corresponds to the year of beginning, Layer B = corresponds to the year of arrival, T = result of the transformation. (This method is used by GIS and RS solutions).

3.2.4. Analysis of Climatic Data: Rainfall

To better understand the factors of evolution and transition of agricultural land in the Groundnut Basin, we have analyzed the inter-annual evolution of rainfall over the period from 2009 to 2018. Indeed, Senegal has two main seasons that mark the climatic regime: a dry season from November to April–May, and a rainy season from May–June to October, depending on the geographical location [41]. Accordingly, the rainfall values used in this analysis only concern the period from June to October, which coincides with the rainy season in our study area [71]. The methodology adopted is based on a statistical approach, using the averages of the five months of rainfall for each year in the series.

4. Results

4.1. Analysis of the Land-Use Evolution

4.1.1. Cropland and Grassland Evolution

Agricultural land is becoming increasingly scarce and threatened by several factors. This situation can be fast/slow and differs from one country to another. Therefore, the transition of agricultural land is relatively fast in developed countries due to industrialization but is slow in underdeveloped countries. Tables 2 and 3 shed light on the rate of evolution in agricultural land area, and urbanization has been increased in the last few decades. On the other hand, grassland has decreased.

Table 2. The statistics of the land-use area in years 2009, 2012, 2015 and 2018 (km^2).

Land Use Pattern	Years/Values			
	2009	2012	2015	2018
Cropland	14,569.53	15,541.06	15,913.48	16,687.16
Grassland	18,656.27	17,704.38	17,295.78	16,566.23
Permanent wetland	758.80	778.41	779.76	768.63
Urban and built up	172.49	174.85	175.41	177.32
Other	807.27	765.66	799.93	765.02
The total surface of the study area	34,964.36	34,964.36	34,964.36	34,964.36

Table 3. Dynamic land-use evolution in the period of 2009–2012, 2012–2015 and 2015–2018 (%).

Number	Periods	Land Use Pattern			
		Cropland	Grassland	Permanent Wetland	Urban and Built Up
P1	2009–2012	+6.67	−5.10	+2.58	+1.37
P2	2012–2015	+2.40	−2.31	+0.17	+0.32
P3	2015–2018	+4.86	−4.22	−1.43	+1.09
The study period	2009–2018	+14.53	−11.20	+1.30	+2.80

(−) Decrease in area; (+) increase in area.

The research revealed that an increase in cropland (Figure 2), at the beginning of the study period, which coincides with the year 2009, the cropland represented around 14,569.53 km^2. On the other hand, at the end of the study in 2018, the cropland is around 16,687.16 km^2. Therefore, a difference of 2117.63 km^2 was noted, including an increase of 14.53%. The dynamic attitude of cropland amounts to 1.45. This dynamic seems to be important. This means a potential doubling of the cultivated areas around 2028.

On the other hand, our analysis shows a decrease in grassland. The grassland represented respectively 18,656.27 km^2 in 2009 and 16,566.23 km^2 in 2018, showing a decrease of about −2090.04 km^2 (−11.20%). Therefore, the cropland increases as the grasslands decrease.

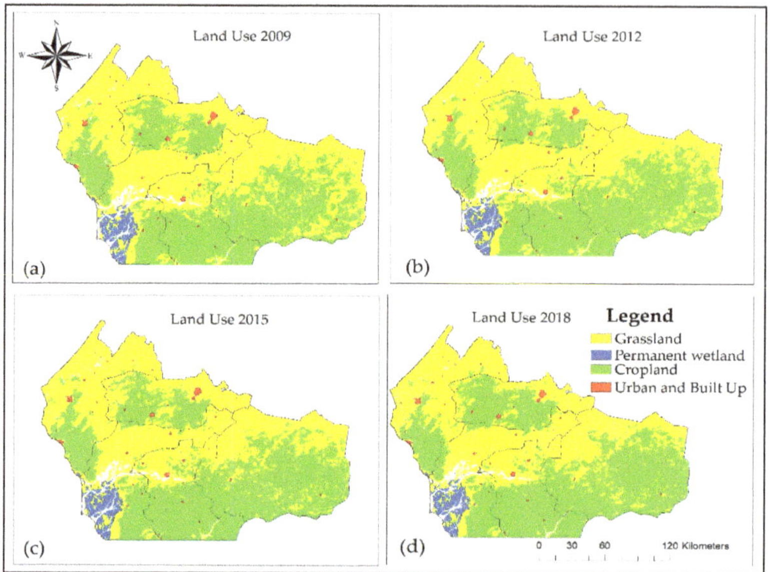

Figure 2. (**a**–**d**) show land use evolution patterns in years of 2009, 2012, 2015, and 2018 in the study area.

As for the results between periods, the cropland also shows fluctuations. Thus, Table 3 reflects an evolution of 6.67% over the period 2009–2012. This result has decreased to 2.40% between 2012 to 2015 with a deviation of 4.27%. Compared to the second period, we have observed in the third period an increase of 4.86%. Therefore, if we compare the first period to the last period, the results reflect that the cropland has not increased significantly. In summary, the cropland has slightly increased by 1.81% between the first and last periods (Table 3). This situation remains the same for the grassland. Respectively, the results are decreasing by −5.10%, −2.31%, and −4.22% and the difference noted between the first and third periods is about 0.88%. In summary, the study highlights important knowledge that an expansion of cropland during the study period (Figure 2). These results are confirmed by the data of the World Bank site (2020). Indeed, they show that the permanent cropland represented 0.30% in 2009 and 0.35% in 2016, increasing 0.05% in Senegal.

4.1.2. Urban and Built Up and Permanent Wetland Evolution

Urbanization is one of the factors that affect agricultural land. Therefore, it appears differently depending on the context and evolution of the population. In our study area, urbanization seems to be slow and occupies the space little by little. The dynamic attitude of urban and built-up turns around 0.28. Between 2009 and 2018, urban and built-up represent 2.80% and permanent wetland 1.30%.

Urban and built-up and permanent wetland are not significantly represented. Indeed, at the beginning of our study in 2009, urban and built-up represents an area of 172.49 km^2 out of 34,964.36 km^2, which corresponds to the total area of the study area. This area has evolved to reach 177.32 km^2 at the end of the study in 2018, including an increase of 2.80%. However, the inter-period results show an increase of 1.37% between 2009 and 2012. This value has changed slightly from 2012 to 2015 with an area of 0.32 km^2, a decrease of about 1.05% less. For the last period, our research reflects an increase of 1.09%. In addition, between the first and the last period, urban and built-up grew by 0.28%.

In summary, urbanization has intensified in the second and last periods. The permanent wetland reflects an increase of 2.58% in the first period, 0.17% in the second period, that is to say, a difference 2.41%. This result continues to decrease, reaching −1.43% in the third period. Finally, between the first and the last period, this variable has decreased by about −1.15%.

4.2. The Regional Difference of the Study Area in Land-Use Transition

Analysis of regional differences is important to understand the dynamics of agricultural land use in each region. We have firstly analyzed the cropland and grassland. Concerning cropland, from 2009 to 2018, an increase is noted in all the five regions. The region of Kaffrine has the highest value with 1337.95 km^2 (22.80%). It is followed by the regions of Diourbel (19.50%) and Fatick (9.71%). The regions of Kaolack and Thiès show 3.83% and 2.53% over the ten years, respectively. This increase hides inter-annual disparities. Between 2009 and 2012, the area of cropland in the Diourbel region decreased by −5.81%. However, over the same period, the area increased in Fatick (4.27%), Kaffrine (13.31%), Kaolack (7.01%), and Thies (1.70%). This approach shows that cropland evolution is not exponential but in a sawtooth pattern. This pattern is similar to grassland. In general, it has decreased. The Kaffrine region occupied first place with −24.81%, followed by the Diourbel (−18.74%) and Kaolack (−7.26%). In contrast, the Thiès region shows an increase of 0.11%.

For urban and built-up and permanent wetland, the situation is the same. Concerning urban and built-up, the region of Thies comes in first place with an evolution of 4.02 km^2 (7.94%) between 2009 and 2018 (Figure 3). This evolution trend remains the same for the Diourbel region with 0.93%. The evolution of urban and built-up is average in the Factick region with 0.77%. In contrast, the regions of Kaolack (−0.17%) and Kaffrine (−0.61%) show a decrease in urban and built-up areas in the same period. Permanent wetland concern largely the Fatick and Thiès regions. In the Thiès region, the evolution is stable. However, in the region of Fatick, it has been recorded an increase of 9.80 km^2 (1.31%). However, this result hides disparities. In the same region shows an increase of 2.60% from 2009–2012, to reach a decline of −1.43% from 2015–2018. Finally, the analysis of regional differences in land use shows several aspects. The first aspect shows that in the regions of Kaffrine and Diourbel, cropland has evolved rapidly, but urban and built-up remains low in Kaffrine (Figure 3). In contrast, urban and built-up expansion is relatively rapid in the Thiès region, and moderate in the Diourbel region. However, in the Thies region, the evolution of cropland and grassland has evolved weaknesses compared with other regions.

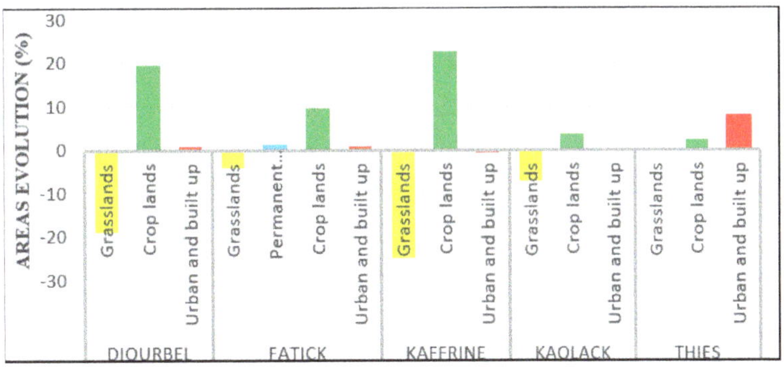

Figure 3. Regional differences in land use evolution from 2009 to 2018 (%).

4.3. *Analysis of Land Use Transitions*

4.3.1. Inter-Period Transition

To understand agricultural land transition dynamics, the series is divided into three equal periods. The first period from 2009–2012 shows a decrease in grassland in favor of other variables. 944.58 km^2 of grassland was transformed into cropland. Then, 12.23 km^2 of grassland transformed into a permanent wetland, and 2.12 km^2 became urban and built up. This first approach shows a significant decrease of grassland at the expense of urbanism and permanent wetland indeed. Urbanism and permanent wetland have occupied about 14.35 km^2 of grassland. Urban and built-up alone, records 2.27 km^2 on the grassland.

Transition in the second period (2012–2015) appears to be less intensive than in the first period. The transformation of cropland decreased. The area transformed from grassland to cropland went from 944.58 km^2 to 363 km^2, a decreasing 580.8 km^2. Dynamic continues with the changes from grassland to permanent wetland with 2 km^2. Similarly, 1.15 km^2 of grassland has been converted to urban up and built. This situation shows a relatively slow evolution of urbanization. Therefore, compared to the first period, the results reflect a difference of 1.03 km^2 of grassland converted to urban up. Concerning the last period, it witnesses transition compared to the second period. Indeed, we still note an extension of agricultural land compared to the grassland, representing 772.12 km^2.

In this last part, our analysis points to a reversal of the permanent wetland and grassland. About 10.96 km^2 of permanent wetland have been transformed into grassland. Urbanization as for it evolved. We observed a 2.21 km^2 of grassland are again transformed into urban and built up. The inter-period analysis of land transition shows significant fluctuations. A few areas of cropland were transformed into urban and built up, in other cases, the grassland was transformed into cropland or urban and built up.

4.3.2. Changes That Occurred during the Study Period

Analysis done during the period of the study period highlights some changes between the variables. Among the most significant changes, the results underline an extension of other components on the grassland. 2083.29 km^2 of grassland was transformed into agricultural areas (Table 4). Similarly, 5.42 km^2 of grassland was transformed into urban and built up. In addition, 3.63 km^2 of grassland was transformed into a permanent wetland. We have noticed a significant change in grassland, in the study area which has continued to increase. This change is especially noticeable in the Kaffrine region (Figure 4). The balance of land-use changes observed over the study period shows that the most important relationships are between grassland and the other land use pattern, namely cropland and urban up and built.

Table 4. Land use transition over the study period (2009–2018) in km^2.

	Cropland	Grassland	Permanent Wetland	Urban and Built Up
Cropland	13,288.12	x	x	x
Grassland	2083.29	15,202.92	3.63	5.42
Permanent Wetland	x	x	747.53	x
Urban and built up	x	x	x	166.35

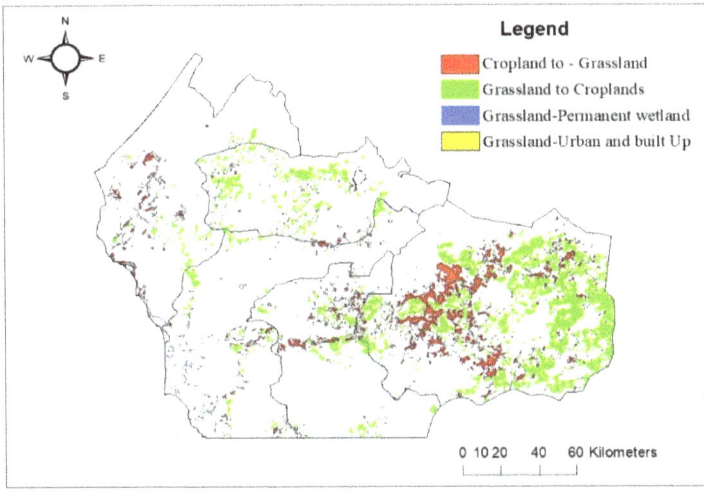

Figure 4. Land use transition throughout the study period (2009–2018).

5. Discussion

5.1. Impacts of Urbanization and Population on Agricultural Land Transition

The National Agency of Statistics and Demography (NASD) site projections of population trends in the study area show a rapid evolution. The population projections data show that there were 5.059,331 million people in 2009 and 6.436,913 million in 2018, with an increase of 27.23% in ten years. A rapid increase in population may be an explanatory factor that is at the origin of agricultural land transition because evolution in population induces demand for housing and occupation of new spaces. On the other hand, the results highlight that urbanization is expanding rapidly in the Thiès area (Figure 3). According to NASD site data, the Thiès region is the most urbanized and populated region after Dakar. This region, whose land area represents less than 2% of Senegal's land area (196,722 km^2), concentrates more than 25% of the national population [45]. As a result, the overcrowding of the capital (Dakar), partly explains the rapid development of urbanization in the Thies region. This region (70 km from Dakar) now serves as a secondary city to correct the territorial imbalance; it has been the area to major state projects such as the new Blaise Diagne International Airport. From this perspective, agricultural land fragmentation and scarcity are still mentioned as of considerable constraints on agricultural modernization. It could be exacerbated in the affected area due to a huge agricultural land acquired to support urbanization and industrialization [72]. Therefore, it is undeniable that the loss of agricultural land to urbanization is a serious threat to food security and poverty alleviation, especially in regions where many people are already poor. Consequently, agricultural development in Senegal has to face many challenges related to good land administration and planning for successful socio-economic development, particular rural economic transformation.

5.2. Climatic Factors That Influence the Evolution of Agricultural Land

The links between the land and the global climate have long been known [73]. Thus, the scientific literature provides positive examples of that problem. It points out that land degradation is a complex process involving the natural ecosystem and the socioeconomic system. Climate and land-use changes are the two predominant driving factors [74]. However, it is clear that climatic factors, including temperature or rainfall, can impact land-use. In this study, the pivotal factors that can influence agricultural land's transition focus on the rainfall.

Generally, the projected changes in climate include recurring climate extremes like droughts, flooding, and outbreaks of pests and diseases exposing the region to the vulnerabilities of the changing environment [75]. The agricultural sector is one of the first affected by this change [64] because rainfall is the main factor affecting agricultural production [76]. Therefore, the erratic spatio-temporal distribution of rainfall can often be the origin of an increase or a decrease in the cropland. Past studies emphasized two normal years with a dryness trend in 2012 and 2013 in West African, particularly Senegal [77]. This dryness can have a negative influence on the area planted. For instance, our study shows a decrease in cropland during the periods of 2012–2015 and 2015–2018. Indeed, the inter-annual evolution of rainfall during the period 1985–2014 in the region of Kaolack shows thirteen years out of thirty that are deficient compared to the average of the series which is 604.0 mm of rain. The most deficient year was 2014 with 423 mm [78]. The above background confirms the rainfall results analyzed for the whole study area. Indeed, the analysis made on the evolution of the rainfall shows that the rainfall varies from one year to another. Indeed, Figure 5 shows three periods. The results of the second period show that the rainfall is decreasing and 2014 is the most deficient year of the whole period. Similarly, after an increase in rain in 2015, the rainfall decreased over the third period. Accordingly, the decrease/fluctuation in rainfall during this period may explain the reduction or increase in cropland in the Groundnut Basin.

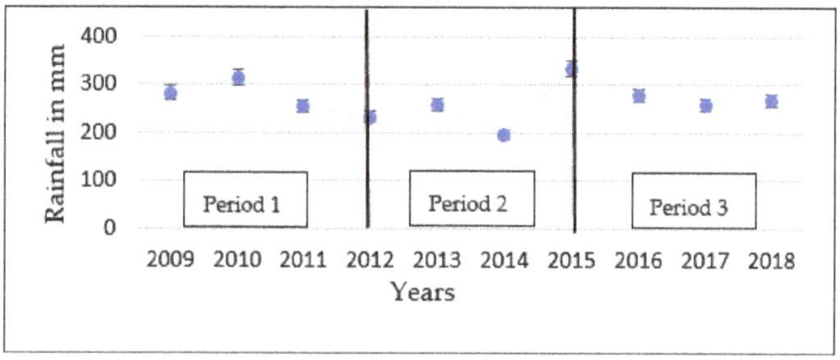

Figure 5. Annual evolution of rainfall: 2009 to 2018. Source CRHS 2021.

On the other hand, our analysis reflects the decline of urban and built-up in the regions of Kaolack and Kaffrine. Formerly inhabited areas are abandoned, leaving the place to fallow. These two regions are predominantly agricultural (Figure 2). Therefore, this situation has been proven in some studies in the past. Indeed, the decline of groundnut cultivation, coupled with the disappearance of certain industrial facilities for processing this raw material, has aggravated the situation, leading to a massive displacement of populations from the former urban centers of the Groundnut Basin to the metropolis of Dakar [65]. Therefore, many socio-economic and climatic factors can influence agricultural land, and some studies demonstrate about 74% of farmers perceived that erratic rainfall seasonality contributes significantly to the land-use change and agricultural land abandonment [79].

5.3. The Means of Financial Has Effect on Agricultural Land Evolution in Senegal

This link between agricultural land transition and financial means has been highlighted in the literature. For instance, cropland increases in the United States, and pastureland decreases when government payments go up [80]. The same observation is noted in China, where economic restructuring also influences the overall evolution of farmland areas [25]. Despite a shortfall in rainfall in recent years, Senegal has significantly improved its results due to the selection of seeds and strong mechanization, which have positively impacted agricultural yields. Yields have witnessed a dramatic increase [81]. To understand this phenomenon, we have analyzed the evolution of agricultural investments. This analysis shows an identical correlation between the variables. According to data from the World Bank site, investments in the agricultural sector in Senegal have evolved considerably during the study period. They represented 16,323 million CFA ($30,471.27) in 2009 and 51,585.6 million CFA ($96,297.26) in 2018, including an increase of 35,262.6 million CFA ($65,826). Accordingly, we found that the agricultural investments for the Groundnut Basin represent 17.3% [82]. Meanwhile, fertilizer use has doubled. The use of fertilizer has evolved from 18,489,000 kg in 2008 to 37,000,000 kg in 2018, including an increase of 18,511,000 kg. These huge investments in the sector can justify the evolution of agricultural land in the area. The evolution of agricultural land in the Groundnut Basin depends more on financial means.

6. Conclusions and Recommendations

The findings of this study reveal that the agricultural land has not yet been subjected to real human pressure. From 2009–2018, urban and built-up occupies only 177.5 km^2 of the total area (34,964.36 km^2) and increases at 2.80%, with a dynamic attitude of 0.28. This situation is because the fact that it is an under-populated area with little urbanization. The density represented 184.10 people per km^2. Therefore, basic needs such as housing, infrastructure, and services are poorly developed in the area. Analysis of the regional difference shows that the Thiès region alone occupies 54.61 km^2. Today, this region is considered an integral part of the Dakar region (25% of the country's population), which is the most urbanized in the country [83]. The most visible case in the study is the extension of cropland on the grassland. It represents 14,569.53 km^2 in 2009 and 16,687.16 km^2 in 2018, including an increase of 2117.63 km^2 (14.53%). The region of Kaffrine alone recorded an increase of 1337.95 km^2 (22.80%) and followed by Diourbel (19.50%). The justifying factors can be related to agricultural investments and climatic performances such as rainfall or fertilizer. The use of fertilizer is increase reaching about 18,511,000 kg over the period. For the transition of areas is relatively intense, and the results reflected that 2083.29 km^2 of the grassland was transformed into cropland; 5.36 km^2 of the grassland was transformed into urban and built up. In general, the agricultural land in the study area has not yet undergone a major transition.

According to these findings, recommendations are necessary to ensure efficient and balanced management of agricultural land in the future. First, agricultural land use planning is essential. Given the increasing urbanization and large-scale agriculture, the establishment of an agricultural nature protection zone is necessary, to develop and seek consensus on rules guiding the sustainable utilization of agricultural resources. Second, the agricultural land is disappearing faster than population growth. It is therefore imperative to move towards zero lands "artificialization". That is to say that we must resort to the restoration of degraded land to compensate for the land newly occupied by urban and built up. Third, land tenure management. Secure land and property rights are critical for reducing poverty and for enhancing economic development, gender equality, social stability, and sustainable resource use. We propose a restructuring of the basics of law No. 64–46 of 17 June 1964 to facilitate access to and proper management of agricultural land.

Thus, the restructuring of this law will strive to put in the place a legal system that will facilitate access to and management of land in general and agricultural land in particular. However, it is relevant to integrate all stakeholders in the reform process, strengthen the existing land access procedure, and to further include land issues in the decentralization and agricultural development policy laws. In addition, access and control over land are problematic in Senegal, especially under customary rule [84]. Therefore, a more equitable redistribution of access to land, especially to those who can invest in agricultural development, might be an important point in restructuring this law.

Furthermore, land issues are becoming increasingly complex due to economic development and population growth. Thus, the lack of coordination between socio-economic development laws can be seen as a blocking factor in the reform. Similarly, customary laws on land rights are increasingly challenged in the context of globalization. However, the lack of a clear delineation between the state and private domain may be a limiting factor in the reconstruction of this law. Therefore, it is necessary to formulate rural (such urban) spatial planning promotes; promote the implementation of comprehensive land consolidation projects throughout the region; and optimizes agricultural, ecological, and construction space [85].

Author Contributions: Conceptualization, B.F. and G.D.; methodology, B.F.; validation, G.D., and B.F.; formal analysis, G.D; resources, B.F.; data curation, B.F.; writing—original draft preparation, B.F.; writing—review and editing, G.D.; visualization, B.F.; supervision, G.D.; project administration, G.D.; acquisition of funding, G.D. Both authors have read and agreed to the published version of the manuscript.

Funding: This research was funded by the National Natural Science Foundation of China, Grant Number 41571167.

Institutional Review Board Statement: Not applicable.

Informed Consent Statement: Informed consent was obtained from all subjects involved in the study.

Data Availability Statement: The data presented in this study are available online. Kindly check the Table 1.

Acknowledgments: We would like to express our gratitude to the professionals of the Northeast Agricultural University who encouraged us to make this project a success.

Conflicts of Interest: The authors declare no conflict of interest.

References

1. Vanbergen, A.J.; Aizen, M.A.; Cordeau, S.; Garibaldi, L.A.; Garratt, M.P.; Kovács-Hostyánszki, A.; Lecuyer, L.; Ngo, H.T.; Potts, S.G.; Settele, J.; et al. Transformation of agricultural landscapes in the Anthropocene: Nature's contributions to people, agriculture and food security. *Adv. Ecol. Res.* **2020**, *63*, 193–253. [CrossRef]
2. Sachs, J.; Schmidt-Traub, G. *The Sustainable Development Goals Report 2020*; Cambridge University Press: Cambridge, UK, 2020.
3. Bolca, M.; Turkyilmaz, B.; Kurucu, Y.; Altinbas, U.; Esetlili, T.; Gulgun, B. Determination of Impact of Urbanization on Agricultural Land and Wetland Land Use in Balçovas' Delta by Remote Sensing and GIS Technique. *Environ. Monit. Assess.* **2007**, *131*, 409–419. [CrossRef] [PubMed]
4. Holman, I.; Brown, C.; Janes, V.; Sandars, D. Can we be certain about future land use change in Europe? A multi-scenario, integrated-assessment analysis. *Agric. Syst.* **2017**, *151*, 126–135. [CrossRef]
5. Al Tarawneh, W.M. Urban Sprawl on Agricultural Land (Literature Survey of Causes, Effects, Relationship with Land Use Planning and Environment) A Case Study from Jordan (Shihan Municipality Areas). *J. Environ. Earth Sci.* **2014**, *4*, 97–124.
6. Rubiera-Morollón, F.; Garrido-Yserte, R. Recent Literature about Urban Sprawl: A Renewed Relevance of the Phenomenon from the Perspective of Environmental Sustainability. *Sustainability* **2020**, *12*, 6551. [CrossRef]
7. Duranton, G.; Puga, D. The Growth of Cities. 2013. Available online: https://www.oecd.org/economy/growth/Growth_of_cities_Duranton.pdf (accessed on 29 August 2021).
8. Eurostat Regional Yearbook. Land Cover and Land Use. 2011. Available online: https://ec.europa.eu/eurostat/documents/3217494/5728541/KS-HA-11-001-12-EN.PDF.pdf/88b77a74-2628-4ed7-bb9f-24244d6c78a1?t=1414775978000 (accessed on 10 August 2021).
9. Sroka, W.; Mikolajczyk, J.; Wojewodzic, T.; Kwoczynska, B. Agricultural Land vs. Urbanisation in Chosen Polish Metropolitan Areas: A Spatial Analysis Based on Regression Trees. *Sustainability* **2018**, *10*, 837. [CrossRef]
10. United Nations. *Department of Economic and Social Affairs—Population Division. World Urbanization Prospects. The 2014 Revision*; ST/ESA/SER.A/366; UN: New York, NY, USA, 2015. Available online: https://population.un.org/wup/Publications/Files/WUP2014-Report.pdf (accessed on 27 July 2021).
11. Food and Agriculture Organization (FAO). *The Future of Food and Agriculture Trends and Challenges*; Food and Agriculture Organization: Rome, Italy, 2017.
12. Ambros, P.; Granvik, M. Trends in Agricultural Land in EU Countries of the Baltic Sea Region from the Perspective of Resilience and Food Security. *Sustainability* **2020**, *12*, 5851. [CrossRef]
13. György, K.T.; Sadowski, A.; Bandlerova, A. Land Use and Land Reform in former Central and East European Countries. 2007. Available online: https://www.researchgate.net/publication/23508765_Land_use_and_land_reform_in_former_Central_and_East_European_countries (accessed on 27 July 2021).
14. Cook, E. *Agriculture, Forestry and Fishery Statistics 2018 Edition*; European Commission: Luxembourg, 2018; pp. 28–31. Available online: https://ec.europa.eu/eurostat (accessed on 16 September 2021). [CrossRef]
15. Popovici, E.A.; Balteanu, D.; Kucsicsa, G. Assessment of changes in land-use and land-cover pattern in Romania using CORINE Land Cover database. *Carpathian J. Earth Environ. Sci.* **2013**, *8*, 195–208. Available online: https://www.researchgate.net/publication/272161616_Assessment_of_changes_in_land-use_and_land-cover_pattern_in_Romania_using_CORINE_Land_Cover_database (accessed on 28 August 2021).

16. De Arriba Bueno, R. Evaluating land reform and market in Bulgaria. *J. Econ. Bus.* **2007**, *10*, 11–13.
17. Csaki, C.; Lerman, Z. Agricultural Transition Revisited: Issues of Land Reform and Farm Restructuring in East Central Europe and the Former USSR. 1996. Available online: https://www.researchgate.net/publication/290007608_Agricultural_transition_revisited_Issues_of_land_reform_and_farm_restructuring_in_East_Central_Europe_and_the_former_USSR (accessed on 27 July 2021).
18. Izakovičová, Z.; Mederly, P.; Petrovič, F. Long-Term Land Use Changes Driven by Urbanisation and Their Environmental Effects (Example of Trnava City, Slovakia). *Sustainability* **2017**, *9*, 1553. [CrossRef]
19. European Commission. Urban Sprawl in Europe—The Ignored Challenge—European Environment Agency. 2006. Available online: https://www.eea.europa.eu/publications/eea_report_2006_10/eea_report_10_2006.pdf/view (accessed on 29 July 2021).
20. Roy, J.; Bose, C.; Bose, R.; Das, S.; Dhakal, S.; Dasgupta, M.; Ghate, R.; Roy, S.S.; Konar, M.; Wickramasinghe, A.; et al. Development Pathway. In *Global Environmental Changes in South Asia*; Springer: Berlin/Heidelberg, Germany, 2000; pp. 14–53. [CrossRef]
21. Liu, Y.; Li, J.T.; Yang, Y. Strategic adjustment of land use policy under the economic transformation. *Land Use Policy* **2018**, *74*, 5–14. [CrossRef]
22. Satterthwaite, D.; McGranahan, G.; Tacoli, C. Urbanization and its implications for food and farming. *Philos. Trans. R. Soc. B Biol. Sci.* **2010**, *365*, 2809–2820. [CrossRef]
23. Long, H.; Li, Y.; Liu, Y.; Woods, M.; Zou, J. Accelerated restructuring in rural China fueled by 'increasing vs. decreasing balance' land-use policy for dealing with hollowed villages. *Land Use Policy* **2012**, *29*, 11–22. [CrossRef]
24. Liu, Y.; Yang, R.; Long, H.; Gao, J.; Wang, J. Implications of land-use change in rural China: A case study of Yucheng, Shandong province. *Land Use Policy* **2014**, *40*, 111–118. [CrossRef]
25. Zhou, X.; Li, X.; Song, W.; Kong, X.; Lu, X. Farmland Transitions in China: An Advocacy Coalition Approach. *Land* **2021**, *10*, 122. [CrossRef]
26. Boudreaux, K.; Sacks, D. Land Tenure Security and Agricultural Productivity. MERCATUS ON POLICY No.57. 2009. Available online: https://www.mercatus.org/publications/development-economics/land-tenure-security-and-agricultural-productivity (accessed on 28 August 2021).
27. Bambio, Y.; Agha, S.B. Land tenure security and investment: Does strength of land right really matter in rural Burkina Faso? *World Dev.* **2018**, *111*, 130–147. [CrossRef]
28. Segura Warnholtz, G.; Fernández, M.; Smyle, J.; Springer, J. *Securing Forest Tenure Rights for Rural Development: Lessons from Six Countries in Latin America*; PROFOR: Washington, DC, USA, 2017; ISBN 9780991040780.
29. Kasimbazi, E. Global Land Outlook Working Paper Land Tenure and Rights for Improved Land Management and Sustainable Development. 2017. Available online: https://knowledge.unccd.int/sites/default/files/2018-06/5.Land%2BTenure%2Band%2BRights__E_Kasimbazi.pdf (accessed on 29 August 2021).
30. Singirankabo, U.A.; Ertsen, M.W. Relations between Land Tenure Security and Agricultural Productivity: Exploring the Effect of Land Registration. *Land* **2020**, *9*, 138. [CrossRef]
31. Nasrin, M.; Uddin, M.T. Land Tenure System and Agricultural Productivity in a Selected Area of Bangladesh. *Progress. Agric.* **2013**, *22*, 181–192. [CrossRef]
32. International Fund for Agricultural Development. Improving Access to Land and Tenure Security. 2008. Available online: https://www.ifad.org/documents/38711624/39417918/land_e.pdf/99f1a767-4ed1-41fc-a341-9bbd7fd2fe7f (accessed on 29 August 2021).
33. International Fund for Agricultural Development. Land Tenure Security and Poverty Reduction. 2015. Available online: https://www.ifad.org/documents/38714170/39148759/Land+tenure+security+and+poverty+reduction.pdf/c9d0982d-40e4-4e1e-b490-17ea8fef0775 (accessed on 29 August 2021).
34. Byamugisha, F. Securing Land Tenure and Easing Access to Land. 2016. Available online: www.acetforafrica.org (accessed on 29 August 2021).
35. Secretary-General of the OECD. Cities in the World: An New Perspective on Urbanization. 2020. Available online: www.oecd.org/regional/regional-statistics (accessed on 28 August 2021).
36. UN-DESA (United Nations Department of Economic and Social Affairs). World Population Prospects 2019: Highlights. Available online: https://population.un.org/wpp/Publications/Files/WPP2019_Highlights.pdf (accessed on 3 April 2021).
37. Yawson, D.O.; Mulholland, B.J.; Ball, T.; Adu, M.O.; Mohan, S.; White, P.J. Effect of Climate and Agricultural Land Use Changes on UK Feed Barley Production and Food Security to the 2050s. *Land* **2017**, *6*, 74. [CrossRef]
38. Hinz, R.; Sulser, T.B.; Huefner, R.; Mason-D'Croz, D.; Dunston, S.; Nautiyal, S.; Ringler, C.; Schuengel, J.; Tikhile, P.; Wimmer, F.; et al. Agricultural Development and Land Use Change in India: A Scenario Analysis of Trade-Offs Between UN Sustainable Development Goals (SDGs). *Earth's Futur.* **2020**, *8*, e2019EF001287. [CrossRef]
39. Verburg, P.H.; Van Berkel, D.; Van Doorn, A.M.; Van Eupen, M.; Heiligenberg, H.A.R.M.V.D. Trajectories of land use change in Europe: A model-based exploration of rural futures. *Landsc. Ecol.* **2009**, *25*, 217–232. [CrossRef]
40. Jepsen, M.R.; Kuemmerle, T.; Müller, D.; Erb, K.; Verburg, P.H.; Haberl, H.; Vesterager, J.P.; Andric, M.; Antrop, M.; Austrheim, G.; et al. Transitions in European land-management regimes between 1800 and 2010. *Land Use Policy* **2015**, *49*, 53–64. [CrossRef]
41. Centre de Suivi Ecologique. *Annuaire sur l'environnement et les Ressources Naturelles au Sénégal*, 4th ed.; Centre de Suivi Ecologique: Dakar, Senegal, 2018; pp. 12–19. Available online: https://www.cse.sn/index.php (accessed on 28 August 2021).

42. Bourgoin, J.; Diop, D.; Dia, D.; Sall, M.; Zagré, R.; Grislain, Q.; Anseeuw, W. Regard sur le modèle agricole sénégalais: Pratiques foncières et particularités territoriales des moyennes et grandes exploitations agricoles. *Cah. Agric.* **2020**, *29*, 18. [CrossRef]
43. Faye, N.F.; Gérard, F.; Sall, M.; Affholde, F.; Roudier, P. Poverty and Inequality in Rural Areas of Central Senegal: Status, Causes and Consequences. 2021. Available online: https://agritrop.cirad.fr/598426/1/Ndeye%20et%20al.%20SFER%202021.pdf (accessed on 25 July 2021). (In French)
44. Toure, A.K.; Diakhate, M. Descriptive Analysis of The Influence of Rainfall and Temperature Indicators on Agricultural Yields in Senegal. 2020. Available online: https://www.researchgate.net/publication/338698842_Analyse_descriptive_de_l%27influence_des_indicateurs_de_pluie_et_de_la_temperature_sur_les_rendements_agricoles_au_Senegal (accessed on 10 August 2021). (In French)
45. National Agency Statistics and Demography. General Census of Population and Housing Senegal. 2013. Available online: https://satisfaction.ansd.sn/ (accessed on 30 April 2021). (In French)
46. New Partnership for Africa's Development-FAO. Project to Restore the Agricultural Production Base in The Groundnut Basin. 2006. Available online: http://www.fao.org/3/ah215f/ah215f00.htm (accessed on 30 April 2021). (In French)
47. Ricome, A.; Affholder, F.; Gérard, F.; Muller, B.; Poeydebat, C.; Quirion, P.; Sall, M. Are subsidies to weather-index insurance the best use of public funds? A bio-economic farm model applied to the Senegalese groundnut basin. *Agric. Syst.* **2017**, *156*, 149–176. [CrossRef]
48. United States Agency for International Development. Feed the Future Senegal Naatal Mbay Project: Women's Economic Empowerment Strategy. 2016. Available online: https://pdf.usaid.gov/pdf_docs/PA00MMW4.pdf (accessed on 24 May 2021).
49. Sanogo, D. The Living Hedge in the Southern Groundnut Basin of Senegal: Adoption and Agro-Ecological Consequences. Ph.D. Thesis, Cheikh Anta Diop University of Dakar, Dakar, Senegal, 2000.
50. Ndour, M. *Investing in Rural People in Senegal*; International Fund for Agricultural Development: Rome, Italy, 2015. Available online: https://www.ifad.org/documents/38714170/39972302/Investir+dans+les+populations+rurales+au+Sénégal.pdf/60b009ed-ef32-45f5-ab1d-35fe04674825 (accessed on 5 July 2021).
51. Mbow, M. The Challenges of Senegalese Agriculture in a Climate Change Perspective. Master's Thesis, University of Sherbrooke, Sherbrooke, QC, Canada, 2017.
52. Gueye, G.; Fall, M.; Edwige, C.; Louhounghou, R. Characterization and Typology of Family Farms in Senegal: Volume 3 Groundnut Basin. 2008. Available online: https://www.bameinfopol.info/IMG/pdf/Expl._Fam_2.pdf (accessed on 30 April 2021). (In French)
53. Ramde, F.; Lo, B.S. The Role of The Agricultural Sector in The Senegal Economy. 2015. Available online: https://mpra.ub.uni-muenchen.de/81906/ (accessed on 11 May 2021). (In French)
54. National Agency Statistics Demography. Republic of Senegal Provisional 2017 and Final 2016 National Accounts (2014 Base). 2019. Available online: http://www.ansd.sn/ressources/publications/NOTE%20ANALYSE%20COMPTES%20NATIONAUX%202017_.pdf (accessed on 11 May 2021). (In French)
55. Ambiente, Y.M. *Les Paysages de l'Afrique de l'Ouest: Une Fenêtre sur un Monde en Pleine Évolution*; U.S. Geological Survey EROS: Garretson, SD, USA, 2016; pp. 279–280.
56. Ndong, J.-B. L'évolution de la pluviométrie au Sénégal et les incidences de la sécheresse récente sur l'environnement/The evolution of rainfall in Senegal and the consequences of the recent drought on the environment. *Rev. De Geogaphie Lyon* **1995**, *70*, 193–198. [CrossRef]
57. Faye, M.; Fall, A.; Faye, G.; Van Hecke, E. La variabilité pluviométrique et ses incidences sur les rendements agricoles dans la région des Terres Neuves du Sénégal oriental. *Belgeo* **2018**. [CrossRef]
58. D'Alessandro, S.; Fall, A.A.; Grey, G.; Simpkin, S.; Wane, A. Senegal Agricultural Sector Risk Assessment. 2015. Available online: Agritrop.cirad.fr/583362/ (accessed on 29 July 2021).
59. Rowland, J.; Funk, C. Famine Early Warning Systems Network-Informing Climate Change Adaptation Series A Climate Trend Analysis of Senegal. 2012. Available online: Fews.net/sites/default/files/documents/reports/FS12-3123.pdf (accessed on 24 May 2021).
60. Mechiche-Alami, A.; Abdi, A.M. Agricultural productivity in relation to climate and cropland management in West Africa. *Sci. Rep.* **2020**, *10*, 1–10. [CrossRef]
61. Faye, C. Analysis of Drought Trends in Senegalese Coastal Zone on Different Climatic Domains (1951–2010). *An. Univ. Din Oradea Ser. Geogr.* **2018**, *28*, 231–244. Available online: https://www.researchgate.net/publication/329351771_ANALYSIS_OF_DROUGHT_TRENDS_IN_SENEGALESE_COASTAL_ZONE_ON_DIFFERENT_CLIMATIC_DOMAINS_1951-2010 (accessed on 29 July 2021).
62. Food and Agriculture Organization of the United Nations and African Development Bank Group. Agricultural Growth in West Africa as a Market and Policy Driver. 2015. Available online: http://www.fao.org/3/i4337f/i4337f.pdf (accessed on 30 April 2021). (In French)
63. Niang, A.; Knapman, C. *Land Access for Senegal's Small Producers under Threat*; International Institute for Environment and Development: London, UK, 2017. Available online: https://pubs.iied.org/17375iied (accessed on 29 April 2021).
64. The Republic of Senegal. National Land Reform Commission Land Policy Document. 2016. Available online: http://www.hubrural.org/IMG/pdf/version_preliminaire_du_document_de_politique_fonciere.pdf (accessed on 30 April 2021). (In French)

65. Oya, C.; Ba, C.O. Agricultural policies 2000–2012: Between Voluntarism and Incoherence. 2013. Available online: https://eprints.soas.ac.uk/16799/1/04Oya-Ba130131%20proof.pdf (accessed on 29 April 2021). (In French)
66. Faroukh, A.T.; Tellal, R.; Qarro, M. Analysis of socio-economic mutations in the Benslimane forest Lever for sustainable development (Province of Benslimane, Morocco). *J. Mater. Environ. Sci.* **2017**, *8*, 4415–4425. [CrossRef]
67. Agro-Innov, R. Voluntary Cooperation Program for Innovative and Sustainable Agricultural Entrepreneurship Senegal Portrait-Diagnosis of Agricultural Soil Health. 2017. Available online: https://docplayer.fr/158367777-Programme-de-cooperation-volontaire-pour-un-entrepreneuriat-agricole-innovant-et-durable-senegal-portrait-diagnostic-de-la-sante-des-sols-agricoles.html (accessed on 30 April 2021). (In French)
68. Díaz-Pacheco, J.; van Delden, H.; Hewitt, R. The importance of scale in land use models: Experiments in data conversion, data resampling, resolution and neighborhood extent. In *Geomatic Approaches for Modeling Land Change Scenarios*; Springer: Madrid, Spain, 2018; pp. 163–186. [CrossRef]
69. Guoming, D.; Jingpan, M.; Xiang, C. Study on the transformation of arable land use patterns in modern agricultural areas. *Angew. Chemie Int.* **2018**, *6*, 951–952. [CrossRef]
70. Thierry, A.; Martin, P.; Ismaïla, T.I.; Brice, T. Modelisation des Changements D'occupation des Terres en Région Soudanienne au Nord-Ouest du Benin. *Eur. Sci. J. ESJ* **2018**, *14*, 248. [CrossRef]
71. Sagna, P.; Ndiaye, O.; Diop, C.; Niang, A.D.; Sambou, P.C. Les variations récentes du climat constatées au Sénégal sont-elles en phase avec les descriptions données par les scénarios du GIEC? *Pollut. Atmos.* **2015**, *227*. [CrossRef]
72. Thi, N.P.; Kappas, M.; Faust, H. Impacts of Agricultural Land Acquisition for Urbanization on Agricultural Activities of Affected Households: A Case Study in Huong Thuy Town, Thua Thien Hue Province, Vietnam. *Sustainability* **2021**, *13*, 8559. [CrossRef]
73. Faye, A.; Lejeune; Neya, O. The IPCC Special Report on Climate Change and Land What Impacts for Africa? 2019. Available online: www.climateanalytics.org/publications (accessed on 11 August 2021).
74. Li, Z.; Deng, X.; Yin, F.; Yang, C. Analysis of Climate and Land Use Changes Impacts on Land Degradation in the North China Plain. *Adv. Meteorol.* **2015**, *2015*, 1–11. [CrossRef]
75. Nhemachena, C.; Nhamo, L.; Matchaya, G.; Nhemachena, C.R.; Muchara, B.; Karuaihe, S.T.; Mpandeli, S. Climate Change Impacts on Water and Agriculture Sectors in Southern Africa: Threats and Opportunities for Sustainable Development. *Water* **2020**, *12*, 2673. [CrossRef]
76. Jarju, A.M.; Solly, B. Analysis of the Efficiency of Precipitation on the Evolution of Agricultural Production in Upper-Casamance (South Senegal) between 1985 and 2018. 2020. Available online: www.researchgate.net/publication (accessed on 3 September 2021).
77. Nouaceur, Z.; Murarescu, O. Rainfall Variability and Trend Analysis of Rainfall in West Africa (Senegal, Mauritania, Burkina Faso). *Water* **2020**, *12*, 1754. [CrossRef]
78. Faye, B. The Problems of the Agricultural Sector in the Face of Rainfall Variability in the Commune of Dara Mboss, Kaolack Region from 1980 to 2014. Marster's Thesis, Cheikh Anta Diop University of Dakar, Dakar, Senegal, 2016; pp. 28–47.
79. Rajpar, H.; Zhang, A.; Razzaq, A.; Mehmood, K.; Pirzado, M.B.; Hu, W. Agricultural Land Abandonment and Farmers' Perceptions of Land Use Change in the Indus Plains of Pakistan: A Case Study of Sindh Province. *Sustainability* **2019**, *11*, 4663. [CrossRef]
80. Mu, J.E.; Sleeter, B.M.; Abatzoglou, J.T.; Antle, J.M. Climate impacts on agricultural land use in the USA: The role of socio-economic scenarios. *Clim. Chang.* **2017**, *144*, 329–345. [CrossRef]
81. Davies, M.; Camara, T.; Diop, S. *Invest in Senegal: A Competitive Investment Destination in West Africa*; Deloitte: London, UK, 2017; p. 28.
82. Senegal Emergent Plan. National Agricultural Investment Program for Food Security and Nutrition. PNIASAN Senegal 2018–2022, Final Report. 2016. Available online: https://www.dapsa.gouv.sn/sites/default/files/publications/PNIASAN%20v4%20D%c3%a9c_1.pdf (accessed on 30 August 2021).
83. World Bank. *Revue de l'Urbanisation: Villes Emergentes pour un Sénégal Emergent*; World Bank: Washington, DC, USA, 2017. [CrossRef]
84. Pronat, E.; LANdac. Securing Women Land Rghts in Africa Working paper 3: Securing Women Land Rights in Africa-Senegal. 2018. Available online: http://www.landgovernance.org/assets/20181127-A4-Working-paper-03_Senegal.pdf (accessed on 12 August 2021).
85. Lyu, L.; Gao, Z.; Long, H.; Wang, X.; Fan, Y. Farmland Use Transition in a Typical Farming Area: The Case of Sihong County in the Huang-Huai-Hai Plain of China. *Land* **2021**, *10*, 347. [CrossRef]

land

Article

Land-Use Change and Health Risks in the Process of Urbanization: A Spatiotemporal Interpretation of a Typical Case in Changzhou, China

Dongyang Yang [1], Chao Ye [2,*] and Jianhua Xu [2,3]

1 Key Research Institute of Yellow River Civilization and Sustainable Development & Collaborative Innovation Center on Yellow River Civilization, Henan University, Kaifeng 475004, China; yangdy@lreis.ac.cn
2 School of Geographic Sciences & Institute of Eco-Chongming, East China Normal University, Shanghai 200241, China; jhxu@geo.ecnu.edu.cn
3 Research Center for East-West Cooperation in China, East China Normal University, Shanghai 200241, China
* Correspondence: cye@geo.ecnu.edu.cn

Abstract: China has undergone rapid urban expansion in recent decades. At the same time, environmental pollution and its risk to public health have increased. However, the relationship between urban land-use changes and health is ambiguous and insufficiently understood. Based on a typical city-scale case—namely, Changzhou, China—this research aimed to interpret the evolution of health risks alongside land-use change during the process of urbanization. We gathered data from multiple sources, including population mortality data, socioeconomic data, remote-sensing images, data for the points of interest of enterprises, and relevant information on environmental health events and cancers. The results showed that Changzhou's urbanization was typical insofar as it was characterized by massive growth in industry, a rapid increase in the urban population, and urban land expansion. Health risks related to environmental pollution increased considerably with urban land expansion over time, and they increased with proximity to the pollution. The results from a generalized linear model confirmed that Changzhou's urbanization triggered increasing health risks. Our study interpreted the relationship between urban land expansion and health risks from a spatiotemporal perspective. It can be used as a reference for urban planning and policymaking with regard to urban environmental health.

Keywords: land use change; urbanization; environmental pollution; health risks

Citation: Yang, D.; Ye, C.; Xu, J. Land-Use Change and Health Risks in the Process of Urbanization: A Spatiotemporal Interpretation of a Typical Case in Changzhou, China. *Land* **2021**, *10*, 820. https://doi.org/10.3390/land10080820

Academic Editor: Shiliang Su

Received: 2 July 2021
Accepted: 3 August 2021
Published: 5 August 2021

1. Introduction

Morbidity and mortality from chronic diseases related to environmental pollution are increasing alongside urbanization in China [1,2]. Urbanization is a spatiotemporal process by which urban components are transformed. These components include land use, industry, population, and many other natural and socioeconomic factors within the urban territorial system. These transitions inevitably affect public health. In this regard, there is an urgent need to understand the relationship between urban land expansion, urbanization, and environmental health issues in China.

In recent decades, China has seen unparalleled speed in rural–urban transition. The urban population rose from 191 million in 1980 to 831 million in 2018, and the urbanization rate reached 59.58% [3]. In general, urbanization improves living conditions, including education, employment, services, income, and health [4]. However, unplanned urban expansion often causes environmental hazards and health risks within the developing world [5–7]. China's urbanization, both historical and present, was accompanied by the process of industrialization. With rapid growth in industrialization, China's urbanization led to widespread environmental pollution and damage, especially in terms of land pollution and degradation [8–10]. Decommissioned industrial sites and current industrial parks have contaminated and seriously deteriorated the land [11,12]. The fast pace of urban development left little time for careful planning and design. Having experienced 30 years of rapid growth and urbanization, China now faces unique environmental health challenges [13,14].

There are numerous studies on environmental health issues caused by urbanization [2,13,15]. Different aspects of urban living, including income, healthcare, and education, were considered determinants that might affect public health [15–17]. In particular, environmental pollution has long been identified as a major risk factor [2,18,19]. Brownfields and industry parks often use hazardous substances and pollutants, and these have been a major concern of researchers [20–26]. Many studies have focused on pollution exposure and health risk assessments based on environmental pollution around brownfields and industry parks [22,27–29]. Simultaneously, scholars have documented that the proportion of mortality caused by chronic diseases and cancer increased obviously in urban China [30], and chronic disease incidence rates were higher in big cities than in small cities and rural areas in China [13]. Meanwhile, the dynamic changes in health risks during the urbanization process and the evolutionary relationship between urbanization, urban land change and health risks has rarely been studied, especially at the scale of a city. Hence, there is a need to adopt a spatiotemporal perspective to urbanization in order to understand environmental health issues in China.

Urbanization is a dynamic spatiotemporal process during which natural and human factors radically change from rural to urban. Different patterns of urbanization differ in how quickly they develop, and with regard to their effects on land and public health. Considering the spatiotemporal transformations of urban components—including land use, industry, and population—we offer an interpretation of the relationship between land-use change and health issues in Changzhou, China. In what follows, we present our findings and discuss policy implications. We expect that our results will contribute to effective policymaking regarding urbanization in order to abate urban land pollution and health risks for the sake of urban sustainability.

2. Materials and Methods

2.1. Study Area

Changzhou, located on the southern bank of the Yangtze River, is a prefecture-level city in southern China's Jiangsu province. Changzhou borders the provincial capital of Nanjing to the west, Zhenjiang to the northwest, Wuxi to the east, and Zhejiang province to the south (Figure 1). Changzhou is part of the highly developed Yangtze Delta region of China, extending northwest from Shanghai. In 2018, its total population was 4.74 million, and its urban population was 3.85 million. The urbanization rate reached 73.3%, 13.72% higher than the average in China.

Data published by the Ministry of Environmental Protection of China show that most chemical and petrochemical construction projects are distributed near rivers and lakes in highly populated areas [31]. Changzhou is located in the Yangtze River Delta close to many water bodies and dense populations. It also has a large chemical industry, with hazardous chemical substances.

2.2. Multi-Source Data

We gathered data on Changzhou from multiple sources: population mortality data, socioeconomic data, point-of-interest (POI) data, remote-sensing images, and high-resolution images. The population mortality data (from 2001 to 2015) and the socioeconomic data, including data on demographics, economy development, and pollution emissions, were obtained from the Changzhou Statistical Yearbook (from 1990 to 2015) on China National Knowledge Infrastructure (http://tongji.cnki.net/). The urbanization rate was calculated as the proportion of urban population in the total population. The POI data were obtained from Baidu Maps (http://map.baidu.com/). Given that thermal power, steel, cement, petrifaction, chemical engineering, non-ferrous smelting, textile, and paper-making are high-polluting industries, the POI data for these industries were extracted. Remote-sensing images (Landsat TM and OLI_TIRS) from 1995 and 2015 were downloaded from the Geospatial Data Cloud (http://www.gscloud.cn/). Data on cancer were gathered from published literature [32–37]. In addition, national and local authoritative media have reported massive environmental health events in Changzhou. A briefing on these data and their sources is presented in Table 1. Relevant information, such as the type and location of these events, was collected from the reports and verified through a field survey.

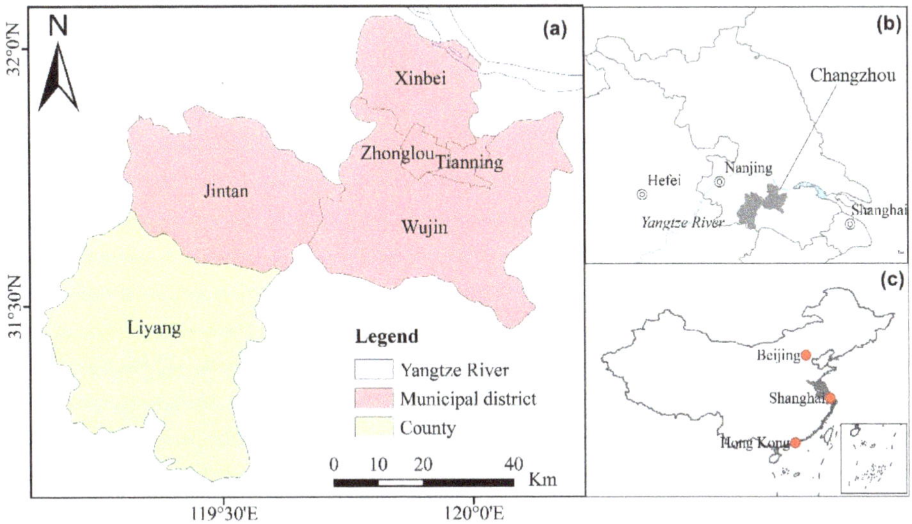

Figure 1. Study region (**a**) and its location in Jiangsu (**b**) and China (**c**).

Table 1. Data and sources.

Data	Source
Population mortality data	Changzhou Statistical Yearbook
Socio-economic data	Changzhou Statistical Yearbook
POI	Baidu map
Remote-sensing images	Geospatial Data Cloud
High-resolution images	Baidu map
Cancer data	Published literature [32–37]

Construction land in 1995 and 2015 was interpreted based on remote-sensing images, and the spatial distribution and time variation of the newly increased construction areas and enterprises could be examined by applying overlay analysis and point density analysis using ArcGIS software. Environmental health events with locations were mapped to show their spatial distribution.

2.3. Linear Weighting Method

To measure the comprehensive pollution from three major industrial polluting sources—wastewater, waste gas, and solid waste—we used the linear weighting method to calculate the comprehensive pollution index (F) based on data for pollution emissions:

$$F_i = \sum_{j=1}^{n} x_{ij} \times w_j$$

where x_{ij} is the normalized value of polluting emission j in year i, and w_j represents the weight of polluting emission j, calculated as follows:

$$w_j = \frac{\sigma_j}{\sum_{j=1}^{n} \sigma_j}$$

where σ is the standard deviation of observations x_{ij}.

2.4. Bivariate Spatial Association

Lee (2001) developed a bivariate spatial association based on univariate spatial correlation analysis to examine the spatial association between bivariate observations [38]. The bivariate spatial association is defined as follows:

$$I_{kl}^{i} = Z_k^i \sum_{j=l}^{n} W_{ij} Z_l^i$$

where W_{ij} denotes the spatial weight matrix, $Z_k^i = \left[x_k^i - \overline{x_k}\right]/\sigma_k$, $Z_k^i = \left[x_k^j - \overline{x_l}\right]/\sigma_l$, x_k^i is the observation k at location i, x_l^j is the observation l at location j, and σ_k and σ_l denote the variance of x_k and x_l, respectively.

As with spatial correlation analysis, the bivariate spatial association visualizes the results using a Moran's I scatterplot and the bivariate LISA cluster map [39]. We used this method to examine the spatial association between newly increased construction land and newly increased polluting enterprises.

2.5. Generalized Additive Model (GAM)

Health outcomes can exhibit linear or nonlinear variation, as can urbanization, land-use expansion, and industrial development. The GAM model provides primary functions (linear, polynomial, or spline) to fit the variation of dependent variables [40], such as health outcomes, and independent variables. Given the complexity of the relationship between urbanization and environmental health, we constructed the following GAM model:

$$Y = \beta_0 + \sum_{j=1} f_j(X_j) + \epsilon_j$$

where Y is a dependent variable (such as total mortality), $f_j(\cdot)$ is a random univariate function for independent variable X_j (such as the urbanization rate), and ϵ_j is a normal random error term.

The model determines the relationships between the dependent variable and independent variable according to the degree of freedom (DF): if DF = 1, the relationship is linear; if DF > 1, the relationship is nonlinear; and a higher DF indicates a more significant nonlinear relationship.

We included the urbanization rate (UR, the percentage of the urban population in the total population), the construction land area (CLA, km^2), and the proportion of heavy industrial output to total industrial output (PHI, %) as independent variables. We included total mortality (2001–2015), the number of cancer cases (2002–2011), and the mortality from cancer from 2009–2015 in Changzhou as dependent variables into the GAM, to explore the relationship between urbanization and health risks. Mortality from cancer in 2009 and 2010 was estimated according to its linear relationship with that of Wujin district, with a R^2 of 0.98.

3. Results

3.1. Temporal Interpretation of Land-Use Change and Environmental Health Risks in Changzhou

3.1.1. Evolution of Urbanization and Land Use

In the past decades, Changzhou experienced rapid urbanization (Figure 2). In 2003, the urbanization rate (the proportion of the urban population) was 56.45%. In 2018, it reached 72.5%. The built-up area also increased rapidly during the period. The built-up area was 92.28 km^2 in 2003, and 300 km^2 in 2018. There was massive growth in industrial land along with Changzhou's urban expansion: the industrial land area increased from 25.91 km^2 to 95.64 km^2 during the study period, with a growth rate of 269.12% (Figure 2).

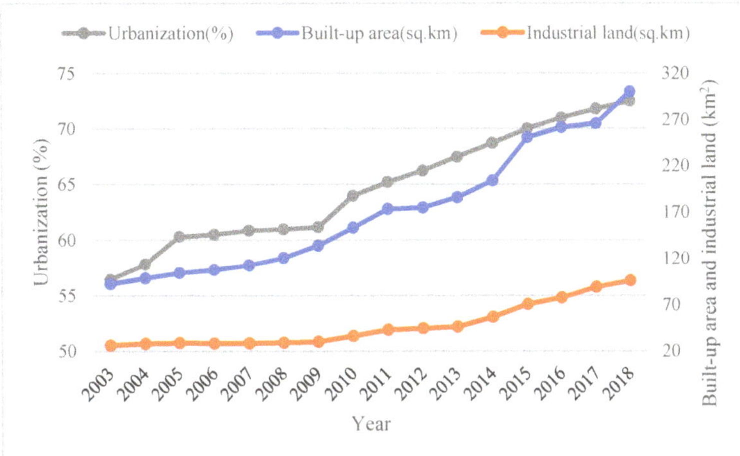

Figure 2. Urbanization evolution and land use change of Changzhou from 2003 to 2018.

3.1.2. Evolution of Population and Pollution

However, Changzhou's urbanization had a profound effect on the environment. Figure 3 shows that the pollution index—calculated using the linear weighting method based on the three industrial waste sources, namely, waste gas, wastewater, and waste solids—increased from 0.13 to 0.66, indicating that pollution emissions increased greatly from 1990 to 2015 (Figure 3). The cumulative emissions per unit area also increased. In particular, the cumulative emissions of industrial solid waste per km^2 increased from 1.94 to 60.23 million kg. As industrial waste emissions were new pollutants produced each year, both cumulative pollution emissions and pollution exposure led to increased health risk. Risk to health increased with more pollutants and pollution exposure. Furthermore, population density increased, although there was a small fluctuation before 2008. Thus, the environment was deteriorating while population increased. In other words, more and more people were exposed to an increasingly polluted environment.

3.1.3. Variation Profile of Health Risks

Pollution in Changzhou had conspicuous effects on health. Total mortality was $5.5/10^3$ in 2001, but reached $7.0/10^3$ in 2015. The number of cases and mortality from cancer increased markedly in the past decades in Changzhou according to the data gathered from published literature. The number of inpatient cases of cancer in Changzhou was 1995 in 2002, and reached 6649 in 2011 (Figure 4). The mortality from cancer in Changzhou was $208.98/10^5$ in 2011 and increased to $222.08/10^5$ in 2015 [33]. Mortality was $214/10^5$ in Wujin district in 2009 and increased to $242.6/10^5$ in 2014 [36]. This was higher than the mortality in the whole city. The mortality from cancer in Liyang county was the lowest, although it increased from $151.88/10^5$ in 2010 to $171.4/10^5$ in 2014 [37].

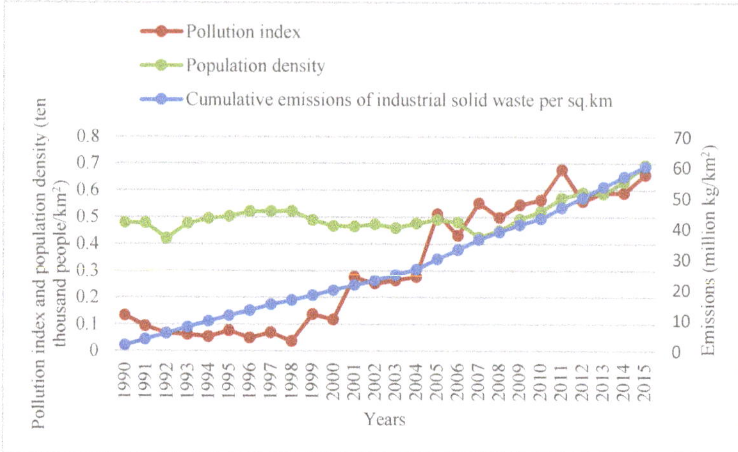

Figure 3. Time variations of environmental pollution and population density.

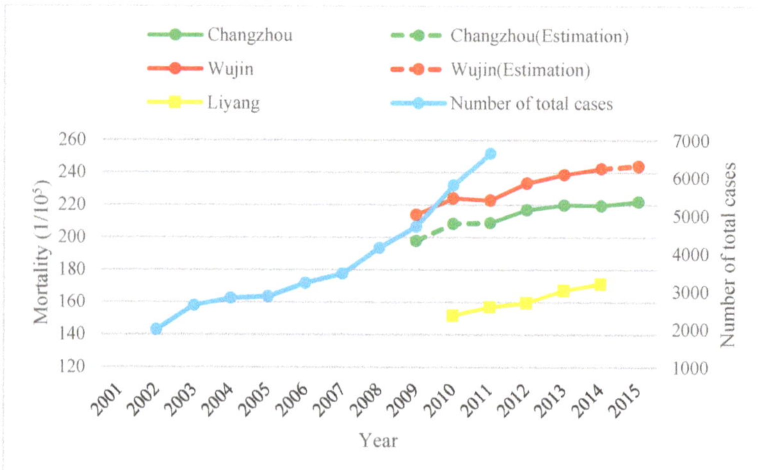

Figure 4. Time variations of the case number and mortality of cancers in Changzhou.

3.2. Spatial Interpretation of Land Use and Environmental Health in Changzhou

3.2.1. Bivariate Spatial Association of Urban Expansion and High-Polluting Industries

The area of construction land in 1995 and 2015 was interpreted based on remote-sensing images (Figure 5a,b) to examine urban expansion in Changzhou. Construction land in 2014 totaled 1075.77 km^2, a significant increase compared with that in 1995 (618.63 km^2). There were a total of 2802 high-polluting enterprises located in the coverage of the construction land in 2015. The number of enterprises located in the coverage of the construction land in 1995 was 1150, and the number of enterprises located in the coverage of the increased construction land was 1404. Errors in image interpretation suggested that some enterprises were not located in the coverage of construction land. In general, the results indicated that high-polluting enterprises significantly increased along with urban expansion during the period from 1995 to 2015 (Figure 5c,d).

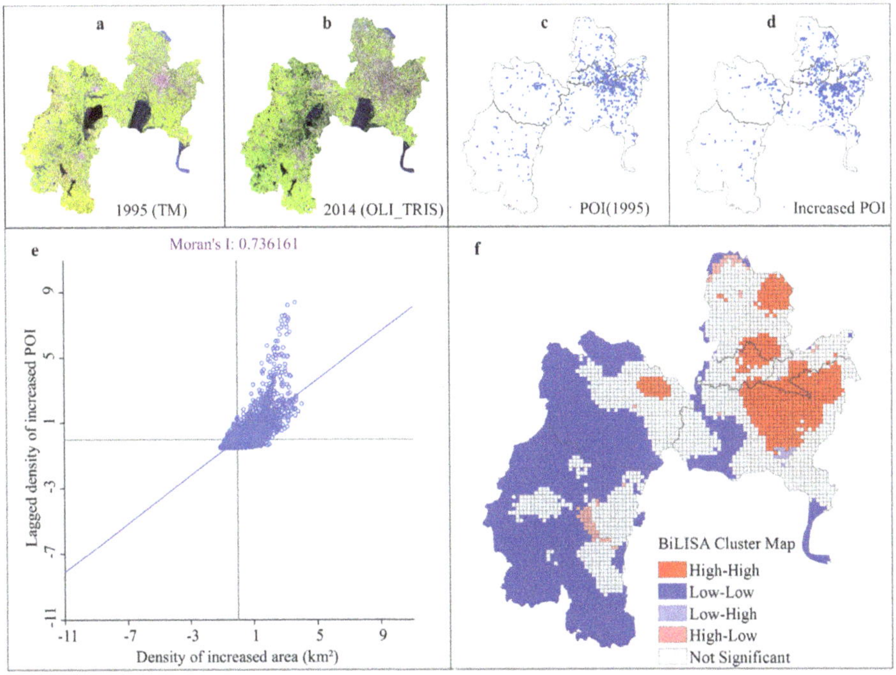

Figure 5. Spatial evolution of construction land and high-polluting industries from 1995 to 2014: (**a**,**b**) is remote-sensing image in 1995 and 2014 respectively; (**c**,**d**) is POI of high-polluting enterprise in 1995 and 2014 respectively; (**e**,**f**) is Mora's *I* scatter plot and LISA cluster map respectiavely.

We calculated the number of newly increased enterprises in each 1 km × 1 km grid using point density analysis with ArcGIS software. Further, we examined the bivariate spatial associations between the area of newly increased construction land and the number of newly increased enterprises in each 1 km × 1 km grid. The bivariate global Moran's *I* was approximately 0.74, indicating that newly increased enterprises had a strong positive spatial association with newly increased construction land (Figure 5e). Figure 5f shows the bivariate local spatial clusters of newly increased enterprises and newly increased construction land. Clusters of High-High were mainly distributed in the suburban areas of Changzhou, including Wujin, Tianning, Zhonglou, and Xinbei. An exurb area in Xinbei and the northeast area of Jintan were also found with clusters of High-High, indicating that there were significantly high increases in both construction land and high-polluting enterprises. In contrast, rural areas far away from urban areas were found with significantly low increases in both construction land and high-polluting enterprises (clusters of Low-Low).

3.2.2. Spatial Distribution of Environmental Health Events

A series of environmental health events caused by environmental pollution occurred during the urbanization of Changzhou. These environmental health events were mainly scattered in 77 places, among which 11 places suffered the occurrence of cancers, according to web media reports. Most of these environmental health events (62.67%) were distributed in the urban districts of Changzhou. Jintan also saw multiple occurrences of severe environmental health events. Field surveys and high-resolution image interpretation showed that all of the places that experienced environmental events were adjacent to high-polluting enterprises or brownfields. Several significant environmental health events and their locations are shown in Figure 6, revealing the spatial relationship between industrialized urban expansion and environmental health. Locations (1) and (2) show that high-polluting enterprises in the exurb area of Changzhou were correlated with the occurrence of cancer events. Xinbei first reported the occurrence of cancer events in 2003. In 2008, it had one of the highest rates of cancer. Reports indicated that 90 in 750 people were afflicted with cancer in Henshanqiao village in Wujin (location (2)) in March 2013. Location (3) shows that the high-polluting enterprises in the suburban areas of Changzhou caused residents to feel unwell frequently. (4) shows the location of a commercial residential building built on the site of several decommissioned chemical plants. (5) shows the location of student poisoning events at the Changzhou foreign language school. These generated a lot of press coverage in April 2016.

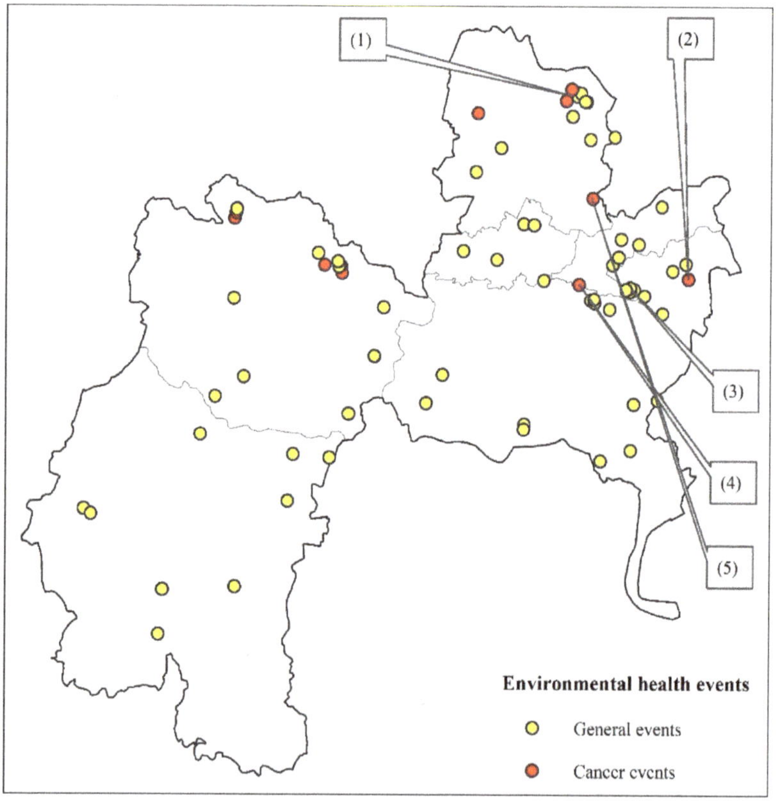

Figure 6. Locations with environmental health event reports. Note: the points do not represent accurate geographical coordinates of the events, but refer to the approximate locations and their adjacent regions according to the media reports; general events refer to the events which cause uncomfortableness or health ailments in humans, and cancer events refer to the events which are mainly related to cancer and cancer risks.

3.3. *Further Evidence of the Relationship between Land-Use Change and Environmental Health Risks*

Transformations in population, land use, and industry are essential features of urbanization. Thus, UR, CLA, and PHI were considered as proxy variables and were included as independent variables in the GAM model, to further verify the impact of urban expansion or urban land-use change on health risks. The results are shown in Table 2. UR and PHI were found to have a linear relationship with total mortality ($p < 0.01$), and CLA presented a nonlinear relationship with total mortality. The explained deviances and the determination coefficients were low (Dev < 55%; $R^2 < 0.50$). However, UR, CLA, and PHI showed significant nonlinear relationships with the number of cancer cases ($p < 0.001$), and the models had good explanatory power and goodness of fit (Dev > 95%; $R^2 < 0.70$). In addition, UR showed a linear relationship with the mortality of cancer, and CLA presented a nonlinear relationship with the mortality of cancer ($p < 0.01$). The models had good explanatory power and goodness of fit. PHI also showed a linear relationship with the mortality of cancer, but the model was non-significant and had poor explanatory power and goodness of fit ($p = 0.379$, adjusted $R^2 = -0.01$).

Table 2. Results of the GAM model and its comparison with the SLR.

Pairs of Variables		GAM				SLR		
		eDF	Dev (%)	p	Adjusted R^2	β	p	R^2
Total mortality	UR	1	48.4	0.0038	0.45	0.0413	0.0040	0.44
	CLA	2.008	54.3	0.0198	0.47	0.0091	0.0215	0.29
	PHI	1	44.3	0.0065	0.40	0.0822	0.0067	0.40
Number of cancer cases	UR	3.108	97.3	<0.001	0.96	113.64	<0.001	0.83
	CLA	1.794	98.8	<0.001	0.98	41.68	<0.001	0.98
	PHI	3.231	95.8	<0.001	0.94	179.72	<0.001	0.74
Cancer mortality	UR	1	95.9	<0.001	0.95	1.909	<0.001	0.95
	CLA	1.811	92.1	0.0047	0.89	0.2381	0.0028	0.83
	PHI	1	15.7	0.379	−0.01	−1.853	−0.012	0.38

Note: eDF donates the estimated degree of freedom, Dev donates the explained deviance, and β is the regression coefficient of SLR.

The specific effects of these urbanization factors on health are shown in Figure 7. The total mortality grew linearly with increases in UR and PHI (Figure 7(a1,a3)), and generally showed nonlinear growth along with an increase in CLA (Figure 7(a2)). The number of cancer cases nonlinearly increased along with the three factors, but the nonlinear relationships were more significant between the number of cancer cases and UR and PHI (Figure 7(b1–b3)). The mortality of cancer exhibited linear growth and nonlinear growth with increased UR and CLA, respectively (Figure 7(b1,b2)). Figure 7(b3) shows that the mortality of cancer linearly decreased with an increase in PHI, but the explained deviance (15.7%) and adjusted R^2 (−0.01) proved that the relationship was non-significant and unreliable. In addition, the simple linear regression (SLR) was calculated for comparison with the GAM. The estimated coefficient generally confirmed the relationships identified by the GAM, but the SLR showed disadvantages when the relationship between dependent and independent variables was complicated or nonlinear, because they yielded a lower R^2 than that of the GAM (Table 2).

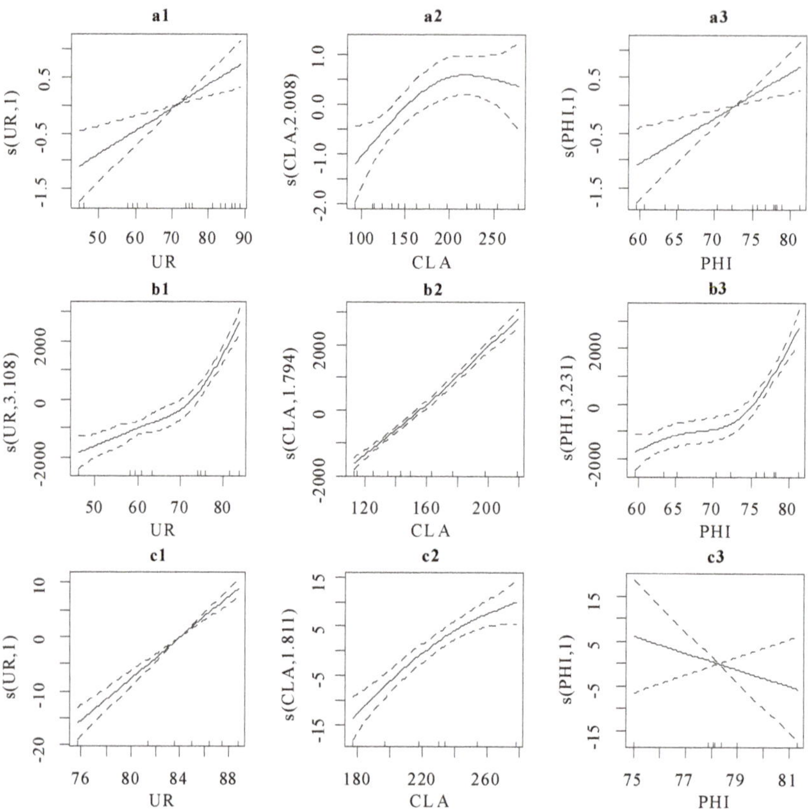

Figure 7. Effect of urbanization on health: (**a1–a3**), total mortality; (**b1–b3**), number of cancer cases; (**c1–c3**), mortality of cancer. Note: the vertical axis donates the fitted function value, the horizontal axis is the observations of the independent variable; solid line represents the fitted line (or curve) of the dependent variable, and the dotted line represents the confidence intervals.

4. Discussion

This paper interpreted the evolution of health risks associated with rapid urban expansion from a spatiotemporal perspective based on the case of Changzhou, China. The results showed that the pattern of Changzhou's urbanization is a typical case of industrial land expansion, which aggravated environmental pollution. Health risks in Changzhou tended to increase over time and with proximity to pollution. This kind of urbanization exposed the population to a polluted environment, and was related to increased health risks and the frequent occurrence of environmental health events.

With rapid urbanization, urban residents face decreasing environmental benefits, and even suffer a poisoned environment when pollutant emissions proliferate in the urban space. Furthermore, the development of industry increases pollution sources and extends the spatial distribution of pollutants, especially with people living close to past or current industrial sites [29,41–43]. In general, industrial urbanization exposes an increasing number of people to a polluted environment, and long-term pollution exposure is related to increased health risks. The spatiotemporal perspective here adopted can help facilitate an understanding of the mechanism and spatiotemporal patterns of health risks, and risk assessments of environmental health.

High-resolution spatiotemporal data can mitigate uncertainties and bias in exposure assessments [44,45]. However, there are few monitoring platforms and databases for gathering spatiotemporal data on chronic non-communicable diseases and human mortality in China. The spatiotemporal discontinuity of environmental health events and privacy issues related to diseases caused by pollution present challenges for quantitative analyses of the relationship between land-use change and health. In our study, however, we collected multi-source data in an effort to explain this relationship. Our results showed that Changzhou's urbanization featured rapid development, rapid population growth, extensive urban land expansion, and industrialization dominated by heavy industry (Figure 2). Industrial waste emissions increased alongside increasing population levels in a worsening environment (Figure 3). Further, increases in the total death rate of the city's population, the number of cancer cases, and cancer mortality support the pattern of exacerbated health risks (Figure 4).

Increases in new construction land and high-polluting enterprises showed clusters of High–High in the inner suburban areas of Changzhou, an exurb area of Changzhou, and a suburban area in Jintan district, indicating that urban expansion presented a high spatial association with variations in enterprises of high-polluting industries (Figure 5e,f). The spatial distribution of environmental health events presented spatial adjacency with enterprises of high-polluting industries or brownfields (Figure 6). This indicated that urbanization exposed people to polluted environments and caused environmental health events. In particular, cancer events in Xinbei and Wujin were both adjacent to high-polluting enterprises.

The results of the GAM analysis also revealed that rapid urban land expansion resulted in increased health risks. The total mortality and the number of cancer cases generally increased along with the development of urbanization factors, including population, land use, and industry. The variation in total mortality showed a nonlinear response to the construction land area, and even transformed when the construction land area reached 200 km^2 (Figure 7(a2)). Larger construction land or urban space may disperse concentrated populations to avoid gathering around pollution sources, and restrain increases in environmental health risks. Increases in the urbanization rate and construction land area were also found to be related to increases in cancer mortality. According to the data on health risks, these increased along with the development of urbanization.

There were differences in the time ranges for several kinds of data. Changzhou's administrative division has changed since 2002. Thus, we selected the period from 2003–2018 when calculating the urbanization rate. The total population and industrial waste emissions are measured for the entire city, and thus cannot be divided into administrative division. Therefore, we selected a longer time range. There were no long-term data on cancer because monitoring platforms on chronic non-communicable diseases were only recently introduced in China. Thus, we extrapolated relevant health data and events from literature and media coverage. When we conducted statistical analyses, mainly in the GAM model, the common time range was used to treat the discrepancies.

Our results have some policy implications. First, more attention should be paid to environmental health. The government should inspect, analyze, and transform current urbanization patterns with rapid industrial land expansion. Urbanization plans should consider the increase in population, industrial development, and urban expansion, and environmental assessments are needed for redevelopment plans and new extensions of land. Second, monitoring databases and platforms for land pollution and public health are urgently needed and should be accessible to the public. Complete real-time data can help interpret the dynamics of pollution and health and assist with risk warning. Besides, environmental impact assessment should be paid sufficient attention to support urban functional zoning in future. Strict and scientific criteria related to environmental health should be set and strictly enforced, including safe distances between urban land space and industrial sites or brownfield land, and safe pollutant emission standards. Precaution is more important than governance.

5. Conclusions

This paper investigated the evolution of health risks with rapid urban expansion from a spatiotemporal perspective. The results help clarify the relationship between health risks and land-use change during the process of urbanization. Urban expansion, massive growth in industry, a rapid increase in population, and large-scale urban land expansion expose populations in polluted environments to environmental health risks. Our study confirmed that inner suburbs close to decommissioned industrial sites and outer suburbs close to industrial parks are generally high-risk areas in cities with industrial urbanization. Multifaceted changes, such as land use change, urban environment change, and residents' lifestyle change during urbanization process, may independently or synergistically affect health. More attentions on the influences of these factors and the dominant factor that influences health, and their regional differences are recommended in future studies. Accurate spatial assessment and pre-warning for health risks also deserve further investigation.

Author Contributions: C.Y. conceived and initiated the study, J.X. provided useful suggestions. D.Y. mainly undertook statistical analysis and manuscript writing. All authors have read and agreed to the published version of the manuscript.

Funding: This study was sponsored by the National Natural Science Foundation of China (Grant No.41571138) and Natural Science Foundation of Henan (No. 202300410076), China.

Data Availability Statement: The data presented in this study are available on request from the corresponding author. The raw data are not publicly available, except memberships.

Conflicts of Interest: The authors declare that they have no competing interests.

References

1. Talukder, S.; Capon, A.; Nath, D.; Kolb, A.; Jahan, S.; Boufford, J. Urban health in the post-2015 agenda. *Lancet* **2015**, *385*, 769. [CrossRef]
2. Gong, P.; Liang, S.; Carlton, E.J.; Jiang, Q.; Wu, J.; Wang, L.; Remais, J.V. Urbanisation and health in China. *Lancet* **2012**, *379*, 843–852. [CrossRef]
3. Changzhou Staticstical Yearbook 2019. Available online: http://tongji.cnki.net/ (accessed on 18 November 2020).
4. Bloom, D.E.; Canning, D.; Fink, G. Urbanization and the Wealth of Nations. *Science* **2008**, *319*, 772–775. [CrossRef]
5. Alirol, E.; Getaz, L.; Stoll, B.; Chappuis, F.; Loutan, L. Urbanisation and infectious diseases in a globalised world. *Lancet Infect. Dis.* **2011**, *11*, 131–141. [CrossRef]
6. Song, W.; Liu, M. Farmland Conversion Decreases Regional and National Land Quality in China. *Land Degrad. Dev.* **2017**, *28*, 459–471. [CrossRef]
7. Zhou, C.; Li, S.; Wang, S. Examining the impacts of urban form on air pollution in developing countries: A case study of China's megacities. *Int. J. Environ. Res. Public Health* **2018**, *15*, 1565. [CrossRef]
8. Yang, H.; Huang, X.; Thompson, J.R.; Flower, R.J. China's Soil Pollution: Urban Brownfields. *Science* **2014**, *344*, 691–692. [CrossRef]
9. Bian, J.; Ren, H.; Liu, P.; Zhang, Y. Sustainable Urbanization Performance Evaluation Based on "Origin" and "Modernization" Perspectives: A Case Study of Chongqing, China. *Int. J. Environ. Res. Public Health* **2018**, *15*, 1714. [CrossRef]
10. Wang, D.; Kwan, M. Selected studies on urban development issues in China: Introduction. *Urban Geogr.* **2017**, *38*, 360–362. [CrossRef]
11. Cheng, F.; Geertman, S.; Kuffer, M.; Zhan, Q. An integrative methodology to improve brownfield redevelopment planning in Chinese cities: A case study of Futian, Shenzhen. *Comput. Environ. Urban Syst.* **2011**, *35*, 388–398. [CrossRef]
12. Liu, Y.; van Oort, F.; Geertman, S.; Lin, Y. Institutional determinants of brownfield formation in Chinese cities and urban villages. *Habitat Int.* **2014**, *44*, 72–78. [CrossRef]
13. Zhu, Y.G.; Ioannidis, J.P.; Li, H.; Jones, K.C.; Martin, F.L. Understanding and harnessing the health effects of rapid urbanization in China. *Environ. Sci. Technol.* **2011**, *45*, 5099–5104. [CrossRef]
14. Thompson, A.L.; Houck, K.M.; Adair, L.; Gordon-Larsen, P.; Popkin, B. Multilevel examination of the association of urbanization with inflammation in Chinese adults. *Health Place* **2014**, *28*, 177–186. [CrossRef]
15. Whiting, D.; Unwin, N. Cities, urbanization and health. *Int. J. Epidemiol.* **2008**, *38*, 1737–1742. [CrossRef] [PubMed]
16. Vlahov, D.; Freudenberg, N.; Proietti, F.; Ompad, D.; Quinn, A.; Nandi, V.; Galea, S. Urban as a determinant of health. *J. Urban Health* **2007**, *84*, 16–26. [CrossRef]
17. Kwan, M.P.; Wang, J.; Tyburski, M.; Epstein, D.H.; Kowalczyk, W.J.; Preston, K.L. Uncertainties in the geographic context of health behaviors: A study of substance users' exposure to psychosocial stress using GPS data. *Int. J. Geogr. Inf. Sci.* **2018**, *33*, 1176–1195. [CrossRef]

18. De Hollander, A.E.M.; Staatsen, B.A.M. Health, environment and quality of life: An epidemiological perspective on urban development. *Landsc. Urban Plan* **2003**, *65*, 53–62. [CrossRef]
19. Zhang, J.; Mauzerall, D.L.; Zhu, T.; Liang, S.; Ezzati, M.; Remais, J.V. Environmental health in China: Progress towards clean air and safe water. *Lancet* **2010**, *375*, 1110–1119. [CrossRef]
20. Ren, W.; Xue, B.; Geng, Y.; Sun, L.; Ma, Z.; Zhang, Y.; Mitchell, B.; Zhang, L. Inventorying heavy metal pollution in redeveloped brownfield and its policy contribution: Case study from Tiexi District, Shenyang, China. *Land Use Policy* **2014**, *38*, 138–146. [CrossRef]
21. Raymond, C.M.; Gottwald, S.; Kuoppa, J.; Kyttä, M. Integrating multiple elements of environmental justice into urban blue space planning using public participation geographic information systems. *Landsc. Urban Plan* **2016**, *153*, 198–208. [CrossRef]
22. Li, P.; Wu, J.; Qian, H.; Lyu, X.; Liu, H. Origin and assessment of groundwater pollution and associated health risk: A case study in an industrial park, northwest China. *Environ. Geochem. Health* **2014**, *36*, 693–712. [CrossRef] [PubMed]
23. Li, D.; Zhang, C.; Pizzol, L.; Critto, A.; Zhang, H.; Lv, S.; Marcomini, A. Regional risk assessment approaches to land planning for industrial polluted areas in China: The Hulunbeier region case study. *Environ. Int.* **2014**, *65*, 16–32. [CrossRef] [PubMed]
24. Hou, D.; Li, F. Complexities Surrounding China's Soil Action Plan. *Land Degrad. Dev.* **2017**, *28*, 2315–2320. [CrossRef]
25. Li, F.; Qiu, Z.; Zhang, J.; Liu, W.; Liu, C.; Zeng, G. Investigation, pollution mapping and simulative leakage health risk assessment for heavy metals and metalloids in groundwater from a typical brownfield, middle China. *Int. J. Environ. Res. Public Health* **2017**, *14*, 768. [CrossRef]
26. Cabral Pinto, M.M.S.; Silva, M.M.V.; Ferreira da Silva, E.A.; Marinho-Reis, A.P. The cancer and non-cancer risk of Santiago island (Cape Verde) population due to totential toxic elements exposure from soils. *Geosciences* **2017**, *7*, 78. [CrossRef]
27. Lu, Y.; Song, S.; Wang, R.; Liu, Z.; Meng, J.; Sweetman, A.J.; Jenkins, A.; Ferrier, R.C.; Li, H.; Luo, W.; et al. Impacts of soil and water pollution on food safety and health risks in China. *Environ. Int.* **2015**, *77*, 5–15. [CrossRef]
28. Yousaf, B.; Liu, G.; Abbas, Q.; Wang, R.; Imtiaz, M.; Zia-ur-Rehman, M. Investigating the uptake and acquisition of potentially toxic elements in plants and health risks associated with the addition of fresh biowaste amendments to industrially contaminated soil. *Land Degrad. Dev.* **2017**, *28*, 2596–2607. [CrossRef]
29. Zhao, L.; Xu, Y.; Hou, H.; Shangguan, Y.; Li, F. Source identification and health risk assessment of metals in urban soils around the Tanggu chemical industrial district, Tianjin, China. *Sci. Total Environ.* **2014**, *468–469*, 654–662. [CrossRef]
30. Li, X.; Song, J.; Lin, T.; Dixon, J.; Zhang, G.; Ye, H. Urbanization and health in China, thinking at the national, local and individual levels. *Environ. Health* **2016**, *15*, 32. [CrossRef]
31. MEP; NDRC; MWR. *Planning for the Prevention and Control of Water Pollution in Key River Basins (2016–2020)*; Minstry of Ecology and Environment of the People's Public of China: Beijing, China, 2017; pp. 3–199. Available online: http://www.mee.gov.cn/gkml/hbb/bwj/201710/t20171027_424176.htm (accessed on 18 November 2020).
32. Bu, J.; Suo, L. Incidence trend of malignant tumor from inpatient. *Chin. J. PHM Dec.* **2013**, *29*, 807. (In Chinese)
33. Dong, H.; Yao, Y.; Yao, X. Death causes and life lost from maligant tumor among residents in Changzhou city from 2011 to 2015. *Mod. Prev. Med.* **2017**, *44*, 2512–2516. (In Chinese)
34. Wang, J.; Zhang, W.; Zhang, Y. Analysis of the Death Features of Malignant Tumor Patients in Tianning District of Changzhou. *Occup. Health* **2006**, *22*, 724–726. (In Chinese)
35. Shang, F.; Yu, R. Analysis on epidemic and death situation of malignant tumor among residents of Tianning District of Changzhou city from 2011–2013. *Occup. Health* **2015**, *31*, 2939–2942. (In Chinese)
36. Shi, S.; Xu, M.; Qiang, D. Analysis on death situation and life lost of malignant tumor among residents of Wujin District of Changzhou city from 2009–2014. *Jiangsu J. Prev. Med.* **2017**, *28*, 58–60. (In Chinese)
37. Zheng, G. Malignant Tumor Death Analysis of Liyang City Residents in 2010–2014. *China Health Stand. Manag.* **2016**, *27*, 6–7. (In Chinese)
38. Lee, S.-I. Developing a bivariate spatial association measure: An integration of Pearson's r and Moran's I. *J. Geogr. Syst.* **2001**, *3*, 369–385. [CrossRef]
39. Anselin, L.; Syabri, I.; Smirnov, O. Visualizing multivariate spatial correlation with dynamically linked windows. In Proceedings of the CSISS Workshop on New Tools for Spatial Data Analysis, Santa Barbara, CA, USA, 24–28 June 2002.
40. Song, Y.Z.; Yang, H.L.; Peng, J.H.; Song, Y.R.; Sun, Q.; Li, Y. Estimating PM2.5 concentrations in Xi'an City using a generalized additive model with multi-source monitoring data. *PLoS ONE* **2015**, *10*, e0142149. [CrossRef]
41. Bambra, C.; Cairns, J.M.; Kasim, A.; Smith, J.; Robertson, S.; Copeland, A.; Johnson, K. This divided land: An examination of regional inequalities in exposure to brownfield land and the association with morbidity and mortality in England. *Health Place* **2015**, *34*, 257–269. [CrossRef]
42. Silbergeld, E.K. Managing hazards in place: The risks of residual risks. *Environ. Res.* **2017**, *158*, 806–811. [CrossRef]
43. Liu, Z.; Lu, Y.; Wang, T.; Wang, P.; Li, Q.; Johnson, A.C.; Sarvajayakesavalu, S.; Sweetman, A.J. Risk assessment and source identification of perfluoroalkyl acids in surface and ground water: Spatial distribution around a mega-fluorochemical industrial park, China. *Environ. Int.* **2016**, *91*, 69–77. [CrossRef]
44. Kwan, M.P. The Limits of the Neighborhood Effect: Contextual Uncertainties in Geographic, Environmental Health, and Social Science Research. *Ann. Am. Assoc. Geogr.* **2018**, *108*, 1482–1490. [CrossRef]
45. Yang, D.; Lu, D.; Xu, J.; Ye, C.; Zhao, J.; Tian, G.; Wang, X.; Zhu, N. Predicting spatio-temporal concentrations of PM2.5 using land use and meteorological data in Yangtze River Delta, China. *Stoch. Environ. Res. Risk Assess.* **2018**, *32*, 2445–2456. [CrossRef]

land

MDPI

Article

Values-Led Planning Approach in Spatial Development: A Methodology

Armands Auzins [1,*] and Uchendu Eugene Chigbu [2]

[1] Institute of Civil Engineering and Real Estate Economics, Riga Technical University, Kalnciema Street 6-210, LV-1048 Riga, Latvia

[2] Department of Land and Property Sciences, Faculty of Natural Resources and Spatial Sciences, Namibia University of Science and Technology, 13 Jackson Kaujeua Street, Private Bag 13388, Windhoek 9000, Khomas, Namibia; echigbu@nust.na

* Correspondence: armands.auzins@rtu.lv; Tel.: +371-29-439-004

Abstract: The scope of land management, which includes spatial planning as an activity in the public domain, demands that a planning process that is based on publicly or societally acceptable values is a matter of necessity. This study proposes a methodology for introducing a values-led planning (VLP) approach in spatial development. The motivation of the study is to promote the embrace of assessed values in planning. The study draws from issues evoked in various topical studies on European comparative perspectives. By way of argumentation, the study makes three relevant contributions to the literature and spatial planning and development practice. First, it presents and discusses the essential elements required in the design of methodology. In this way, it figuratively depicts VLP as a consequence of interactions between four key elements of spatial planning. Second, it proposes an actual methodology for action. Third, it discusses the applicability of the methodology. The proposed methodology would be useful for planners, including public authorities, land managers, and community leaders, who make socio-spatial decisions in land management and related activities.

Keywords: values-led approaches; values; methodology; land management; land-use; spatial planning; land administration; values-led planning; spatial development

Citation: Auzins, A.; Chigbu, U.E. Values-Led Planning Approach in Spatial Development: A Methodology. *Land* **2021**, *10*, 461. https://doi.org/10.3390/land10050461

Academic Editor: Adrianos Retalis

Received: 30 March 2021
Accepted: 23 April 2021
Published: 26 April 2021

1. Introduction

Spatial development as a process focuses on decisions and activities related to the coordination and spatial distribution of land-use. A spatial planning system conceptually characterises "the ensemble of territorial governance arrangements that seek to shape patterns of spatial development in particular places" [1]. However, when it comes to spatial development practice, the desire to achieve win-win outcomes from spatial planning poses a substantial challenge [2]. Such outcomes are still rare in the real world of planning where difficult choices have to be made to the detriment of win-win values [3,4]. Many spatial planning studies provide detailed research outcomes in the context of publicly controlled planning [5]. Thus, spatial planning may more appropriately be concerned with the planning of development, involving substantial changes and responding to questions of what changes should be made, how substantial those changes are, and comparing them to an existing situation. Accordingly, planning aims at feasible solutions for identified needs and problems. These solutions should be analysed and evaluated before making binding decisions and engaging in their implementation [5] (pp. 32–38). Rational assumptions in planning support the optimisation of the decision-making process that considers choices between different values and existing uncertainties as well as circumstances and alternatives. However, based on this rational reasoning, the planning process also demands mutual discussions by the involved stakeholders. The strategy of "collaborative rationality" proposes solutions for wicked problems in planning, thus focusing on the characteristics of a collaborative and rational planning process which is inspired more by Habermas's concept of communicative rationality than by scientistic planning [6].

The assessment of shared/social values for ecosystem and landscape services and sustainability has been widely discussed in the scientific literature in recent years [7–15]. An empirical study demonstrates that the boundaries between instrumental and deliberative paradigms are often vague and suggests integrating some qualities of both [7]. Proposing the framework and classification of values, Kenter et al. (2015) conceptualised the dynamic interplay between shared/social and individual values as well as emphasised the importance of shared/social values for decision making [8]. In this regard, VLP primarily contains (1) group values (within valuation) where values are expressed by a group of stakeholders and (2) deliberated values where values are an outcome of a deliberative process. Kenter et al. (2016) provided arguments that values and preferences for ecosystem services need to be generated through a process of deliberation and learning [9]. They also proposed a new theoretical model for deliberative valuation that enables more effective integration of social learning and plural knowledge and values in valuation and decision making. Shared values can be deliberated through formal and informal processes where individuals may separate their personal preferences from a "broader metanarrative about what values ought to be shared" [10]. The proposed framework by Connor and Kenter (2019) provides an opportunity to bridge and reconcile the different types of values through deliberations—intrinsic, instrumental, and relational values [11]. Garcia-Martin et al. (2017) demonstrated a European perspective on landscape values perceived by local stakeholders, the patterns in the spatial distribution of values, and their connection to different socio-economic backgrounds and landscape characteristics [12]. Fagerholm et al. (2019), presenting an assessment of ecosystem services benefits, provided argumentation on the links from services to benefits and from benefits to different types of values [13]. Keller and Backhaus (2020) used the term landscape instead of ecosystem to underline the multiple dimensions of the landscape-services approach besides ecological issues. They defined landscape services, emphasising the benefits of landscape qualities for individuals and society [14]. Exploring drivers and processes of European landscape change, Van der Sluis et al. (2019) referred to the framework that shows the landscape as a social-ecological system providing landscape services for the people [15] (p. 459). Tiboni et al. (2020), promoting and testing a methodology to analyse the effects of urban regeneration, identified how different urban operations may contribute to creating public value. They assessed various possible development scenarios and compared them with the baseline of the current situation [16].

Thus, previous studies provided arguments that a planning process has to be based on assessed values—a reason why this study focuses on a methodology for a values-led planning (VLP) approach to support spatial development. It has been affirmed that the "introduction of VLP approach based on consolidated new knowledge from stakeholders' experience and empirical evidence will help better understand and guide the relevant processes and their effects in specific territories based on (1) the identified values as an outcome of experts' work and (2) the attitudes from stakeholders' preferences concerning these values" [17] (p. 281). Therefore, a VLP approach is concerned with the "evaluation and planning–implementation concept" and consequent principles. It aims to balance mainly the interests of environmental protection and new development. The usefulness of the VLP approach is found in dynamics and potential changes in land-use and its values. Practically, "the potential for further spatial development should be assessed and then supported by binding decisions" [18].

In the European context, research on spatial planning practices have been addressed from different planning cultures. Some scientific contributions clearly distinguish between planning systems and planning cultures [19]. Research on spatial planning systems and planning cultures in Europe distinguished both and associated a planning culture with the "underlying shared values, norms, and beliefs of the planning community or the societal environment that affect planning practices" [20] (p. 26). This study asserts that it is difficult to dissociate shared values entirely from approaches to planning. This assertion is both logical and practical because the cultural features of the country influence the planning system in each country. For instance, village renewal—a methodology-based programme for spatial planning in various European countries—is done differently in Germany compared to other European countries [21]. The difference lies in the way the shared values in different countries are shaped. This is why Auziņš (2018) argued that "planning practices inherent to the system cannot be drawn from a comparison of legal-administrative framework conditions alone. Therefore, the outcome from the comparative analysis of planning practices (changes in cultures) is essential rather than of planning systems, which are only represented by hierarchies, artefacts, and institutional settings" [22] (p. 2). European comparative studies reveal the trends and directions in the evolution of spatial planning systems and territorial governance as well as the design of new typologies in Europe. This has led to the revision of EU policies and national spatial planning and territorial governance, focusing on synergies and contradictions between both [23]. Cross-fertilisation between the EU cohesion policy and spatial planning practice also has recently been on European planning communities' agenda [24].

As the concern of this study is the embrace of assessed values in planning, it proposes a methodology. Methodology development in the context of this study emphasises applicable techniques and guiding recommendations for introducing a VLP approach into land-management practice. This is especially concerned with "the science and practice related to the conceptualisation, design, implementation, and evaluation" of land-based interventions "with the purpose to improve the quality of life and the resilience of livelihoods in a responsible, effective, efficient, consensual, and smart manner" [25] (p. 66). This means engaging in analytical research and considering the consolidated outcome of spatial development case studies as well as focusing on stakeholders' involvement in the planning processes. It also means finding new ways to advance values in planning through proposed techniques. Going forward, this study is organized into five sections. Section 2 specifies the approach to this study. As part of framing the methodology, Section 3 presents the theoretical perspective of values-led approaches. In addition, Section 3 answers the question about why values matter in spatial planning by using Rokeach's theory of values to explain the values-led approach in planning. Section 4 frames the path to a methodology for the VLP approach in spatial development. Finally, Section 5 provides a conclusion on the way forward to support spatial planning practices.

2. Approach to the Study

In this study, a methodology reflects the overall frame which justifies applying particular spatial planning techniques and tools. By proposing and discussing the methodology, this study does not specifically consider the value-related issues to be a wicked problem in spatial planning and development. However, it does have a wicked component because a lack of values in planning poses problems in development outcomes. Yet, there is no immediate and no ultimate test of a solution for it [26]. This makes it pertinent to keep searching and testing techniques for VLP, thus tackling values-related planning concerns.

In devising a methodology for VLP, this study draws from findings from various topical studies on European comparative perspectives (from spatial planning documentary sources) conducted from 2017 to 2020. The study particularly responds to Auziņš's and Viesturs's (2017) call for creating a methodology for the VLP approach to ensure applicable implementation strategies for existing frameworks of spatial planning and development [17]. Hence, this study extends previous research conducted on European-wide comparative studies of spatial planning concepts using comprehensive evidence gathering (CEG) [17,18,22,27]. The methodology is proposed as a result of the synthesis of the key study outcome gathered during both the analytical research (examination of a range of scientific literature and documentary sources) and empirical research by making case studies (exploring the spatial planning practice and territorial governance in selected differently experienced European countries). Therefore, the study aims to show and discuss the issues which are essential for designing the content of methodological solutions to support the implementation of the VLP approach into land-management practice.

Consequently, the approach of the study is fourfold (Figure 1): first, to provide a theoretical understanding of values-led approaches; second, to examine the organisation and peculiarities of the planning process and to identify the best way possible to involve stakeholders in a spatial planning process; third, to analyse the planning environment and shared values of the actors and to characterise and propose a typology of values, conceptualised participation, and evaluation techniques, including a value-causing assessment (VCA); and fourth, based on issues emerging from the synthesis of values-led approaches—including the organisational peculiarities and the shared values of actors in the spatial planning process—the study proposes a methodological framework for the VLP approach to be introduced in spatial development. The first three steps in this methodological approach are based on three knowledge-generating causal steps: (1) the understanding of theoretical orientation; (2) institutional issues and cases; and (3) values considerations in planning that led to the outcome—a methodology for VLP approach to spatial development. Thus, applying the proposed techniques for the integration of the assessed values with stakeholders' preferences essentially presents a novelty of this study.

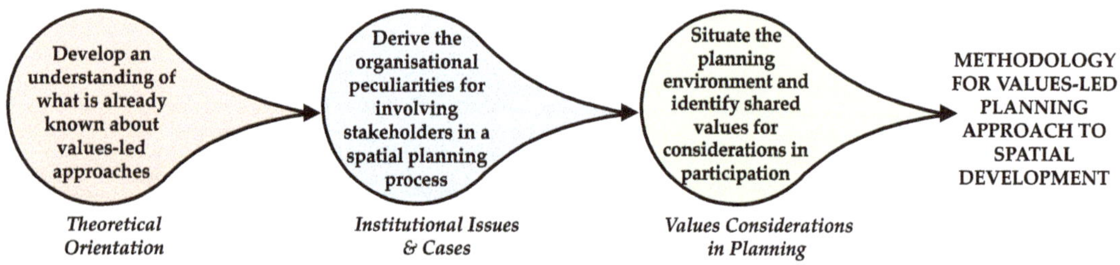

Figure 1. The approach adopted in the study.

3. Values-Led Approaches: A Theoretical Orientation for Spatial Planning

Generally, approaches entail how decisions and activities are carried out by those tasked with leadership in any sector or discipline of life. These decisions and activities also relate to planning, which involves coordinating different or related and interlinked or isolated decisions and activities to achieve envisaged goals. In planning, values form an essential part of making decisions and implementing activities [21–23].

3.1. *Why Do Values Matter in Approaches?*

The importance of values in planning has attracted interest from many scholars. That is why it has become common to read about value-focused decision making [28], values-led conservation [29], values-driven leadership [30], values-led participatory design [31], values-led entrepreneurship [32], value-led management [33], and value-focused approaches [34]. In terms of spatial planning, planners have always operated with values or are aware of the need for values in their work. This is why planning has been done with different sets of human-related values. For instance, Chigbu et al. (2019) called for a tenure security-sensitive approach that would protect landowners' rights in any process of land use and planning in the global south [35]. However, Auziņš's and Viesturs's VLP approach re-evoked the discourse in spatial planning in the context of values-sensitive ways, because they directly argued for the "creation of positive synergy in managing land-related resources" [17] (p. 275).

This study furthers this discourse by basing its argumentations on the premise that all sorts of planning—whether land/spatial planning, natural resource, or human-resource planning—are related because they are all human activities. This premise is not only relevant for grasping values and how they apply in planning, but it is also a truism because planning is about people [36]. That is why the implementation of planning activities either improves or worsens the living conditions of people [17]. Davoudi (2016) argued that there is a "value of planning" and there are "values in planning" [37]. This implies that planning does not only lead to value (e.g., a social value of planning) but depends on values. Hersperger et al. (2017) evaluated outcomes in planning in Swiss landscapes and concluded that the values tied to goals and indicators are linked to an efficient outcome [38]. From Australian experience, Rawluk et al. (2017) noted that concrete and abstract social values influence the success of environmental management planning concerning bushfire mitigation because values relate to natural places and attributes [39]. Ives et al. (2017) recognised the importance of "capturing residents' values for urban green space" in urban land administration [40]. What all these scholars [17,36–40] are alluding to is that values are at the core of any planning process. Hence, values matter in any sort of planning [41]. This is why this study seeks to link values-led approaches to planning from a spatial perspective. However, the study cannot achieve this aim without understanding what values-led approaches entail in spatial planning and development.

3.2. Using Rokeach's Theory of Values to Understand a Values-Led Approach in Planning

To grasp the connotations of values-led approaches, it is necessary to, first and foremost, grasp the meanings embedded in the concept of value. Rokeach's (1973 and 1979) theory of values provides a path to understand values in the context of planning [42,43]. In theory, he defines a value as an "enduring belief that a specific mode of conduct or end-state of existence is personally or socially preferable to an opposite or converse mode of conduct or end-state of existence" [42] (p. 5). Based on this definition, Rokeach developed a value survey (RVS) consisting of values considered as preferable behavioural and terminal values which can apply to "a group, a social organization, a total society, or even an ideal society" [42] (p. 38). It is possible to understand Rokeach's (1973 and 1979) theorisation of values in many ways [42,43]. However, a generic inference from the theory, which can be suitable for planning, is that values can be implicit or explicit. They can relate to individuals, places, and groups of people. As is always the case with many theories, there is room for improvement. In this regard, Schwartz (2012) investigated Rokeach's theory and concluded that it does not adequately explain the underlying structure of values as a system [44]. This cultural perspective of values is relevant in planning either as a method or a system. Hence, the concept of values-led approaches is related to the notion that planning should be based on human values in the form of a vision for action in delivering outcomes of land use and spatial development [45]. However, this does not entirely explain what is meant by a values-led approach to spatial planning.

3.3. (Re)Stating the Concept of Values-Led Approach in Planning

The scope of land management, which includes spatial planning as an activity in the planning domain, demands that a planning process, based on publicly or societally acceptable values, is a matter of necessity. This is because the direct ways to capture and fulfil the interest of the public exist in spatial-development affairs. Evidence from the empirical studies of Reimer et al. (2014) supports that new approaches to spatial planning are needed because various urban and rural areas are facing new challenges [19]. The VLP approach as prescribed by Auziņš and Viesturs (2017) provides an opportunity for testing new values-led tools and techniques. As they put it [17] (p. 281):

> *"Introduction of VLP approach based on consolidated new knowledge from stakeholders' experience and empirical evidence will help better understand and guide the relevant processes and their effects in specific territories based on the identified values (an outcome of expertise) and attitudes (stakeholders' preferences), thus avoiding such problems as, for*

example, unplanned urban sprawl, environmental/landscape fragmentation and damage, unequally populated areas, remarkable differences in income, insolvent territories, etc. It is argued that . . . implementation of the new approach within the proposed framework will lead to improved regional and local land-use policies and thus better territorial governance, developing more inclusive and resilient territories for the benefit of entrepreneurship, society, and nature. Innovation activities will be recognised, for example, when applying developed methodological guidelines in the planning process. Complex yet significant relationships between the values and preferences of the stakeholders concerning land use and development are to be assessed."

The study [17] is concerned with the conceptual background and feasibility aspects of the VLP approach. In contrast, this paper capitalises on the achieved outcome and provided arguments, thus proposing a methodology and pathway for its implementation.

The arguments for the VLP approach reflect the evolution of planning cultures and systems. For instance, during the period of the Ebola pandemic in West Africa, many communities clamoured for the reassessment of their public-health values in planning [46], intending to improve their funerals, cultural festivals, and other celebratory rites that make them vulnerable to pandemics. In the COVID-19 (or Coronavirus) period, radical changes in behaviour and attitudes to public facilities whether in urban, peri-urban, or rural areas have emerged. For instance, the practices of physical/social distancing in public spaces like playgrounds and parks are values-related planning issues that require integration into spatial planning. Values reflect the realities in the organisation of desired activities and well-being. Thus, if spatial planning is about situating people in the most convenient ways to ensure adequate living conditions, then values should form a critical consideration of the spatial planning process.

There are several values to be considered in planning. They all relate to the cultural, institutional, ethical, or ideological (or philosophical) sense of worth, which shapes how people decide, act, and appreciate the manner of life they live or want to live. In broad terms, values can be categorised in many forms. Spatial values can be in the form of intrinsic or extrinsic values. Intrinsic value reflects an ethical and philosophical property. Intrinsic values can manifest in planning because people have to make personal or group choices about their path to development "based only on how things possess inherent worth or satisfy their preferences" [47] (p. 1462). Intrinsic values can also manifest in how the people "consider the appropriateness of how they relate with nature and with others, including the actions and habits conducive to a good life, both meaningful and satisfying" [47] (p. 1462). These intrinsic values are relational because they are mostly based on life principles, preferences between options, and virtues linked with person-to-person and place-to-person relationships. Values can manifest in place-based relationships or a sense of place [48,49]. Such values can also be extrinsic. Thus, they are objectified in their physical worth, and they cannot be avoided since planning is about people and is done by people. Hence, a planning system that works to improve the existing living conditions and livelihood requires a planning approach that allows values-related issues (whether social, economic, or environmental), and their effects in specific territories are urgently needed. This is why a VLP approach is crucial for identifying the synergy that would enhance balanced socio-economic and environmental impact as well as governance in human settlements. Scholars have done preliminary work in defining the framework for understanding the VLP approach [17,18,22,27]. The missing piece is a methodology for making values-based spatial development a reality.

4. Framing the Path to a Methodology for VLP Approach in Spatial Development

Spatial planning is "a multifaceted and highly complex activity, embedded in specific cultural contexts composed of interactive processes among involved actors, their cultural cognitive frames, and their particular social contexts" [50] (p.83). Values which emerge from a culture "play a mediating role between people or society and the environment, influencing people's intentions, way of life, sense of place, practices, norms, and rules" [51]

(p. 25). Therefore, VLP is essential because when communities prioritise their needs around what they value most, their efforts are bound to produce outcomes that matter most to them. In operationalising the EU Spatial Development Agenda, VLP has entailed putting values in place, promoting and integrating values in planning towards sustainable place-making and place-shaping.

The operationalisation of the VLP approach requires identifying those elements of its conceptual framework and reconciling them with existing empirical issues that are commonly accessible from the literature. In this regard, this study identified four key elements. These are values-accommodating and enable leading a spatial planning process to spatial development without negating values. They include (1) the nature of the physical "land-use" and "space" being planned [52], (2) the values-based "function" of spatial planning [53], (3) the values-based "principle" of spatial planning [54], and (4) the nature of "prevailing ideological and belief systems" under which spatial planning operates [55]. The nature of "land-use" and "space" is determined by the actual spatial planning, which considers the land use and spatial networks as means of administering uses to people and services, including nature—embracing all functions of space. However, spatial functions can be non-values-based. Participation is a major way to ensure that values-based functions are actualised in the planning process. However, this depends on the principles applied in spatial planning. Where bottom-up principles are applied, there is a tendency to embrace equity, equality, accountability, participation, and various others. Yet, all of these issues are concerned with specific beliefs and ideologies. It can be observed that ideologies and belief systems drive planning vision, which appears to guide success or failure. Four elements or dimensions of spatial planning (in Figure 2) are recognised as crucial in a spatial planning process because of their interaction that determines a relationship capable to lead a VLP approach in the planning process.

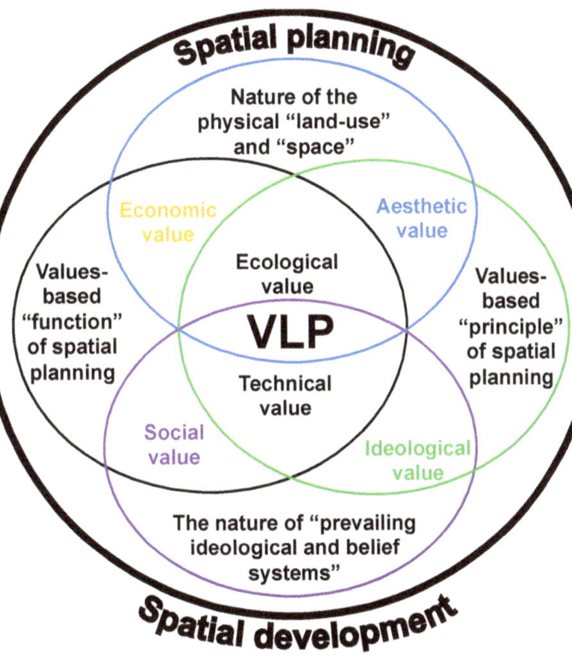

Figure 2. VLP is a consequence of an interaction between four key elements of spatial planning.

4.1. The Typology of Values

Values, apart from being a set of belief systems or culture, are the core instruments in expressing, interpreting, and understanding viewpoints in the planning process. They also represent a core embedment of the planning experience from conceptualisation to the outcome and its monitoring and further alterations in plans.

Further, each of these four main elements or dimensions (in Figure 2) produces particular *core values* that enable the operation of spatial planning (as a comprehensive process) and the creation of preconditions for further changes in land use and space through spatial development. The interrelation between these four dimensions manifests in six types of core values that influence spatial planning. Based on developed useful classifications and assessments' approaches, the typology of *spatial (territorial) values* is proposed (in Table 1).

Table 1. Values and their characteristics.

Typology	Values	Description
Core values	Economic value	Land use and development as they reflect in welfare gains
	Social value	Involvement and cooperation as they reflect on social inclusion and security
	Ideological value	Traditions and cultural matters as they reflect in beliefs
	Ecological value	Environmental quality and ecological liveability as they reflect in acceptability
	Aesthetic value	Emotional perception and critical attitudes as they reflect in designs
	Technical value	Application of technology-based knowledge as it reflects in stimulating planning innovations
Spatial values	The values of land-related resources	Can be seen in landscape as a stock; factual land-use types or geospatial units in the frame of a particular project, e.g., agricultural land, forest, waterbody, roads, built-up land, etc.
	The values of landscape functions	Can be seen and perceived in the landscape as ecosystem and landscape functions, goods, and services—the functions based on de Groot and Hein (2007) [56] when considering the multi-functional nature of ecosystems and the evolvement of the approach of ecosystem services when considering it as a strategy and focusing on ecosystem services as these are related to the activities, decisions, and investments of humans [57], e.g., provisioning, regulation and support, cultural, etc.
	The values of land-use patterns	According to the institutional settings of a particular spatial planning system, these can be identified in spatial development plans from different socio-economic functionalities—functional zones in an area, e.g., rural areas, green/natural areas, transport infrastructure, industrial areas, public areas, family housing, etc.
	Synergetic values of land-use patterns	Can be perceived and analysed through the planning-implementation experience, e.g., green or public areas concerning residential areas and connection to public-transport infrastructure in an area
	Intangible values	Can be recognised through historical evolution and are socially meaningful for the future, e.g., intrinsic, place-based, historical, cultural, etc.

"Spatially allocated values are the ones that may be recognised spatially as both provided by ecosystems and human-made formations. Ecosystems, e.g., agricultural and forest land and ecological landscapes, are an essential part of natural capital. Their quality is the basis for environmental viability. However, the impact of human beings on ecosystems is considerable. Ecosystems provide humans with services that are relevant for survival, health, and welfare/culture. Human-made formations, e.g., developments, improvements, and utility supply, are an essential part of human-made capital. Its quality is the basis for socio-economic equity. Human-made formations provide humans with services that are relevant for dwelling, workplace, and recreation.

4.2. Conceptualised Participation

Public participation and stakeholders' involvement are integrated into the spatial planning process in all observed European countries [18,22]. However, they are approached differently in different countries and even in regions or municipalities within the same countries. The difference also appears in the conditions of particular planning practices. The *conceptualization of participation* captures various approaches and techniques that support the understanding of possible discussions and agreements in the way from a conflict/disagreement towards a consensus-building, which results in the extent to which the outcome of participation and collaboration is considered in a final or binding decision. Conceptualisation allows finding ways to reach sustainability in a decision-making process when assessing stakeholders' preferences in an area. This also means that public participation should be considered as a values-related matter in planning. For instance, in Germany's case, participation is viewed as a cultural value to be embraced in any planning process [21]. Trust is a prerequisite for cooperation, for the resolution of collective action problems, and effective democratic governance. Trust is indeed vital to participants' belief that a meaningful discussion and cooperation with the members of a planning association (committee) is possible [58]. This is especially important if the planning process is driven by an expert approach (often opposite to a deliberative approach). Yet, building and maintaining trust is extremely complex. Trust is deeply connected with risk, power, and modernity. However, citizen engagement and placemaking [59] are crucial processes for improved communication, informed deliberation, and trust-building practice.

The deliberative planning method as a form of participatory planning and an urban planning theory focuses on involving the community in spatial planning and management processes. During the discussions about possible further development, the involved local stakeholders may participate, communicate, and deliberate. Often, participatory planning is debated in contrast to deliberative planning [58]. The first includes all points of view in the decision-making process and gathers them all in one vision. However, the second involves those being motivated by an intended outcome and choosing a vision that is based on sound evidence and arguments [60,61]. Deliberative democracy provides principles for spatial planning. The most important is to determine the beneficial conditions and particular collaborative forms, which provides the best possible outcomes and largely contributes to the decision-making process. Therefore, key questions should be formulated: how may different concepts of stakeholders' participation contribute to the binding decision? To what extent does the introduction of a particular concept improve the implementation of what is decided and the decision-making process on the whole? Legacy et al. (2019) recently discussed new ways of conceptualising participation that "can create new informal spaces where injustices and inequalities are voiced and the structures and hegemonies created" [62]. This study on methodology development assumes the interrelation between three processes: (1) conceptual participation, (2) deliberative planning, and (3) decision-making leading towards a consensus-oriented governance model.

4.3. Value Causing Assessment (VCA)

It is acknowledged in the study on methodology development that the problem structuring methods (PSM) [63] are appropriate to apply in structuring the values and in assisting planners and involved stakeholders. The inclusion of goals and objectives as evaluation criteria and the weighting given to these criteria adjust the quantitative scheme of cost-benefit analysis (CBA) to the needs of planning. However, such quantitative methods and the multi-criteria analysis (MCA) and ordinal ranking methods still use a broad scheme of CBA. PSM represents an alternative set of methods developed to address similar concerns in the operation research field. These methods were developed to address situations where there is no single objective to be achieved and where parameters are contested [63]. The widely used quantitative multi-criteria decision analysis (MCDA) methods require that all dimensions be valued in ordinal terms. Both the valuation and weighting are sensitive to the identity of those performing the tasks. Feitelson (2011) pointed out that an "evaluation process in a communicative setting aims to utilize expertise to raise issues that can be easily described to a relatively wide audience" [64]. PSM makes progress in this direction by engaging experts and stakeholders and by seeking to understand better their perceptions of the issues at hand and of the options for addressing them. The method based on MCDA has been proposed by Feitelson's distinguished study. The method can also be applied for the VCA to the extent that professionals evaluate alternatives according to criteria that are partially derived from the objectives. These objectives can be determined either from existing plans and policy documents or in a collaborative process. The experts are asked to rationalise their evaluation in reporting the outcome. These rationalisations are summarised by criteria and form building blocks of the VCA whose outcomes are a set of values that can be easily understood and discussed by non-experts and people who have not previously engaged in the planning process. For instance, the alternatives (trade-offs of values) are designed/described with main features and evaluated by experts (see step 1.5 in Table 2).

Table 2. Steps of VCA.

1. Identification and assessment of values (professional expertise)
1.1 Determination of territory and setting of preliminary land-use objectives and its criteria (from previous planning practice and policies) by the responsible planner 1.2 Establishment of the board of experts by the responsible planner 1.3 Assignment of tasks for experts lead by the responsible planner 1.4 Identification and structuring of values from field surveys, stakeholders' knowledge and community involvement, plans, documentaries, and other sources by experts 1.5 Evaluation of alternatives (trade-offs of values) by experts 1.6 Consensus-oriented discussion and provision of the outcome as a result of professional expertise by the board of experts 1.7 Summarization of the outcome when considering its contribution to the land-use objectives (by criteria) by the responsible planner
2. Beliefs and preferences of stakeholders (stakeholder deliberation)
2.1 Establishment of the board of stakeholders/steering committee by the responsible planner 2.2 Organisation of stakeholders' deliberations based on the outcome of professional expertise, provided arguments, and discussions, involving community members and experts by the responsible planner 2.3 Selection between possible alternatives regarding decisions by stakeholders and responsible planner 2.4 Organisation of public hearing and discussion by stakeholders 2.5 Consensus-oriented discussion and provision of the outcome on identified alternatives by the board of stakeholders/steering committee 2.6 Summarization of the outcome when considering its contribution to the land-use objectives (by criteria) by the responsible planner

Table 2. *Cont.*

3. Allocation of planning alternatives (assessed decision-making)
3.1 Establishment of the joint working group, involving representatives from both boards, responsible planner, and a representative of the legitimate body (official authority responsible for legally binding decision-making) by the legitimate body
3.2 Synthesis of outcomes from professional expertise and stakeholder deliberation towards making an assessed and well-coordinated decision, considering that possible feedbacks and mitigation measures (trade-offs) are solved before providing outcomes (1.7 and 2.6) by the joint working group
3.3 Approval of the synthesis report (a decision) on the allocation of planning alternatives by the legitimate body
3.4 Proper arrangements into spatial/land-use plans (land-use patterns) and regulations on land use and development by the responsible planner
3.5 Monitoring of the implementation of the decision by stakeholders

The VCA is proposed based on the identified "gap between post-modernist planning theory (communicative collaboration) and largely modernist planning practice (rational-technocratic process)" [64]. Thus, in the light of the advanced systematic qualitative approach for evaluating planning alternatives, as is emphasised by Feitelson (2011), VCA will also lead to the identification of particular values that should be deliberated by stakeholders rather than to choose among the alternative options. Accordingly, the VCA uses the expertise of professionals to focus stakeholder deliberations on spatial planning and local development. To employ VCA, various known evaluation techniques have been reviewed [18,27]. Evaluation techniques, represented through case studies and the outcome of applied research, are considered relevant for the VLP approach. At the same time, the advantages of qualitative evaluation methods for VLP over some widely used quantitative ones (e.g., MCDA, CBA) should be considered. For instance, CBA often does not address all the facets of complex tasks with multiple externalities and wide-ranging distributional implications due to democratic decision-making and orientation towards consensus, and it also does not reflect social welfare preferences or assure quality among all involved stakeholders. In this respect, for instance, it is suggested that to facilitate and set the agenda for discussions in spatial planning, a qualitative yet systematic method is necessary. However, the effective combination of both qualitative and quantitative approaches in communicative planning is encouraged, especially where there is a need to process gathered data sets. From the discursive perspective of planning practice, the deliberation process and decision-making focus on using mainly qualitative evaluation methods. The application of VCA should be based on the philosophy that policy for the people needs policy with the people. Hereafter, the steps from 1.1 to 3.5 of VCA are subsequently proposed in Table 2.

4.4. Preconditions for Ensuring a Values-Enabling Planning Environment

By developing the methodology, it is acknowledged that three essential preconditions are necessary to promote the environment in which spatial planning is taking place. This implies that it has to be a values-enabling planning environment. This is essential for values-based spatial development to emerge. If an environment is not supportive of values but rather impedes values, it cannot lead to a values-based spatial development. The planning environment involves conditions that surround and enable the planning process. For instance, laws, policies, professional ethics, human behaviours, and capacity, among others. The planning environment should play a fundamental role in shaping the nature of spatial planning practices within geographies, legal jurisdictions, or administrative boundaries. Three essential elements that characterise the planning environment based on the shared values of stakeholders include: (1) the organization of spatial planning and involvement in it; (2) a planning process and tools; and (3) the relationship of planning activities to fostering sustainable communities. These are further explained below.

4.4.1. The Organization of Spatial Planning and Involvement in it

The organization of the planning process and stakeholders' involvement allow for a values-enabling environment if the administrative structures, policy frameworks, institutional and social settings, collective actions, and social learning are collaborative in their design. Hence, when developing a values-enabling or supportive environment in spatial planning, the following issues are relevant to consider for applying the methodology:

- The organisation of planning in practice when identifying a locus of power, hierarchies, institutions, etc.;
- The recognition of the most influential actors in planning, including the organisational structures, authorities, partnerships, the groups of common interest, and individuals;
- The legal and administrative fundaments that formally support spatial planning;
- Policymaking and implementation that reflects deep-rooted values, e.g., dominant policy style, ascertaining its impact on institutional performance, and social activity looking over last years;
- The existence of the linkages among stakeholders, including collaboration forms/ networks, cross-border relations, integration of sectoral interests, problem-driven cooperation, etc.;
- An assessment of the linkages between places relevant in the planning context, e.g., rural-urban, inner-urban, and peri-urban;
- The identification of key defining moments, events, and people in the evolution of planning practice;
- Methods for how the people may benefit from spatial planning, including informing, learning, collaborating, understanding values, critically acting;
- Dynamic changes if measuring territorial governance, thus ascertaining the movement between both command/control and consensus-oriented governance models.

4.4.2. Planning Process and Tools

Examining both the peculiarities of the planning process and the ways concerning how the planning practice is supported and improved allows focusing more on the deliberative making of plans, planning modes, formal and informal (complementary) planning tools, and project-oriented techniques. To ensure that the planning process and applicable or associated planning tools do not hinder the evolution of values as a part of the methodology for action, the following issues are relevant to consider:

- The determination of key driving forces influencing the evolution of local planning practice over the last decades, including how values have evolved in political, economic, and neo-liberal agendas, and social actions, etc.;
- The recognition of structures and networks with an important influence on the development of local planning practice if considering the ways they changed over time;
- The evolvement of the role of professionals/planners over time and its current status, e.g., an executive-arm, a technician, a consultant, a negotiator, or an assistant;
- An overview of the education for professionals/planners;
- An assessment of the coherence of a planning community in a particular practice, identifying several different planning communities, and ascertaining its variations in urban/rural, regional/local, or other contexts;
- Observation of different instrumental planning tools in a particular practice, e.g., informal modes of operation and planning tools that lie outside the institutionalised planning system;
- The characteristics of emerged planning modes and tools to support spatial planning practice, e.g., general/specific planning regulations, set of planning documents, legally binding/guiding, formal and/or informal arrangements for territorial governance, enhancing multi-sector participation, and networking, more oriented towards strategies or land use;
- The recognition of projects which support and/or provide improvements to formal spatial planning;

- Dynamic changes if measuring a spatial planning-implementation linkage—besides the planning, also considering the implementation of plans in practice through decision-making in land-use management—thus ascertaining the movement from just formal institutionalised planning mode towards complementing informal planning arrangements.

4.4.3. The Relationship of Planning Activities to Fostering Sustainable Communities

For maintaining and repositioning the values in the planning process, it is necessary to analyse the planning environment itself and the shared values of the involved actors, whose preferences and actions may influence the planning outcomes. This allows focusing more on the cultural awareness of stakeholders in the planning process. It would also ensure that the shared assumptions, values, and preferences of the involved parties are put under scrutiny to promote a planning environment in support of generally accepted values. To achieve this, the following issues relevant for applying the methodology should be considered:

- The extent to which a spatial planning in particular practice succeeds in achieving the principles such as sustainability, equal opportunities, public participation, transparency, integrated approach, and coherence;
- The importance of community involvement and activity in spatial planning as well as the social value of planning;
- The promotion of community development and management if considering identified and discussed spatial values and preferences of stakeholders in spatial planning;
- The extent to which the perception, beliefs, shared values, and behaviour of the actors involved can be recognised through the spatial planning;
- The importance of the role of values in spatial planning and the extent to which the planning agendas and discourses, e.g., dominating ideas, views, and styles, substantiate the preference of values;
- The extent to which spatial planning serves different interests, including local communities, the business community, private developers, international investors, etc.;
- The impact of the international planning ideas and knowledge on the evolution of planning practice;
- Reasons for increasing activity of the civil society and identification of bottom-up initiatives and networks mobilising around urban and rural development issues as well as the importance of these networks in current planning debates and agendas;
- An assessment of emerged, distinct approaches of the planning experience and tradition to spatial planning;
- The impact of the evolution of planning education and experience on the VLP approach to planning.

4.5. *Towards an Application of the Proposed Methodology for the VLP Approach in Spatial Development*

To apply the proposed methodology, planners, including public authorities, land managers, and community leaders involved in spatial planning, must understand the diverse characteristics of values and how they can influence development outcomes negatively or positively.

In spatial planning, socio-economic, environmental, and institutional aspects of values and preferences of stakeholders embody *sustainability dimensions*. They can refer to future-oriented or historical changes. Therefore, the planning and implementation should form the necessary preconditions for sustainable use of land-related resources, thus reconciling spatial development interests with all dimensions or elements of values that apply within specific planning environments. For instance, the interests of preservation and revival of natural resources constitute environmentally significant ecological values. They may also be social values for people who are from a very environmentally aware society. These sorts of interests are primary in Europe but can be of secondary concern elsewhere. Irrespective of the geography of planning, some values are generally of basic interest to people globally.

For instance, the values related to housing, work, and place improvements are known to be of primary interest in any form of spatial setting, e.g., urban, rural, or peri-urban. Therefore, it is essential to focus on ensuring that these primarily accepted values are handled as a matter of priority before secondary values. Doing this requires adopting effective strategies for tackling the challenges and approaches usually encountered in identifying, assessing, and discussing these values among involved stakeholders.

Furthermore, the review of governance styles and institutional settings, as well as sustainable development analysis, should be performed to grasp how complementary planning tools can support the process of formal spatial planning in practice without negating the importance of promoting values. Evidence-based knowledge of the "collaborative planning approach" [65] may support making binding decisions to promote consensus-building for the benefit of local society. Hypothetically, sustainable decisions based on harmonized values and preferences lead to "sustainable communities" [66]. Some arguments towards a "sustainable intensification" of land use emphasise the "management of growing pressure of human needs, while at the same time minimizing the impact on the environment" [67]. However, a new paradigm of sustainability towards a sustainable future rather than a sustainable development is becoming necessary. In this light, the sustainability aspects should focus on "how significantly human needs have to be diminished or changed for the impact on the environment and land-related resources to be the smallest possible" [18]. In the context of implementing the developed methodology for the VLP approach, we would point to it as a definition of sustainability challenge.

As the VLP approach does not replace formal (institutionalised) spatial planning but complements it, the implementation measures focus on informal (complimentary) tools, e.g., thematic plans, and its integration with formal planning tools (spatial development plans) through the planning process. *The framework for implementation of the VLP approach* is designed with three key elements of the VLP agenda and particularly targeted measures (in Table 3):

Table 3. The framework for the implementation of the VLP approach.

Key Elements	Measures
The organisation of the planning process and involvement of stakeholders	• Development of institutional settings to soften the normative approach in the way they facilitate deliberative processes • Establishment of administrative structures to support a bottom-up approach in territorial governance [68] and placemaking [59] as well as agglomerative (cross-border and soft-area) cooperation • Provision of policy style to facilitate deliberative democracy and communicative planning • Promotion of social settings to strengthen collective actions and social learning towards consensus-building and participation in decision-making
The planning process, modes, tools, and techniques supporting the planning practice	• Making of deliberative plans to meet demand and supply in spatial planning and development aims • Integration of formal and informal planning tools to make the planning process more flexible • Development of thematic planning tools to provide VCA and the mapping of values • Promotion of project-oriented techniques to support the integration of planning modes

Table 3. *Cont.*

Key Elements	Measures
Planning environment and shared values of stakeholders	• Emphasis on the cultural awareness of stakeholders in planning to support the deliberation • Sharing of assumptions, values, and preferences of involved parties to balance interests and achieve well-informed planning results • Understanding of cooperation, discussions, and gaining new knowledge as beneficial social values to promote the trust-building practice • Gaining confidence that made preferences and actions, including trade-offs, will influence further outcomes in a sustainable way

From the meetings with experts during comprehensive evidence gathering (CEG) and the workshops of stakeholders [27], some relevant *guiding suggestions to the VLP methodology* were derived. Thus, the VLP approach benefits society while the benefits justify the values, but the values ground the decision-making process. The approach should conceptually provide advantages for decision-making in land management. It is necessary to promote integration between sectoral policies, spatial planning, and land use. The introduction of the approach should enhance the cross-scale and cross-sectoral coherence between three interrelated land-use (land-use intensity driven by market mechanisms), spatial planning (land-use objectives and priorities determined through the planning process assessing environmental impact), and sectoral policies (restrictions and compensations due to the assessment of policy impact).

Public participation should be purposefully managed. The involvement of inhabitants should not be organised without a specific purpose and informed/explained agenda. The discussions have to be constructive and provided with arguments. The responsibility of parties (authorities or other bodies) about decision-making also has to be clear and declared, especially when the crucial issue is about the extent to which the outcome of a discussion is considered in a binding decision. The interests of participating parties/stakeholders can be different by their status and competencies. In the process of public participation and discussions, the differences cannot be so vivid. However, it would be interesting (even if it sounds quite utopic) to measure the interests by type of participants. The level of competence looks quite important to understand. However, a great deal depends on the provided information to the participants and its understanding by participants. The level of competence influences the quality of discussions and the ability to trade-offs. The public conflict has to be prevented/avoided due to the discourse of public discussions. The early involvement of participants helps to avoid conflicts. Thus, the participants are informed enough due to the discourse of the planning process but not only at the end of it. Otherwise, during the late phase of the planning process, the participants feel more like they are formally informed and provided with factual information but are not involved purposefully to contribute to the planning. Any participatory method has to fit its place and time. So-called thought leaders also have powerful roles to drive public opinion and changes. The establishment of forums of professionals as capable enough formations to initiate and manage discursive changes into particular fields also should be considered for improved planning and development. Particular tools may support participation and decision-making. However, the competence (knowledge, skills, and attitudes) of participating key stakeholders is very significant.

5. Conclusions

This study has used evidence from the literature and CEG to create an understanding of (1) the relevance of values in spatial planning; (2) the concept of values-led approaches in planning; and (3) framed methodology for introducing a VLP approach in spatial development.

The study strove to implement the methodology because of the growing need in Europe and many parts of the world for new innovative ideas ensuring people-centred planning. The proposed methodology is relevant because it projects the importance of conducting spatial development to appeal beyond the mere coordination of activities and space. It embraces often neglected aspects of spatial planning and development concerning the values and preferences of the people whom the planning outcome should serve. It is a methodology that presents preconditions that, if followed, would ensure that stakeholders in a planning process would be able to decide consciously on how the outcome of planning can reflect their worldviews and their development needs. However, it is important to state that the methodology proposed in this study does not, in any way, suggest that all values are relevant for promotion in a planning process. It acknowledges that values evolve and are always in a state of flux, and not all values would lead to adequate spatial development outcomes. The methodology does not assume that values (as an intricate consideration in planning) are always clearly identified and assessed. Values do come with many problems or even create problems. For instance, values that promote gender inequality, racism, nepotism, spatial inequality, inequality in sexual orientations, and spatial or environmental injustice (to mention a few) would not promote a responsible VLP approach. Hence, while putting into operation the methodology proposed in this article, only values that promote good territorial governance and land management practice should be considered by spatial planners, communities, public authorities, and land managers.

This study is biased in favour of the European tradition of spatial planning. However, it is not peculiarly influenced by any specific country experience but rather by studies emanating from multiple European countries. Thus, the issue of adaptability in its application matters. This is why the study emphatically recommends an assessment of the planning environment and shared values as a core activity in facilitating the development of local communities and capitalising on professionalism (constructive attitudes) rather than general policies and mainstream planning to serve market-driven developments. If this is done in consideration of the objectives of the VLP approach, there is bound to be some successful outcome in spatial development. The key objectives should include but not be limited to: (1) more supportive and collaborative territorial governance as well as promoted informal institutions and organisational forms and (2) the building of trust through balancing of planning interests as well as increased cultural awareness, shared perception, and making the appropriate assumption of values and preferences.

For successful implementation of the proposed methodology for solutions to support further planning practices, the key recommendations should be organised into three directions: (1) improvements in institutional settings, organisation of planning process, and involvement; (2) integration of informal (complementary) planning tools into formal planning agenda; and (3) conceptualisation of public participation, collaboration, and deliberation. With these recommendations in place, the likelihood of embracing beneficial values through the VLP approach would be high.

Author Contributions: For this research article, conceptualization, A.A. and U.E.C.; methodology, A.A. and U.E.C.; validation, A.A. and U.E.C.; formal analysis, A.A.; investigation, A.A.; writing—original draft preparation, AA. and U.E.C.; writing—review and editing, A.A. and U.E.C.; visualization, U.E.C.; funding acquisition, A.A. Authors have read and agreed to the published version of the manuscript. All authors have read and agreed to the published version of the manuscript.

Funding: This work was supported by the European Regional Development Fund within the Activity 1.1.1.2 "Post-doctoral Research Aid" of the Specific Aid Objective 1.1.1 "To increase the research and innovative capacity of scientific institutions of Latvia and the ability to attract external financing, investing in human resources and infrastructure" of the Operational Programme "Growth and Employment" No. 1.1.1.2/VIAA/1/16/161.

Data Availability Statement: Not applicable.

Conflicts of Interest: The authors declare no conflict of interest. The funder had no role in the design of the study; in the collection, analyses, or interpretation of data; in the writing of the manuscript, or in the decision to publish the results.

References

1. Nadin, V.; Stead, D. European spatial planning systems, social models and learning. *Disp Plan. Rev.* **2008**, *44*, 35–47. [CrossRef]
2. Turkelboom, F.; Leone, M.; Jacobs, S.; Kelemen, E.; García-Llorente, M.; Baró, F.; Termansen, M.; Barton, D.N.; Berry, P.; Stange, E.; et al. When we cannot have it all: Ecosystem services trade-offs in the context of spatial planning. *Ecosyst. Serv.* **2018**, *29*, 566–578. [CrossRef]
3. Muradian, R.; Gómez-Baggethun, E. The institutional dimension of "market-based instruments" for governing ecosystem services: Introduction to the special issue. *Soc. Nat. Resour.* **2013**, *26*, 1113–1121. [CrossRef]
4. Gwaleba, M.J.; Chigbu, U.E. Participation in property formation: Insights from land-use planning in an informal urban settlement in Tanzania. *Land Use Policy* **2020**, *92*, 104482. [CrossRef]
5. Larsson, G. *Land Management as Public Policy*; University Press of America: Landham, MD, USA, 2010; p. 246.
6. Innes, J.E.; Boother, D.E. Collaborative rationality as a strategy for working with wicked problems. *Landsc. Urban Plan.* **2016**, *154*, 8–10. [CrossRef]
7. Raymond, C.M.; Kenter, J.O.; Plieninger, T.; Turner, N.; Alexander, K. Comparing instrumental and deliberative paradigms which underpin the assessment of social values for cultural ecosystem services. *Ecol. Econ.* **2014**, *107*, 145–156. [CrossRef]
8. Kenter, J.O.; O'Brien, L.; Hockley, N.; Ravenscroft, N.; Fazey, I.; Irvine, K.N.; Reed, M.S.; Christie, M.; Brady, E.; Bryce, R.; et al. What are shared and social values of ecosystems? *Ecol. Econ.* **2015**, *111*, 86–89. [CrossRef]
9. Kenter, J.O.; Reed, M.; Fazey, I. The Deliberative Value Formation model. *Ecosyst. Serv.* **2016**, *21*, 208–217. [CrossRef]
10. Irvine, K.N.; O'Brien, L.; Ravenscroft, N.; Cooper, N.; Everard, M.; Fazey, I.; Reed, M.S.; Kenter, J.O. Ecosystem services and the idea of shared values. *Ecosyst. Serv.* **2016**, *21 Pt B*, 184–193. [CrossRef]
11. O'Connor, S.; Kenter, J.O. Making intrinsic values work; integrating intrinsic values of the more-than-human world through the Life Framework of Values. *Sustain. Sci.* **2019**, *14*, 1247–1265. [CrossRef]
12. Garcia-Martin, M.; Fagerholm, N.; Bieling, C.; Gounaridis, D.; Kizos, T.; Printsmann, A.; Müller, M.; Lieskovský, J.; Plieninger, T. Participatory mapping of landscape values in a Pan-European perspective. *Landsc. Ecol.* **2017**, *32*, 2133–2150. [CrossRef]
13. Fagerholm, N.; Torralba, M.; Moreno, G.; Girardello, M.; Herzog, F.; Aviron, S.; Burgess, P.; Crous-Duran, J.; Fereiro-Domínguez, H.; Graves, A.; et al. Cross-site analysis of perceived ecosystem service benefits in multifunctional landscapes. *Glob. Environ. Chang.* **2019**, *56*, 134–147. [CrossRef]
14. Keller, R.; Backhaus, N. Integrating landscape services into policy and practice—A case study from Switzerland. *Landsc. Res.* **2020**, *45*, 111–122. [CrossRef]
15. van der Sluis, T.; Arts, B.; Kok, K.; Bogers, M.; Gravsholt Busck, A.; Sepp, K.; Loupa-Ramos, I.; Pavlis, V.; Geamana, N.; Crouzat, E. Drivers of European landscape change: Stakeholders' perspectives through Fuzzy Cognitive Mapping. *Landsc. Res.* **2019**, *44*, 458–476. [CrossRef]
16. Tiboni, M.; Botticini, F.; Sousa, S.; Jesus-Silva, N. A Systematic Review for Urban Regeneration Effects Analysis in Urban Cores. *Sustainability* **2020**, *12*, 9296. [CrossRef]
17. Auziņš, A.; Viesturs, J. A Values-led Planning Approach for Sustainable Land Use and Development. *Balt. J. Real Estate Econ. Constr. Manag.* **2017**, *5*, 275–286. [CrossRef]
18. Auziņš, A. Capitalising on the European Research Outcome for Improved Spatial Planning Practices and Territorial Governance. *Land* **2019**, *8*, 163. [CrossRef]
19. Reimer, M.; Getimis, P.; Blotevogel, H. *Spatial Planning Systems and Practices in Europe: A Comparative Perspective on Continuity and Changes*; Routledge: New York, NY, USA, 2014; p. 336.
20. Getimis, P. Comparing spatial planning systems and planning cultures in Europe. The need for a multi-scalar approach. *Plan Pract. Res.* **2012**, *27*, 25–40. [CrossRef]
21. Chigbu, U.E. Village renewal as an instrument of rural development: Evidence from Weyarn, Germany. *Community Dev.* **2012**, *43*, 209–224. [CrossRef]
22. Auziņš, A. Key Trends and Aspects Influencing Changes into Spatial Planning Systems and Practices in Europe. In Proceedings of the International Conference "Economic Science for Rural Development" No. 48, Jelgava, Latvia, 9–11 May 2018; pp. 26–35. [CrossRef]
23. ESPON EGTC. Comparative Analysis of Territorial Governance and Spatial Planning Systems in Europe. *ESPON*. 2020. Available online: https://www.espon.eu/planning-systems (accessed on 9 November 2017).

24. *Spaces of Dialog for Places of Dignity: Fostering the European Dimension of Planning*; AESOP, Book of Proceedings (E-Book); University of Lisbon: Lisbon, Portugal, 2017; p. 3327. Available online: https://aesop2017.pt/images/Congresso/proceedings/BookofProceedings20171215.pdf (accessed on 9 November 2017).
25. De Vries, W.T.; Chigbu, U.E. Responsible Land Management—Concept and application in a territorial rural context. *FUB Flächenmanag. Bodenordn.* **2017**, *79*, 65–73.
26. Chigbu, U.E.; Onyebueke, V.U. The COVID-19 pandemic in informal settlements: (Re)considering urban planning interventions. *Town Plan. Rev.* **2021**, *92*, 115–122. [CrossRef]
27. Auziņš, A. Comparative Analysis of Spatial Planning-Implementation Practices and Territorial Governance. In *Methods and Concepts of Land Management. Diversity, Changes and New Approaches*; EALD; Hepperle, E., Paulsson, J., Maliene, V., Mansberger, R., Auziņš, A., Valciukiene, J., Eds.; Vdf Hochschulverlag AG an der ETH: Zürich, Switzerland, 2020; pp. 23–38. [CrossRef]
28. Keeney, R.L. Value-focused thinking: Identifying decision opportunities and creating alternatives. *Eur. J. Oper. Res.* **1996**, *92*, 537–549. [CrossRef]
29. Jepson, P.; Canney, S. Values-led conservation. *Glob. Ecol. Biogeogr.* **2003**, *12*, 271–274. [CrossRef]
30. Gold, A. Principled principals? Values-driven leadership: Evidence from ten case studies of 'outstanding' school leaders. *Educ. Manag. Adm.* **2003**, *31*, 127–138. [CrossRef]
31. Iversen, O.S.; Halskov, K.; Leong, T.W. Values-led participatory design. *CoDesign* **2012**, *8*, 87–103. [CrossRef]
32. Tennant, M.G. Values-led entrepreneurship: Developing business models through the exercise of reflexivity. *Local Econ.* **2015**, *5*, 520–533. [CrossRef]
33. Artelle, K.A.; Stephenson, J.; Bragg, C.; Housty, J.A.; Housty, W.G.; Kawharu, M.; Turner, N.J. Values-led management: The guidance of place-based values in environmental relationships of the past, present, and future. *Ecol. Soc.* **2018**, *23*, 35. [CrossRef]
34. Coss, D.L.; Smith, K.; Foster, J.; Dhillon, S. Big data in auditing: A value-focused approach to cybersecurity management. *J. Inf. Syst. Secur.* **2019**, *15*, 77–100.
35. Chigbu, U.E.; Alemayehu, Z.; Dachaga, W. Uncovering land tenure insecurities: Tips for tenure responsive land-use planning in Ethiopia. *Dev. Pract.* **2019**, *29*, 371–383. [CrossRef]
36. Chigbu, U.E.; Antonio, D. Country-level strategy for tenure responsive land-use planning: Questions to ask and actions to take. In Proceedings of the World Bank Conference on Land and Poverty, Washington, DC, USA, 16–20 March 2020.
37. Davoudi, S. The value of planning and the values in planning. *Town Plan. Rev.* **2016**, *87*, 615–618. [CrossRef]
38. Hersperger, A.M.; Mueller, G.; Knöpfel, M.; Siegfried, A.; Kienast, F. Evaluating outcomes in planning: Indicators and reference values for Swiss landscapes. *Ecol. Indic.* **2017**, *77*, 96–104. [CrossRef]
39. Rawluk, A.; Ford, R.M.; Neolaka, F.L.; Williams, K.J. Public values for integration in natural disaster management and planning: A case study from Victoria, Australia. *J. Environ. Manag.* **2017**, *185*, 11–20. [CrossRef] [PubMed]
40. Ives, C.D.; Oke, C.; Hehir, A.; Gordon, A.; Wang, Y.; Bekessy, S.A. Capturing residents' values for urban green space: Mapping, analysis and guidance for practice. *Landsc. Urban Plan.* **2017**, *161*, 32–43. [CrossRef]
41. Taneja, R.; Faden, L.Y.; Schulz, V.; Rawal, A.; Miller, K.; Bishop, K.A.; Lingard, L. Advance care planning in community dwellers: A constructivist grounded theory study of values, preferences and conflicts. *Palliat. Med.* **2019**, *33*, 66–73. [CrossRef]
42. Rokeach, M. *The Nature of Human Values*; Free Press: New York, NY, USA, 1973; p. 438.
43. Rokeach, M. *Understanding Human Values: Individual and Society*; Free Press: New York, NY, USA, 1979; p. 322.
44. Schwartz, S.H. An overview of the Schwartz theory of basic values. *Online Read. Psychol. Cult.* **2012**, *2*, 1–20. [CrossRef]
45. Chigbu, U.E.; Kalashyan, V. Land-use planning and public administration in Bavaria, Germany: Towards a public administration approach to land-use planning. *Geomat. Land Manag. Landsc.* **2015**, *4*, 7–17. [CrossRef]
46. Chigbu, U.E.; Ntiador, A.M. Ebola in West Africa: Implications on 'community interaction' in urban Nigeria. *Int. J. Educ. Res.* **2014**, *2*, 329–346. Available online: http://www.ijern.com/journal/2014/October-2014/26.pdf (accessed on 20 May 2020).
47. Chan, K.M.; Balvanera, P.; Benessaiah, K.; Chapman, M.; Díaz, S.; Gómez-Baggethun, E.; Gould, R.; Hannahs, N.; Jax, K.; Klain, S.; et al. Opinion: Why protect nature? Rethinking values and the environment. *Proc. Natl. Acad. Sci. USA* **2016**, *113*, 1462–1465. [CrossRef] [PubMed]
48. Chigbu, U.E. Fostering rural sense of place: The missing piece in Uturu, Nigeria. *Dev. Pract.* **2013**, *23*, 264–277. [CrossRef]
49. Chigbu, U.E. Rurality as a choice: Towards ruralising rural areas in sub-Saharan African countries. *Dev. S. Afr.* **2013**, *30*, 812–825. [CrossRef]
50. Othengrafen, F. Spatial Planning as Expression of Cultured Planning Practices: The Examples of Helsinki, Finland and Athens, Greece. *Town Plan. Rev.* **2010**, *81*, 83–110. Available online: www.jstor.org/stable/40541556 (accessed on 26 March 2021).
51. Horlings, L.G. Values in place; A value-oriented approach toward sustainable place-shaping. *Reg. Stud. Reg. Sci.* **2015**, *2*, 257–274. [CrossRef]
52. Chigbu, U.E.; Schopf, A.; de Vries, W.T.; Masum, F.; Mabikke, S.; Antonio, D.; Espinoza, J. Combining land-use planning and tenure security: A tenure responsive land-use planning approach for developing countries. *J. Environ. Plan. Manag.* **2017**, *60*, 1622–1639. [CrossRef]
53. Tulumello, S.; Cotella, G.; Othengrafen, F. Spatial planning and territorial governance in Southern Europe between economic crisis and austerity policies. *Int. Plan. Stud.* **2020**, *25*, 72–87. [CrossRef]
54. Gentry, R.R.; Lester, S.E.; Kappel, C.V.; White, C.; Bell, T.W.; Stevens, J.; Gaines, S.D. Offshore aquaculture: Spatial planning principles for sustainable development. *Ecol. Evol.* **2017**, *7*, 733–743. [CrossRef]

55. Friedrich, L.A.; Glegg, G.; Fletcher, S.; Dodds, W.; Philippe, M.; Bailly, D. Using ecosystem service assessments to support participatory marine spatial planning. *Ocean Coast. Manag.* **2020**, *188*, 105121. [CrossRef]
56. De Groot, R.; Hein, L. The Concept and Valuation of Landscape Goods and Services. In *Multifunctional Land Use. Meeting Future Demands for Landscape Goods and Services*; Mander, Ü., Wiggering, H., Helming, K., Eds.; Springer: Berlin/Heidelberg, Germany; New York, NY, USA, 2007; pp. 15–36. [CrossRef]
57. Burkhard, B.; de Groot, R.S.; Costanza, R.; Seppelt, R.; Jørgensen, S.E.; Potschin, M. Solutions for sustaining natural capital and ecosystem services. *Ecol. Indic.* **2012**, *21*, 1–6. [CrossRef]
58. Laurian, L. Trust in Planning: Theoretical and Practical Considerations for Participatory and Deliberative Planning. *Plan. Theory Pract.* **2009**, *10*, 369–391. [CrossRef]
59. Beza, B.B. The role of deliberative planning in translating best practice into good practice: From placeless-ness to placemaking. *Plan. Theory Pract.* **2016**, *17*, 244–263. [CrossRef]
60. Sager, T. Deliberative Planning and Decision Making: An Impossibility Result. *J. Plan. Educ. Res.* **2002**, *21*, 367–378. [CrossRef]
61. Legacy, C.; Curtis, C.; Neuman, M. Adapting the deliberative democracy 'template' for planning practice. *Town Plan. Rev.* **2014**, *85*, 319–340. [CrossRef]
62. Legacy, C.; Metzger, J.; Steele, W.; Gualini, E. Beyond the post-political: Exploring the relational and situated dynamics of consensus and conflict in planning. *Plan. Theory* **2019**, *18*, 273–281. [CrossRef]
63. Rosenhead, J.; Mingers, J. A New Paradigm of Analysis. In *Analysis for a Problematic World Revisited: Problem Structuring Methods for Complexity Uncertainty and Conflict*; Rosenhead, J., Mingers, J., Eds.; Wiley: Chichester, UK, 2001; pp. 1–19.
64. Feitelson, E.I. Issue Generating Assessment: Bridging the Gap Between Evaluation Theory and Practice. *Plan. Theory Pract.* **2011**, *12*, 549–568. [CrossRef]
65. Randolph, J. *Environmental Land Use Planning and Management*, 2nd revised ed.; Island Press: Washington, DC, USA, 2012; p. 704.
66. Edge, S.; McAllister, M. Place-based local governance and sustainable communities: Lessons from Canadian biosphere reserves. *J. Environ. Plan. Manag.* **2009**, *52*, 279–295. [CrossRef]
67. Petersen, B.; Snapp, S. What is sustainable intensification? Views from experts. *Land Use Policy* **2015**, *46*, 1–10. [CrossRef]
68. ESPON COMPASS. *Comparative Analysis of Territorial Governance and Spatial Planning Systems in Europe. Applied Research 2016–2018. Final Report*; ESPON EGTC: Luxembourg, 2018. Available online: https://www.espon.eu/planning-systems (accessed on 20 May 2020).

land

MDPI

Article

Analyzing the Effects of Institutional Merger: Case of Cadastral Information Registration and Landholding Right Providing Institutions in Ethiopia

Solomon Dargie Chekole [1], Walter Timo de Vries [2], Pamela Durán-Díaz [2,]* and Gebeyehu Belay Shibeshi [1]

1 Institute of Land Administration, Bahir Dar University, Bahir Dar 5001, Ethiopia; solomon.dargie@bdu.edu.et (S.D.C.); gebeyehu.belays@bdu.edu.et (G.B.S.)

2 Chair of Land Management, Department of Aerospace and Geodesy, Technical University of Munich, 80333 Munich, Germany; wt.de-vries@tum.de

* Correspondence: pamela.duran@tum.de; Tel.: +49-89-289-25789

Abstract: Strong national institutional arrangements in the geospatial information management are essential for successful implementation of sustainable land administration system. However, it is not only the existence of institutions but also their effectiveness that leads to the intended goals and reaching of objectives. There are international calls to merge highly related land administration institutions, yet Ethiopia executes two related land administration tasks (landholding right provision and cadastral registration) by two different institutions, the Urban Land Development and Management Bureau, and the Urban Land Adjudication and Information Registration Agency. Thus, the objective of this article is to analyze the effect of merging cadastral information registration and urban landholding right providing institutions lead to effective and strong national land institution. To achieve this, we had a qualitative approach analysis based on desk review and case study research methods. We conducted semi-structured interviews with the directors of the two institutions, and a group discussion with professional experts from both institutions. The findings of this study show that institutional merger between the two institutions believed to unravel the challenges of failing to achieve institutional goals. Although these institutions design strategic plans every year, the level of achievement or operational performance is low. The major cause for this problem is the poor coordination between the institutions. In view of this issue, we recommend merging the two institutions in one since it: reduces the effects of data duplication; provides one-window services; reduces operational costs; fills communication gaps among the staff; reduces time of operation; improves customer service; increases efficiency within processes; and provides a more efficient operation of land markets.

Keywords: cadastral system; cadastral information; institutional merger; land right; institutions; Ethiopia

Citation: Chekole, S.D.; de Vries, W.T.; Durán-Díaz, P.; Shibeshi, G.B. Analyzing the Effects of Institutional Merger: Case of Cadastral Information Registration and Landholding Right Providing Institutions in Ethiopia. *Land* **2021**, *10*, 404. https://doi.org/10.3390/land10040404

Academic Editors: Ruishan Chen, Uchendu Eugene Chigbu and Chao Ye

Received: 23 February 2021
Accepted: 5 April 2021
Published: 13 April 2021

1. Introduction

Land is a source of material wealth; it provides people with resources to sustain on. It is also a major economic asset from which people and nations get significant profit. In many developing countries, land is considered as an important economic and social asset where the status and prestige of people is determined. Because of such a high importance given to land, as compared to other properties, the legal protection accorded to land is al-ways strict in nature [1]. Due to this, strong institutions are established to administer and manage land. According to Holden, institutions are defined as systems of established and prevalent social rules that structure social interactions [2]. Institutions play an important role in shaping socio-economic outcomes [3]. For this to happen, institutions must be ef-fective and functional [2]. The change in institutional arrangements, in many cases, was made in anticipation of a better land administration system [4]. The United Nations

Global Geospatial Information Management (hereinafter UN-GGIM) has affirmed in the Addis Ababa declaration for "Good Land Governance for Agenda 2030" that strong land administration institutions are required to support effective and efficient land administration and management to address the need to secure land and property rights for all [5]. According to UN-GGIM institutional arrangement is defined as the formal and informal structures that support organizations to establish the legal, organizational and productive frameworks. This type of arrangement establishes the sustainable management of geospatial information, inclusive of its creation, updating and dissemination of geospatial data. This in turn provides an authoritative, reliable and sustainable geospatial information base for all users [5]. Relevant institutional arrangements contribute to the strengthening of geospatial information management.

Theoretically, there are three types of institutional arrangements [6]: hierarchy based, market based, and network based. Each type of institutional arrangements has its own mechanism to understand the causes of problems experienced, the gains to be achieved, and the processes through which better institutional arrangements can be achieved. In hi-erarchy-based institutional arrangements, patterns of interaction have two main drivers: (1) authority, operationalized in administrative orders, rules and planning, and (2) dominance and authority as the basic control system on the other. Market-based institutional arrangements are based on competition, bargaining and exchange between actors. Network-based institutional arrangements take the form of cooperation between actors, where inter-organizational relations are ruled by the acknowledgement of mutual interdependencies, trust, and the responsibilities of each actor. In this paper when we talk about insti-tutional arrangement, we are referring to the hierarchy-based type of arrangement. Institu-tional arrangements may be realized by creating new or changing existing structures or management forms within the government.

The importance of institutional arrangements in geospatial information management was recognized by the UN-GGIM at its third session in July 2013, when it identified the need for countries to examine institutional arrangements in geospatial information management, and thereby provide governments with options on how best to create strong na-tional geospatial entities [7]. One of the mechanisms to (re)arrange institutions is through merger, which assembles institutions having related tasks so that resources could be uti-lized on its best use [8].

1.1. Rational Justification for Mergers

The rational justification and benefits of merger arises from different perspectives such as increasing institutional performance, improvement of customer service, efficient operation of land market [9]. A merger is an agreement, based on voluntarily, which unites two or more existing institutions into one [10]. There are several reasons why organizations merge. When looking at mergers, it is important to look at the subject on a case by case basis as each merger has both merits and demerits [11]. Mergers are mainly done to avoid duplication of responsibility across institutions [9]. Avoiding duplication would have economic benefits and help reduce bureaucracies. This in turn increases in-stitutional performance and reduces misuse of resources. Due to this, mergers have re-ceived frequent and thorough study by researchers [12]. The initiative to merge land regis-tration and cadaster institutions coined from FIG commission 7 where the research find-ings of 1994 show that the strategic management and operations of cadastral systems are vested in different organizations [13]. Pursuing these research findings, cadaster 2014 vi-sion was articulated by directing emphasis on merger of the land registration and cadaster institutions. Their findings show (1) improvement of customer service with increased efficiency (2) provision of more data in better quality (3) provision of data that are sufficiently accurate (4) provision of data to the government and citizens at the right time. The most reiterated overarching importance of cadastral institutional rearrangement is to pro-vide efficient operation of the land market. Following this, many developed countries (e.g., Sweden, Finland, Netherlands, Turkey, and Belgium) followed the principle of merger.

For instance, Sweden started the cadaster and registry merging in 2008. Following this, a study conducted to evaluate pre-merger and post-merger institutional performance was undertaken and the finding shows that an institutional merger brought institutional effec-tiveness [14]. What is emerging is that while developed countries have moved on to mod-ern cadastral systems, majority of the countries in Africa are still stuck in the old tradi-tional systems although a few countries have embarked on the process of modernizing their cadaster in line with the internationally recognized benchmarks. These include; Rwanda, South Africa, Botswana, Lesotho, Ghana, and Ethiopia just to mention a few [15,16]. Among these, Rwanda is the only country in Africa that has succeeded in documenting all rights to land based on the Torrens System [17]. Responsibility of all cadastral system activities are given to a single institution, called Office of the Registrar of Land Ti-tles. The system follows the international rules and standards such as cadaster 2014, and LADM [18]. According to GoR Land Administration Manual [18], the implementation of modern cadastral system in Rwanda has been started in 2007. In the first two years, prep-aration works, including public consultations and the development of the legal, institu-tional, and policy frameworks were embarked. On 2009, the actual a fully-fledged Sys-tematic Land Registration (SLR) work were started and completed in 2013 through nine stages: (1) Notification of LTR area and local information campaign; (2) Recruitment and training of local staff; (3) Parcel demarcation; (4) Land adjudication; (5) Data entry and checking; (6) Parcel digitization; (7) Objections and corrections; (8) Lease preparation; (9) Lease issuance. In realizing this registration, Rwanda has engaged the private sector in some specified cadastral activities. The cadastral system has been a very ambitious but also very successful. With this trend, all parcels that account 10.67 million are registered in less than 5 years. For this successful achievement, the London-based trade organization British Expertise awarded certificate of recognition being an Outstanding International Development Project. In this regard, the 2020 BD [19] report ranked Rwandan cadastral system third in the world. In the case of cadastral system of South Africa, it is one of the best and most reliable systems. It accurately defined the boundaries of properties, and the positions of rights affecting those properties [20]. Responsibility for cadastral system falls under the Department of Land Affairs [15]. The primary function of the cadastral system in South Africa is to define (delineate and document) ownership rights. According to the Land Audit Report [20] of South Africa, 94% of the total land in the country is registered digitally in the Deeds Office. In this regard, it can be said that much efforts has been in-vested in to overcoming the problems experienced in many developing countries where lack of cadastral information hinders development. Those illustrative cases show that the theory of merger is becoming acceptable for its effective performance.

1.2. Background and Context of the Study

The origins of the cadastral system in Ethiopia dated back to 1907 when Menilek II signed Ethiopia's first decree related to urban land administration [21–23]. It was by a French company that the cadastral system was introduced but remained without maintenance for more than 80 years [24]. Even though, there was a sporadic effort to es-tablish a cadastral system some time ago in Addis Ababa, as it is not updated for a long time, it is almost non-existent as a system until very recently. Hence, the Addis Ababa City Administration around 1995 decided to launch a cadastral project initially intended to register all property owners liable for property taxation and collect data that would enable the city administration to assess property tax. However, the attempt was not effective which then changed in to establishing a multipurpose cadaster [21]. Even though a step forward was achieved and some experiences were gained as a result of launching the ca-dastral project of 1995 of Addis Ababa, it was not possible to evaluate this project as a successful one due to varieties of reasons, which among others, lack of cooperation be-tween these cadastral system and land right creation institutions [25].

Currently in Ethiopia, urban land is under the responsibility of Ministry of Urban Land Development, Housing and Construction (MULDHC). Within this institution, urban land right provision is undertaken by the Urban Land Development and Management Bureau (hereinafter ULDMB) while the cadastral system is overseen by the Urban Land Adjudication and Registration Information Agency (hereinafter ULARIA). These two institutions are working autonomously though their task are related and could be performed through a single institution. The major activities performed by ULARIA are, according to Proc. No. 81872014 [26], Strategic Plan [25], and Annual report [27], the legal recording of rights including the spatial boundaries [28]. The task of allocating (creating) landholding rights belongs to ULDMB. Land right in this case refers to a set of legally guaranteed enti-tlements or privileges associated to land ownership. They may also be expressed as bun-dle of rights or attributes of ownership. A bundle of rights is a term for the set of legal priv-ileges that is generally afforded to a real estate buyer with the transfer of the title. It in-cludes right of possession, right of control, right of exclusion, right of enjoyment, and right of disposition [1].

The legal protection to land is implemented through land management that changes the resources of land into good effect [29]. Land management encompasses all activities associated with the management of land that are required to achieve sustainable development. Under the umbrella of land management, cadastral system plays major role by ensuring that land is secured [23]. Cadastral system serves as a fundamental source of data infrastructure for land management, as stated in the work by Enemark et al [30]. Spe-cifically, it provides parcel information for the use of land, value, tenure, and development. In this regard, Ethiopia has formulated urban land policies related to land right and cadastral registration systems under two different urban land policies. Proclamation No. 721/2011 [31] dictates about modality of urban land acquisition, while Proclamation No. 818/2014 [26] dictates about urban land adjudication and registration that must be im-plemented all over the country. In line with these proclamations, regulations, directives and manuals have been prepared. The overall objective of the scheme is to accelerate the social-economic and environmental development of urban centers through providing land right and ensuring landholders' security of holding and recognition of title to im-movable property by certifying their right, restriction and responsibility through adjudica-tion and registration.

The organizational structures for landholding right providing and cadastral registration institutions differ widely between countries throughout the world, and reflect local cultural and judicial settings. In some regions these organizations are merged and other countries unmerged [21]. Compared to merged institutions, unmerged type of institutional arrangements exploits resources such as human resource, and leads to the creation of in-compatible parcel information by creating variations of the same object: parcel location, ownership, use and value. Such variations are costly and create a void since there are lim-ited mechanisms that link the management of ownership. Likewise, Ethiopia operates these two related activities by two different institutions, called land right providing institution and urban cadastral system registration institution. In other words, merger did not take place between these two related institutions.

Cadastral system in conjugation with landholding right providing institutions are the building block in land administration and management system. The land manage-ment paradigm endorses modern land administration system through four key functions: land tenure, land valuation, land use and land development. These functions are inte-grated through cadastral system [32]. In this regard, cadastral system and landholding right providing institutions have allowed modern operations for the land administration and management sector.

Cadastral System is an institutional framework varied and complicated by the tasks they must perform, by national cultural, political and judicial settings, and by technology [33]. In this paper, cadastral system is defined as a formal sub-system of land administration that includes the organizational system (a set of professional actors with responsibilities to carry out cadastral activities and maintain cadastral information systems), procedures, and regulations, which altogether ensure that the cadastral system is kept up-to-date. In short, a cadastral system is an organizational system usually referring to the operations that a cadastral institution is conducting [23]. Landholding right provision (creation) institution is a government organ that is responsible for creating and allocating right, restriction and responsibility on the land.

Land right creation and Cadastral system tasks are concerned about the management of spatial and legal data on land. In many countries these tasks are executed by a single institution. Despite international literatures; Cadaster 2014 [13], Cadaster 2034 [34], de Vries et al. [8], Koroso et al. [4] claiming that merging such agencies is more effective for the data handling, Ethiopia executes these two related tasks by two different and independent institutions: land right creation by the ULDMB, while cadastral registration tasks are overseen by ULARIA. In some countries these institutions are merged while in some other countries are not [8]. In countries where the merger did not take place, such as Ethiopia, the content of both institutions is very similar [35]. This type of arrangement misuses resources (such as human resources) and leads to the creation of duplicated and sometimes incompatible parcel information by creating variations of the same object: parcel location, ownership, use and value. Such variations are costly and create a void since there are limited mechanisms that link the management of ownership [33].

As in most developing countries, Ethiopia is challenging by issues related to institutional coordination within the urban land sector. Chekole et al. [23,35] and Tigistu [36] revealed that the historic failure of urban cadastral system pilots in Ethiopia in general, and Addis Ababa in particular remains a concern. It is estimated that over 6.5 million parcels are found within the urban jurisdiction of Ethiopia, and out of this, approximately 680,000 parcels are believed to exist in Addis Ababa [25]. Although a number of cadastral pilot projects were launched in Addis Ababa, there have been many challenges to succeed: - poor cooperation between the two institutions (landholding right provision and cadastral information registration), lack of legal framework to fill the gap between these two institutions, use of different sources of data by the two institutions (the right providing institution uses Arial image Addis Ababa taken in 1996, while the cadastral registration institution uses the Arial image acquired in 2010), the system lacks procedure to eliminate this gap, silence of the law on how the status of parcels placed in the dispute register where there is no response from the right creating institution within 15 days. As a result, most urban people do not enjoy secure land and property rights. This can be evidenced from the report of MoUDC [27], Chekole et al. [23], Chekole et al. [35] and Doing Business Assessment [19], in which, Ethiopia is ranked at 142th out of 192 countries in the world with respect to cadastral system.

Therefore, the main objective of the research is to analyze the effect of institutional merger between ULDMB and ULARIA on institutional effectiveness. Guided by theory of merger, the paper employs desk review and case study research strategies to uncover the research problem. Within the realm of this thematic scope, the following research question is endorsed: does institutional merger between ULDMB and ULARIA improve institutional performance? By unraveling this question, the paper may contribute an alternative solution that can improve performance of urban cadastral system institutions. Furthermore, quality service delivery may be improved through avoiding bureaucratic processes, and reducing time and cost of services.

The paper is organized in to five sections: introduction: in this section brief information are provided to let readers understand the problem, objective, and scope of the paper. The second section endorsed the theoretical framework to give information on how and with which theory the research is guided. Methodology of the research is presented in the third section, whereas results and discussions, and conclusion are presented in the fourth and the fifth sections respectively.

2. Theoretical Framework

This research used the theory of merger as a theoretical framework. The concept of merger was primarily, in the geospatial sector, developed by the International Federation of

Surveyors (FIG) initiative in the form of "Cadastral Statements" among which, "separation between cadastral maps and registers will be abolished" [13]. In support of Cadaster 2014 initiative, the demand for a widely accepted standardized domain model in land administration emerged [13]. The concept of Land Administration Domain Model (LADM) was developed and introduced after assessing the strengths and weaknesses of cadaster 2014 [37,38]. From the assessment, Lemmen et al. [37] have realized existence of different and inconsistent land administration systems across and with countries. Most countries develop their own unique land administration systems; some cadastral system institutions are centralized while others decentralized [29,39]. In the centralized system, institutional arrangement of cadastral system is undertaken by a single institution whilst the decentralized system follows separate institutional arrangements. The different implementations of the various cadastral systems do not make meaningful communication across borders easily [38]. Due to this, LADM has come to tackle these differences. Similarly, the concept developed by Enemark [40] Fit-For-Purpose (FFP) suggests the use of merged institutions rather than separate institutions in the land administration and management systems. One of the concerns for the emergence of FFP is to simplify bureaucratic institutional arrangement barriers in implementing sustainable land administration system. According to Sułkowski et al. [41], merger refers to the formal union of two or more organizations into a single organization usually designed to deliver a more effective operation to meet external challenges and opportunities. Mergers could be classified into horizontal mergers, vertical mergers and conglomerate mergers. A horizontal merger involves the merger of two or more institutions operating related activities. A vertical merger is a combination of two or more institutions involved in different stages of production. Conglomerate merger is a combination in which an institution combines with unrelated institution [41]. This paper follows the horizontal type of merger for the case of Ethiopia, since ULDMB and ULARIA are two institutions arranged at the same structural hierarchy. For the merger to be successful, there are key components required [42]. (1) Smooth communication: - there should be a need to have completely open and direct lines of communication with the key players of the institutions to be merged. (2) Win-win strategy: - in both sides, institutions need to be improving their situation in some way. (3) Shared vision: - a clear set of goals and objectives will keep institutions focused throughout every decision. (4) Well-planned: clearly plan out many of the critical details of merging processes including contingencies. (5) Integration: establish an integration team that is entirely dedicated to executing and implementing the merger. Thus, these components guide how the merger process can take place.

3. Methodology

In order to provide an appropriate framework for the study, we employed descriptive type of research design through semi-structured interviews and focus group discussions (FGD). Semi-structured interviews are made with directors from two urban land administration institutions; ULDMB and ULARIA across all ten sub-cities of Addis Ababa. The aim of the interview is to know how the directors feel about institutional merger between the two institutions. The second research design, FGD, is aimed to collect experts' perception related to institutional merger in these two institutions. Figure 1 illustrates the overall research design. Populations of the study are all professionals who are working in urban land administration across ten sub-cities of Addis Ababa. Our target populations are selected purposely since the study requires professionals who are working directly in the urban land administration and management.

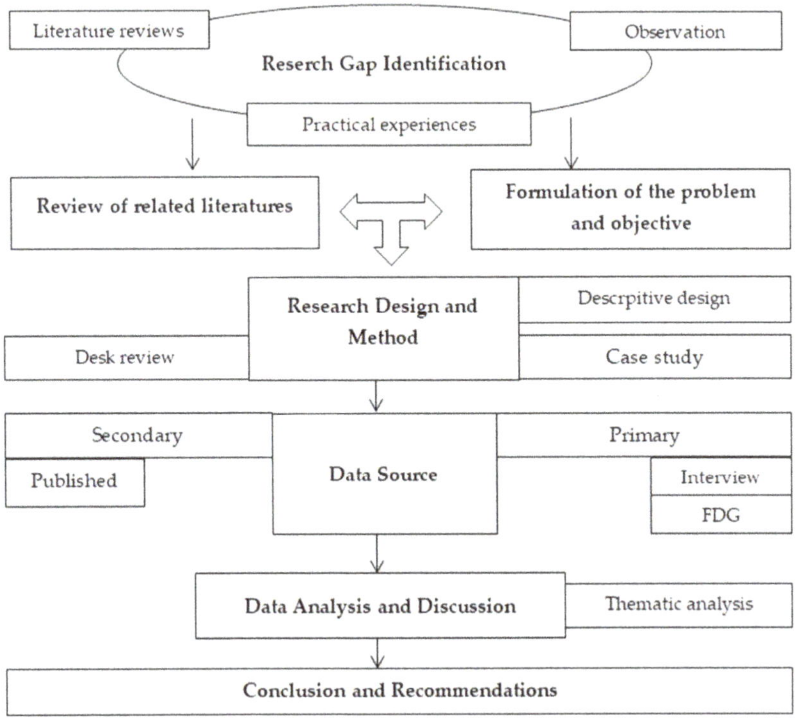

Figure 1. Overall research design.

To analyze the effect of institutional merger between ULMNB and ULARIA, we used desk review and case study research methods. The desk review helped to extract relevant and useful information related to whether merging or separate institutions are effective. The emphasis of case study was to understand existing phenomena. Figure 2 presented location map of the study area, Addis Ababa. Hence, from the study area primary data are collected through semi-structured interview and FGD. These type of data collection instruments believed to provide advantages of accurate screening, capture verbal and non-verbal questions, keep focus, and capture emotions and behaviors [43]. For this purpose, we conducted semi-structured interviews with the directors of ULDMB and ULARIA institutions in all ten sub-cities of Addis Ababa, and did a focus group discussion (FGD) with professional experts. In order to get detail information from our respondents, semi-structured interview tool was selected purposely since interview is a flexible approach, allowing for posing of new questions or check-questions if such a need arises. In conducting FGD, the paper follows the recommendation by Krueger and Casey [44] that indicates a well-designed FGDs should consist of 6–12 participants for the rationale that focus groups should include enough participants to yield diversity information. Focus groups are less threatening to many research participants, and this environment is helpful for participants to discuss perceptions, ideas, opinions, and thoughts [44,45].

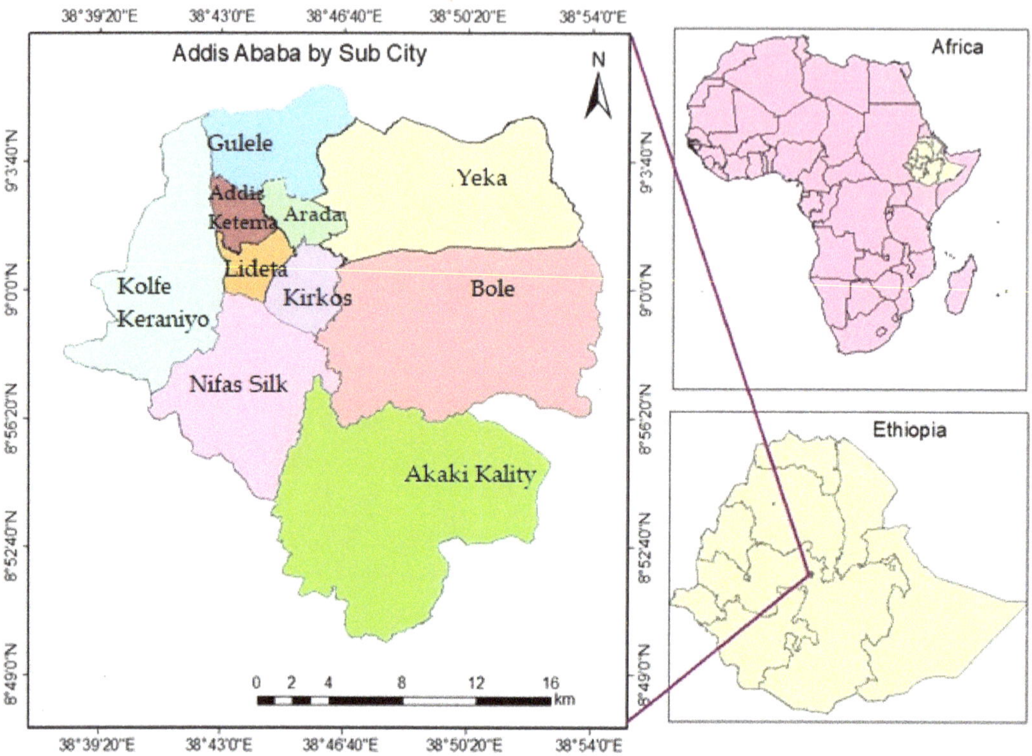

Figure 2. Location map of the study area (Author developed).

With this justification, required data were collected from all institutional directors of ULDMB and ULARIA across the ten sub-cities, which is in total 20 interviews. Apart from this, to understand the views and perceptions of professional experts working in these two institutions, FGD was made in a group of six for these two institutions across all ten sub-cities, which in total 20 FGDs (see summarized information in the Table 1).

Table 1. Summary of Planned and Actual data collected through Interview and Focus Group Discussion.

Data Collection Instruments	Planned		Actual Coverage	
	Sector	Participants	Frequency	Percentage
Interview	ULDMB ULARA	10 Directors from all 10 sub-cities 10 Directors from all 10 sub-cities	10 10	100 100
Focus Group Discussion (FGD)	ULDMB ULARA	10 FDGs (6 professionals per group) 10 FDGs (6 professionals per group)	10 FDGs (6 professionals per group) 10 FDGs (6 professionals per group)	100 100

The collected data were transcribed from Amharic to English. The responses are categorized and written based on their thematic similarity: responses supporting merging the two institutions improve in achieving institutional goals, or the opposite. In addition to this, institutional documents (unpublished) such as: Growth and Transformation Plan II (GTP II) and annual reports were analyzed in order to synthesize the result. Based on the GTP II and annual evaluation reports, a 1–5 Likert scale rating, in the form percent, (refer Table 2) used to evaluate institutional goal achievements.

Table 2. Likert scale rating in Percentage.

Likert Scale Rating in Percentage					
Rate	**Very Low**	**Low**	**Moderate**	**High**	**Very High**
Scale	1	2	3	4	5
Percentage	0–20	21–40	41–60	61–80	81–100

Finally, these data were analyzed and interpreted through qualitative type of data analysis being thematic analysis is at the center of interpretation. Thematic analysis is convenient to interpret the data since the collected data are organized on the basis of thematic similarities [45]. Basically, it emphasizes on identifying, analyzing and interpreting patterns of meaning (or "themes") for qualitative data such as interview and FGD. Finally, these data are presented qualitatively supporting with tables.

4. Results and Discussion

Secure, legally enforceable and marketable land rights are critical for cities in developing countries to be able to ensure that urban land is allocated to its most productive use [17]. The Federal Urban Land & Real Property Registry & Information Agency (FULRPRIA) is an institution established with the Proc. No. 251/2011 [46], and given responsibility of registering all urban lands in accordance with Proc. No. 818/2014 [26]. In order to implement this proclamation, cadastral surveying regulation, landholding adjudication and registration regulation, and enforcement guidelines have also been prepared and approved. In addition to these laws, five universities have developed curricula to address the challenges of urban land administration and management. Apart from this, the urban land sector developed 12 occupational standards (OS) to help build the competency of the land administration professionals. In this regard, the sector recruited 4192 professionals from four regions (Tigray, Amhara, Oromiya, and Southern nations) trained and assessed them starting from Level II-IV professions [25]. All these interventions are made to develop and strengthen the knowledge, skills, and attitudes of land administration professionals, which in turn foster performance of land administration institutions. Despite these efforts and interventions, as stated in the annual report [27], institutional performance of the cadastral system was not possible to evaluate as successful. In connection with this, the report provides the following challenges for its poor performance; silence of the law about sporadic registration procedure, low emphasis by the right provider for cadastral related activities, delayed responses from the right provider for the cases registered in the dispute registry book, low commitment and engagement by the upper management, and existence of incomplete title evidence for the stated owned lands. Hence all these challenges delayed the smooth implementation of urban cadastral system. In the same manner, these issues have been mentioned in the works of Likinaw [47], Chekole [35], Burns [24], and Daniel [21] as challenges for the hindrance of successful cadastral system implementation.

The results from the semi-structured interviews, guided by a previously designed questionnaire (see Table A1), indicate that the achievement of institutional objectives are delayed by varieties of problems, which among others, lack of commitment by the leadership, lack of consistent and organized land information documentation, and lack of proper follow-ups, lack of comprehensive performance evaluation in order to identify its strengths and weaknesses and to propose the redesign strategies, lack of comprehensive review of the cadastral system experiences, lack of adequate study to identify appropriate strategies, poor cooperation between land right providing and registering institutions. According to the directors of ULDMB and ULARIA, despite the importance of follow-ups, leaders' commitment, institutional collaboration, and proper geospatial documentation to ensure sustainability of the cadastral system, in most cases these components are neglected in the strategic plan. Due to this, they are not performing according to their strategic plans. Since the natures of the two institutions' operational tasks are interlinked, independent working could not lead them to achieve their targets. Among the challenged stated above, most are directly related to problems associated with poor institutional collaboration. In connection to this, Auzins [48], asserts that the major source of problem in malfunctioning land administration and management system is miss-arrangement of institutions [48]. As a result, each institution could not achieve the targeted objectives rather they are repeating the same institutional strategic plans every year. In this regard, a study by Wayumba [49] indicated that any cadastral system requires a comprehensive evaluation in order to identify its strengths and weaknesses and to propose the redesign strategies. In the same manner, it is possible to consider available resources during the development of strategic plans. Table 3 presents response rate from both interview and focus group discussions.

Table 3. Interview and FGD results from ULDMB and ULARIA Institutions.

Urban Land Administration Institutions	Merge		Unmerge		Total
	Frequency	Percentage	Frequency	Percentage	
ULDMB (Directors)	8	80	2	20	10
ULARIA (Directors)	10	100	0	0	10
ULDMB (FGDs with experts)	10	100	0	0	10
ULDMB (FGDs with experts)	10	100	0	0	10

In order to verify and validate results of interviews and FGDs, strategic plan (GTP II) and annual performance reports were assessed. According to the strategic plan (GTP II) of Addis Ababa, 430,000 parcels were planned to be adjudicated and registered within 5 years from 2015–2019. However, only 149,584 parcels are adjudicated and registered, which amounts 34.78% performance achievement. This performance level is rated as low based on the scale level indicated in Table 2.

According to the institutional directors, even if the strategic plan formulated for five years, there is an update every year since the strategic plan by itself is an ambitious, which cannot be realized. It was planned to reach on the target by recording 430,000 parcels within five years. Despite these ambitious plans, the reality on the ground was otherwise. They believe that the main challenges for this to happen are lack of proper consideration of available resources: such as human resource, and money during the planning process; communication gaps among land administration stakeholders; commitment, and poor institutional coordination within the land sector; irregular assignment and replacement of institutional leaders. Hence, they believe these are the major causes for the low institutional performance. Similarly, the Kenyan cadastral system has been hindered by the same challenges mentioned above though they could able to solve the problem through redesigning the overall processes, including strategic plans, and institutional arrangements [49]. In this regard, studies; Carlos, et al. [50], and Nicholas [51] suggest that strategic planning process requires considerable thought and planning on the part of the institution's upper-level management. Before setting a strategic plan and then determining how to strategically implement it, the upper management of the respective institutions first needs to take into account available resources. In contrast to the cases of Ethiopia and Kenya, the most exemplary experience of Rwanda has laid remarkable story of success across African cadastral system. Across the country, 10.67 parcels have been completed in less than five years with an average of USD7 cost per parcel [17]. This achievement is attributable to nine years of dedicated reform efforts, which started with a comprehensive review of Rwanda's policy legal and institutional framework, now regularly updated. Preparations for implementation immediately followed, starting first with piloting to identify scalable approaches for achieving the government's ambitious targets, with concurrent monitoring and impact evaluation helping to identify problems that could then be discussed by policy makers [18]. This exceptional experience of Rwanda gave lessons to other countries. In line with this, the findings by Enemark et al. [52], and UN-GGIM [7] indicate that the process of formulating strategic plan first needs to assess its current situation by performing an internal and external audit to identify the institution's strengths and weaknesses, as well as opportunities and threats (SWOT analysis). Based on this analysis, the institutional directors decide on which priority areas they should focus on, how to best allocate the institution's resources, and whether to take actions such as expanding operations through merger. After setting the strategic plan in this way, it is important for the upper management to evaluate the effectiveness of the strategic plan after the implementation phase [7]. Strategic plan evaluation involves three crucial activities: reviewing the internal and external factors affecting the implementation of the strategy, measuring performance, and taking corrective steps to make the strategy more effective [25]. The findings by de Vries et al. [8] give a good insight in how administrators that encourage and intensify collaboration and integration among public agencies could derive an organizational transformation. In this regard, a framework for effective land administration, FELA [7], indicates the role of establishing partnerships and collaboration in enhancing effective land administration system. According to this framework, partnerships and collaborations bring different but complementary skills, experiences, knowledge, and resources altogether to improve institutional goals. Likewise, one of the seven underpinning principles of Integrated Geospatial Information Framework (IGIF) is collaboration and cooperation between land administration institutions. In this regard, IGIF [53] asserts that at the national the level, there needs to be more institutional collaboration and integration across the various land administration institutions. However, in situations where the level of collaboration between related institutions is very weak, according to de Vries et al. [54] one possible solution is to merge those institutions to effectively provide collective services, based on a single institutional and organizational framework [8]. From the land management perspective, Yin and Shanley [55] reveal that operational efficiency is a rational justification for merger. The primary discourse of mergers is rationalist, emphasizing economic gains of increasing efficiency and technical gains of standardization and reduction of redundancy [54]. In addition, mergers fundamentally change the individual organizations, the responsibilities and accountabilities, the work practices and conventions, the physical location of people and resources, and augment effective authority of the merged organization [56].

According to the legal framework of the two institutions, they clearly need to cooperate and, even in some way, they need to be integrated. This is mentioned in Art. 14 sub-article of 2 and 3 of Proc. No. 818/2014 [26], and it reads, in cases where there is inconsistency between the evidences from the empowered right providing institution and the landholder, the issue shall be referred to the right providing institution for verification. In response to this, the right providing institution shall notify its decision relating to the issue raised to the registering institution within fifteen working days. Despite this dictate, survey results show that in most cases, the requested information is not responded within the stated time limit. According to the directors, one of the possible reasons for this to happen is poor coordination between these two institutions. In fact they have a common goal that is securing urban lands in support of sustainable land administration system. However, the reality on the ground seems competing institutions for individual profits. In addition to this, the professionals working in these two institutions are paid differently, regardless of their expertise and experience. This difference came from the fact that ULARIA is established at agency level and supported with project funds. In this case, the professionals working in this institution are paid better than the professionals working in ULDMB. Justified with the weak cooperation and integration, the directors suggested institutional merger between the two institutions and governing with the same legal and institutional framework leads to a better operational performance. According to their perception, providing responsibility to a single institution would be effective than two or more institutions. In this regard, Yin and Shanley [55] argue that institutional mergers lead to a more effective and transparent processes and to similar services than if only collaboration and integration would increase. This argument is also supported by the works of de Vries et al. [54], de Vries et al. [8], Koroso et al. [4] and Wang et al. [57] that confirmed single institution can perform operations more efficiently than multiple organizations. Lessons from best experience in land administration show that institutional merger achieves the highest, effective, and sustainable service delivery to the customer. In support of this, countries with merged institutions rank, on average, 38 out of 189 countries in Doing Business [19]. Countries with separate institutions rank, on average, 47. The report show that merged organizations work better than unmerged organizations [19].

The other question provided to the directors was concerning the delivery of services in the land administration sector through one window in one place. Among the total 20 interviewees, 80% of the respondents described that the major challenge for the customer is getting land related services in different institutions. For instance, one investor may ask for a plot of land for real estate development. The procedure to obtain this land starts in ULDMB specifically under the directorate for land right creation (provision). This institution creates interests (bundle of rights attached to the land, such as right, responsibility, and restriction. In short, interest in land refers to the right, responsibility, and restriction attached to the landholder). The investor then goes to the ULARIA and request for cadastral surveying and registration of the plot. According to the Proc. No. 818/2014 [26], it is only when this plot of land is registered by ULARIA said to be legal owner of the parcel. However, these two institutions are physically located in different addresses, which mean the customer has no access for one window service. In this regard, the investor is obliged to visit both institutions to get the service. Although the requested services are interrelated and could be performed by a single institution, the system does not provide one-window service to the customer. This bureaucratic procedure, according to DB [19], Turisova et al. [58], and Carlos et al. [59], creates inconvenience to the customer since it consumes time, energy and cost compared to a single entry point to get the service. Therefore, merging these institutions would provide better services to the customer through one-window service. Jouni [60] and the World Bank Group [61] support one-window services in the land administration sector to create simple and smooth processes in executing the tasks of land administration.

In relation to the merits and demerits of institutional merger, institutional directors were asked if institutional merger provides advantages of cost reduction and increase efficiencies. In this regard, the directors (80% of interviewee) strongly believe that merger plays vital role in reducing the costs to be incurred, time to be consumed, and increase efficiencies of the merged institution. On top of this, the resource (e.g., human resource) to be deployed would be used efficiently and wisely. Based on counties experience, for instance Rwanda, the success behind their cadastral system relies mainly on the merged institutional arrangement, which is one of the principles in the cadastral statements. However, Jouni [60] and the World Bank Group [61] highlight that no institutional merging can be established overnight, rather it is important to take concrete steps to the right direction of merging. Coherently, the directors of ULDMB and ULARIA in Addis Ababa also believe that urban land right provision and registration activities are closely related fields, and should not be treated nor operated under separate institutions. The targets intended, the resources deployed, the processes involved, and the outcomes provided are more or less the same. The target of urban development and management office is providing landholding right, which needs to be ascertained through cadastral registration. In this case, they believe that land administrators working on the ULDMB can also be deployed in the ULARIA, so that human resources are managed properly. On the basis of these arguments, investigating these two related activities under two different and independent institutions seems an inefficient use of resources.

Finally, regarding the areas for institutional performance improvement to satisfy the needs of their customers, the directors believe that the two institutions first need to be collaborated to execute their planned tasks on the basis of clear guideline that shows their responsibilities and duties through, for instance, signing Memorandum of Understanding (MoU). This mechanism may increase the relation between the two institutions, which in turn positively affect their individual performance. After getting into deep and effective relationships, the cooperation may develop into merging into single institution. However, this does not mean that it can be done in a single stage; rather it should be through some time since the issue needs discussion between the staffs of both institutions. In support of this, de Vries and Miscione [62] state that in a situation where mergers took place, the root of the transformation is embedded in gradual adaptation and so a merger gradually crystallizes instead of being established by decree. Meanwhile, the two institutions need to put a binding agreement that enforces them to execute the tasks they intended.

Merging cadastral information and land right providing institutions is normally supported from the administrative and legal point of view. The issue of institutional merging is directly linked to cost, time and quality, since time is important to deliver the service, cost determines the amount of money or resources available, and quality represents the fit-to-purpose that the service must be provided. In doing so, the components of merging processes: smooth communication, win-win strategy, achievable plan, common objective, and integration should be placed at the center of the merging process. Unless these elements incorporated and considered during merging process, the envisaged plan may not be achieved. Thus, all these components need to be discussed in front of the staffs who are working at both institutions. This process in turn fosters the initiative to commence the process of merging.

The second type of primary data was collected through focus group discussions (FGD) with experts working in both institutions to understand their perception about the current institutional performance, as well as the possibility to merge the ULDMB and ULARIA. The same questions (refer to Table A1) were also provided to them, so that their perceptions could be evaluated. All participants of the focus group discussion elaborated the procedures to execute their own tasks, and they all felt challenged by the poor coordination between the two institutions. Ferro and Sorrentino [63] indicate that institutional strategic partnership between institutions that have related goals is crucial in order to achieve the intended goals. Suitably, all respondents agreed on the basic ideas of merging, and they are trying to emphasize on the need for institutional merger, though they preferred to work gradually. Meanwhile, they emphasized on the use of collaborative mechanisms that can solve their immediate tasks. Accordingly, some kind of arrangements (written and legally binding agreement), which explains the duties and responsibilities between the two institutions may be needed to improve institutional performances. When compared with the insights of the institutional directors, individual staff members of these institutions perceive the merits of institutional merger the same way.

From the result obtained, it can be concluded that all respondents perceive that the cadastral registration institution demands the strong support (e.g., providing landholding related information) from the right providing institution to execute its tasks properly. Apart from this, the landholder must provide his/her holding document evidencing that he is the rightful claimant. It is this document which must be adjudicated with the copy document found in the right providing institution. In this regard, the right providing institution plays major role in providing and confirming that the land holder is the right claimant. This is also supported by the legal framework that dictates about urban land adjudication and registration. In connection with this, the law (Proclamation No. 818/2014) [26] coerces the right creating institution to provide the requested information within fifteen days. It is the failure to comply with this law that resulted in the poor institutional performance. The findings by, Muparari [14] asserts that overlaps in organizational functions and processes leads to duplication of data and efforts. Due to this, international research studies; Cadaster 2014 [13], Sulkowski et al. [41], Bogaerets et al. [39], FIG [60], the World Bank Group [61], and Peter et al. [64] advise to consider merging the two as one in order to: reduce the effects of duplication; provide services through one-window; reduce the cost of operations; fill the gaps in communication among the staffs; reduce time of operation; improvement of customer service; increase of efficiency within processes; and provide a more efficient operation of land markets. Accordingly, this study confirms that institutional merger between the two institutions may be a solution to the challenges faced by them.

5. Conclusions and Recommendation

The main research objective of this paper was analyzing the effects of merging cadastral information registration and urban landholding right proving institutions, ULARIA and ULDMB in Ethiopia, respectively. In order to achieve this objective, primary and secondary data sources were used. Primary data were collected through semi-structured interviews and focused group discussion from the case study area, Addis Ababa, whereas secondary data from internet sources, strategic plans, and institutional annual performance reports. Desk review and case study research methods were used with a qualitative analysis approach. Although the findings confirm that many challenging issues such as: poor institutional coordination within the land sector, lack of proper consideration of available institutional resources while planning strategies, communication gaps among land administration stakeholders, irregular assignment and replacement of institutional leaders, and leaders' commitment are facing, the most challenging one is basically emanates from the poor cooperation between ULARIA and ULDMB, and poorly designed strategic plans. Practically, urban land right provision and registration activities are closely related fields in that the targets intended, the resources deployed, the processes involved, and the outcomes provided are more or less the same. Despite their closeness, these activities are organized and executed by two different and independent institutions. Due to this arrangement, institutional performance could not be successful. Based on the result from both semi-structured interviews and focused group discussion, both ULDMB and ULARIA institutions design strategic plans every year, with low level of achievement or operational performance. This is because of strategic plans are not formulated in consideration with institutional resource capacities.

Therefore, this paper strongly recommends the following two directions in order to improve institutional performance in the urban cadastral system. (1) Strategic plan is a document that guides how specific activities are executed within various hierarchical management levels (upper, middle, lower). Thus, it is imperative to foster communication and interaction among employees and managers at all levels, so as to help the institution to operate as a more functional and effective team. A well-designed strategic plan considers the current situation of the respective institution. In doing so, available resources, money, and time are considered, and priority activities are identified that avoid ambitious plan. (2) Results have shown that the major challenging issue is the poor institutional coordination between ULDMB and ULARIA institutions. For the meantime, these institutions may come to cooperation through signing Memorandum of Understanding (MoU). This arrangement, bounded with clear duties and responsibilities, may reduce their gaps, and increases institutional performances. However, in the long run, we suggest merging these two institutions can: reduce the effects of data duplication; provide services through one-window; reduce the cost of operations; fill the gaps in communication among the staffs; reduce time of operation; improve of customer service; increase efficiency within processes; and provide a more efficient operation of land markets. It should be noted that no institutional merging can be established overnight; rather it is important to take concrete steps to the right direction of merging.

This research focuses on practical application of the theory of merger only in urban land administration processes. Hence, other researchers may conduct a research if the same theory can bring the urban and rural land administration sectors in one umbrella so as to increase efficient operation of land market.

Author Contributions: Conceptualization, S.D.C., W.T.d.V., P.D.-D., G.B.S.; methodology S.D.C.; validation, S.D.C., W.T.d.V., P.D.-D. and G.B.S.; formal analysis, S.D.C.; investigation, S.D.C.; resources, S.D.C.; data curation, S.D.C.; writing—original draft preparation, S.D.C., W.T.d.V., P.D.-D. and G.B.S.; writing—review and editing S.D.C., W.T.d.V., P.D.-D. and G.B.S.; visualization, S.D.C.; supervision, W.T.d.V., P.D.-D. and G.B.S. All authors have read and agreed to the published version of the manuscript.

Funding: This research received external funding from DAAD, in the form of a Short Term Research Scholarship for In-Country/In region Scholarship holders, grant number 57520399. The Article Processing Charges of this research were funded by the Technical University of Munich (TUM).

Data Availability Statement: The data presented in this study are available on request from the first author.

Acknowledgments: The authors would like to thank the Technical University of Munich (TUM), German Academic Exchange Service (DAAD), and the Institute of Land Administration (ILA) of Bahir Dar University (BDU) for providing supporting materials.

Conflicts of Interest: The authors declare no conflict of interest.

Appendix A

Table A1. Semi-structured interview and focused group discussion questions provided to ULDMB and ULARIA institutions.

1. Is your institution working well to achieve the intended objectives?
2. Which intended goals and targets are met and which are not, and why?
3. Do ULDMB and ULARIAs work in close relation to satisfy the needs of their customers?
4. Do you depend on each other to complete your tasks, and share information and documents between the two institutions?
5. One-window service system provision is getting acceptance by many institutions since it improves customers' satisfaction by providing services at one place. In light of this, does your institution provide one window service to customers? If not why?
6. Does your institution work to improve the relation between the institution and customers? In what ways?
7. How does a better efficient operation of land markets could be achieved in the land sector?
8. What are the strengths of your organization in improving service delivery to the customer?
9. What are the areas for improvement of your organization in delivering services to the customer?
10. Do you think that mergers provide advantages of cost reduction and increase of efficiencies for the merged institution?

References

1. Ambaye, D.W. Expropriation in Amhara Region. In *Ethiopia: Law and Practice*; VDM Verlag: Saarbrücken, Germany, 2010; ISBN 978-3639307153.
2. Hodgson, G.M. What Are Institutions? *J. Econ. Issues* **2006**, *40*, 1–25. [CrossRef]
3. Ferrini, L. *The Importance of Institutions to Economic Development*; University of Reading: Reading, UK, 2012.
4. Koroso, N.H.; Zevenbergen, J.A.; Lengoiboni, M. Land institutions' credibility: Analyzing the role of complementary institutions. *Land Use Policy* **2019**, *81*, 553–564. [CrossRef]
5. UN-GGIM. Good Land Governance for the 2030 Agenda. In *The Addis Ababa Declaration*; UN-GGIM: New York, NY, USA, 2014.
6. Reimers, K.; Guo, X.; Li, M. Beyond markets, hierarchies, and hybrids: An institutional perspective on IT-enabled two-sided markets. *Electron. Mark.* **2018**, *29*, 287–305. [CrossRef]
7. UN-GGIM; Framework for Effective Land Administration (FELA). *Reference for Developing, Reforming, Renewing, Strengthening or Modernizing Land Administration and Management Systems*; Expert Group on Land Administration and Management: New York, NY, USA, 2019.
8. De Vries, W.T.; Laarakker, P.M.; Wouters, H.J. Living apart together. *Transform. Gov. People Process. Policy* **2015**, *9*, 545–562. [CrossRef]
9. Investopedia. Guide to Mergers and Acquisitions. What Are Mergers and Acquisitions (M & A)? Available online: https://www.investopedia.com/terms/m/mergersandacquisitions.asp (accessed on 20 March 2021).
10. CFI. What is a Merger; How to Build a Merger Model and the Key Steps to Build a Merger Model? Available online: https://corporatefinanceinstitute.com/resources/knowledge/modeling/build-merger-model/ (accessed on 20 March 2021).
11. Pettinger, T. Helping to Simplify Economics—HELP:—Pros and Cons of Merger, Economics. 2019. Available online: https://www.economicshelp.org/blog/5009/economics/pros-and-cons-of-mergers/ (accessed on 20 March 2021).
12. Jensen, M.C.; Ruback, R.S. The market for corporate control. *J. Financ. Econ.* **1983**, *11*, 5–50. [CrossRef]
13. Kaufmann, J.; Steudler, D. (Eds.) Cadaster 2014. In *A Vision for Future Cadastral System*; Swiss Federal Directorate of Cadastral Surveying: Bern, Switzerland, 1998.
14. Muparari, T.N. *Evaluation of Mergers of Cadastral Systems: A Corporate Cultural Perspective*; MSc; University of Twente Faculty of Geo-Information and Earth Observation (ITC): Enschede, The Netherlands, 2013.
15. Cadastral Template. Cadastral Template 2.0. Available online: http://cadastraltemplate.org/ (accessed on 20 March 2021).

16. Rajabifard, A.; Steudler, D.; Aien, A.; Kalantari, M. The Cadastral Template 2.0, From Design to Implementation FIG. In Proceedings of the Congress 2014 Engaging the Challenges, Enhancing the Relevance, Kuala Lumpur, Malaysia, 16–21 June 2014.
17. World Bank. *How Innovations in Land Administration Reform Improve on Doing Business: Cases from Lithuania, the Republic of Korea, Rwanda and the United Kingdom*; World Bank: Washington, DC, USA, 2015.
18. GoR. *Land Administration Procedure Manual*; Rwanda Natural Resources Authority Office of the Registrar of Land Titles: Kigali, Rwanda, 2016.
19. DB. *Economy Profile of Ethiopia: Comparing Business Regulation in 190 Economies*; The World Bank: Washington, DC, USA, 2020.
20. RoSA. *Land Audit Report: Phase II: Private Land Ownership, by Race, Gender and Nationality*; Rural Development and Land Reform: Johannesburg, South Africa, 2017.
21. Daniel, T. *Reflections on the Situation of Urban Cadaster in Ethiopia: Africa Local Government Action Forum*; Polytechnic of Nambia: Addis Ababa, Ethiopia, 2006.
22. Deininger, K.; Ali, D.A.; Holden, S.; Zevenbergen, J. Rural Land Certification in Ethiopia: Process, Initial Impact, and Implications for Other African Countries. *World Dev.* **2008**, *36*, 1786–1812. [CrossRef]
23. Chekole, S.D.; De Vries, W.T.; Shibeshi, G.B. An Evaluation Framework for Urban Cadastral System Policy in Ethiopia. *Land* **2020**, *9*, 60. [CrossRef]
24. Tony, B.; Kate, F.; Yan, Z.; Gavin, A.; Imeru, T.; Gebeyehu, B.; Shibeshi, A.M.; Solomon, K.; Abebe, Z. Establishing a Legal Cadaster for Good Governance in Ethiopia: Identifying Bottlenecks and Steps towards Scale-Up. In Proceedings of the World Bank Land and Poverty Conference 2017, Washington, DC, USA, 20–24 March 2017.
25. MoUDC. *Strategic Plan of 2020–2025 for Ministry of Urban Development and Construction*; Ministry of Urban Development and Construction: Addis Ababa, Ethiopia, 2020.
26. FDRE. Urban Landholding Adjudication and Registration Proclamation 818/2014, Issued by the House of People Representatives. *Federal Negarit Gazeta*. 2014. Available online: https://chilot.files.wordpress.com/2014/04/proclamation-no-818-2014.pdf (accessed on 20 March 2021).
27. ULARIA. *Annual Performance Evaluation Report on Ethiopia's Land Adjudication and Registration*; Urban Land Adjudication and Registration Information Agency (ULARIA), Ministry of Urban Development Housing and Construction: Addis Ababa, Ethiopia, 2014.
28. Silva, M.A.; Stubkjær, E. A review of methodologies used in research on cadastral development. *Comput. Environ. Urban. Syst.* **2002**, *26*, 403–423. [CrossRef]
29. UNECE. *Land Administration Guidelines: With Special Reference to Countries in Transition*; United Nations: New York, NY, USA, 1996; ISBN 92-1-116644-6.
30. Enemark, S. (Ed.) *Land Administration and Cadastral Systems in Support of Sustainable Land Governance—A Global Approach*, 3rd ed.; Land Administration Forum for The Asia and Pacific Region: Tehran, Iran, 2009.
31. FDRE. Urban Land Lease Holding Proclamation, Federal Democratic Republic of Ethiopia, Proclamation No. 721/2011. *Federal Negarit Gazeta*. 2011. Available online: https://www.ecolex.org/details/legislation/urban-lands-lease-holding-proclamation-no-7212011-lex-faoc169468/ (accessed on 20 March 2021).
32. Rohan, B.; Abbas, R.; Mohsen, K.; Jude, W.; Ian, W. Cadastral Futures: Building a New Vision for the Nature and Role of Cadastres. In *FIG Congress*; FIG: Copenhagen, Denmark, 2011; Volume 15.
33. Enemark, S.; Williamson, I.; Wallace, J. Building modern land administration systems in developed economies. *J. Spat. Sci.* **2005**, *50*, 51–68. [CrossRef]
34. Grant, D.; Haanen, A. (Eds.) Cadaster 2034: A 10–20 Year Strategy for the Development of the New Zealand Cadastral System. In Proceedings of the Engaging the Challenges -Enhancing the Relevance, Kuala Lumpur, Malaysia, 16–21 June 2014.
35. Chekole, S.D.; De Vries, W.T.; Durán-Díaz, P.; Shibeshi, G.B. Performance Evaluation of the Urban Cadastral System in Addis Ababa, Ethiopia. *Land* **2020**, *9*, 505. [CrossRef]
36. Tigistu, G.A. Experience and Future Direction in Ethiopian Rural Land Administration. In Proceedings of the Annual World Bank Conference on Land and Property, Washington, DC, USA, 18–20 April 2011.
37. Lemmen, C.; van Oosterom, P.; Bennett, R. The Land Administration Domain Model. *Land Use Policy* **2015**, *49*, 535–545. [CrossRef]
38. Lemmen, C. A Domain Model for Land Administration. *J. Theor. Biol.* **2012**. Available online: http://www.gdmc.nl/publications/2012/Domain_Model_for_Land_Administration.pdf (accessed on 20 March 2021). [CrossRef]
39. Bogaerts, T.; Zevenbergen, J. Cadastral systems-alternatives. *Comput. Environ. Urban. Syst.* **2001**, *25*, 325–337. [CrossRef]
40. Enemark, S. *Fit-For-Purpose Land Administration*; Joint FIG/World Bank Publication; FIG: Copenhagen, Denmark, 2014; ISBN 978-87-92853-10-3.
41. Sułkowski, L.; Fijałkowska, J.; Dzimińska, M. Mergers in higher education institutions: A proposal of a novel conceptual model. *Manag. Financ.* **2019**, *45*, 1469–1487. [CrossRef]
42. Doida, C. Key Components of a Strong Merger & Acquisition. 2015. Available online: https://www.doidacrow.com/key-components-of-a-strong-merger-acquisition (accessed on 20 March 2021).
43. Onwuegbuzie, A.J.; Dickinson, W.B.; Leech, N.L.; Zoran, A.G. A Qualitative Framework for Collecting and Analyzing Data in Focus Group Research. *Int. J. Qual. Methods* **2009**, *8*, 1–21. [CrossRef]
44. Krueger, R.A.; Casey, M.A. *Focus Groups: A Practical Guide for Applied Research*, 4th ed.; SAGE: London, UK, 2009; ISBN 978-1412969475.

45. Creswell, J.W. *Research Design: Qualitative, Quantitative, and Mixed Methods Approaches*, 2nd ed.; SAGE: Thousand Oaks, CA, USA, 2003; ISBN 9780761924425.
46. FDRE. Federal urban Real Property Registration and Information Agency Establishment Regulation No. 251/2011. *Federal Negarit Gazeta*. 2011. Available online: https://chilot.files.wordpress.com/2012/02/reg-no-251-2011-federal-urban-real-property-registration.pdf (accessed on 20 March 2021).
47. Likinaw, D. Evaluating the Implementation Performance of Legal Cadaster Registration in Addis Ababa City Administration: The Case of Addis Ketema Sub-City. Ph.D. Thesis, Bahir Dar University, Institute of Land Administration, Bahir Dar, Ethiopia, 2020.
48. Auzins, A. Institutional Arrangements: A Gate towards Sustainable Land use. *Nord. J. Surv. Real Estate Res.* **2014**, *1*. Available online: https://journal.fi/njs/article/view/41506 (accessed on 14 February 2021).
49. Gordon, W. An Evaluation of the Cadastral System in Kenya and a Strategy for Its Modernization. Ph.D. Thesis, University of Nairobi, School of Engineering, Department of Geospatial and Space Technology, Nairobi, Kenia, 2013.
50. Llusar, J.C.B.; Tena, A.B.E.; Roca-Puig, V.; Beltrán-Martín, I. An empirical assessment of the EFQM Excellence Model: Evaluation as a TQM framework relative to the MBNQA Model. *J. Oper. Manag.* **2008**, *27*, 1–22. [CrossRef]
51. Nicholas, A. Connecting with Communities: Use of Strategic Planning at the Bureau of Land Management. 2015. Available online: https://www.blm.gov/sites/blm.gov/files/documents/files/2014ConnectingWithCommunities_summary.pdf (accessed on 10 February 2021).
52. Enemark, S.; Hvingel, L.; Galland, D. Land administration, planning and human rights. *Plan. Theory* **2014**, *13*, 331–348. [CrossRef]
53. UN-GGIM. Integrated Geospatial Information Framework (IGIF): A Strategic Guide to Develop and Strengthen. National Geospatial Information Management (NGIM), Part 1: Overarching Strategic Framework. 2018. Available online: https://ggim.un.org/meetings/GGIM-committee/8th-Session/documents/Part%201-IGIF-Overarching-Strategic-Framework-24July2018.pdf (accessed on 20 March 2021).
54. De Vries, W.; Muparari, T.; Zevenbergen, J. Merger in land data handling, blending of cultures. *J. Spat. Sci.* **2016**, *61*, 191–208. [CrossRef]
55. Yin, X.; Shanley, M. Industry Determinants of the "Merger Versus Alliance" Decision. *Acad. Manag. Rev.* **2008**, *33*, 473–491. [CrossRef]
56. Kyvik, S.; Stensaker, B. Factors Affecting the Decision to Merge: The case of strategic mergers in Norwegian higher education. *Tert. Educ. Manag.* **2013**, *19*, 323–337. [CrossRef]
57. Wang, W.Y.C.; Pauleen, D.J.; Chan, H.K. Facilitating the Merger of Multinational Companies. *J. Glob. Inf. Manag.* **2013**, *21*, 42–58. [CrossRef]
58. Turisova, R.; Sinay, J.; Pacaiova, H.; Kotianova, Z.; Glatz, J. Application of the EFQM Model to Assess the Readiness and Sustainability of the Implementation of I4.0 in Slovakian Companies. *Sustainability* **2020**, *12*, 5591. [CrossRef]
59. Bou-Llusar, J.C.; Escrig-Tena, A.B.; Roca-Puig, V.; Beltrán-Martín, I. To what extent do enablers explain results in the EFQM excellence model? *Int. J. Qual. Reliab. Manag.* **2005**, *22*, 337–353. [CrossRef]
60. Anttonen, J. (Ed.) Integrating Generations and FIG/UN-HABITAT Seminar-Improving Slum Conditions through Innovative Financing. In Proceedings of the FIG Working Week 2008, Stockholm, Sweden, 14–19 June 2008.
61. World Bank. Implementation Completion and Results Report on a Credit in the Mount of USD 29.2 Million to the Kingdom of Cambodia for the Land Management and Administration Project. Available online: http://documents1.worldbank.org/curated/en/851281529960662155/pdf/Implementation-Completion-and-Results-Report-ICR-Document-06222018.pdf (accessed on 16 October 2020).
62. de Vries, W.T.; Miscione, G. The Tool That Has to Build Itself: The Case of Dutch Geo-Data. In *Electronic Government*; Springer: Berlin, Germany, 2012; pp. 137–148. ISBN 978-3-642-33488-7.
63. Ferro, E.; Sorrentino, M. Can intermunicipal collaboration help the diffusion of E-Government in peripheral areas? Evidence from Italy. *Gov. Inf. Q.* **2010**, *27*, 17–25. [CrossRef]
64. Laarakker, P.; de Vries, W.; Wouters, R. Land Registration and Cadaster, One or Two Agencies?: Stage 2 of the research. In Proceedings of the 2016 World Bank Conference on Land and Poverty, Washington DC, USA, 14–18 March 2016.

land

MDPI

Article

Digitization as a Driver fur Rural Development—An Indicative Description of German Coworking Space Users

Marco Hölzel * and Walter Timo de Vries

Chair of Land Management, Department of Aerospace and Geodesy, Technical University of Munich (TUM), Arcisstr. 21, 80333 Munich, Germany; wt.de-vries@tum.de
* Correspondence: marco.hoelzel@tum.de; Tel.: +49-89- 289-22565

Abstract: Background: The urban-rural land divide is visible through where people choose to work. This article aims to detect how, where and why people use rural coworking spaces instead of or in addition to working in urban areas. Methods: The research relied on both documented evidence and a structured survey among users of coworking spaces. Results: We found that the choice of working in rural coworking spaces draws on certain benefits and opportunities for its users, such as avoiding social isolation, separating private and professional life, reducing the commuting. An additional benefit for rural towns and villages is that the presence of a coworking space can make the location more vital, lively and attractive. Conclusions (and recommendations): Coworking space could partially bridge the urban-rural land divide. However, understanding this requires more insights in the behavior of rural coworking space users. Further research could look into modelling cause-effect relations and predicting coworking user behavior and the effect on their environment.

Keywords: rural development; rural land use; remote work; coworking spaces; digitization; urban-rural divide; town center revitalization

Citation: Hölzel, M.; de Vries, W.T. Digitization as a Driver fur Rural Development—An Indicative Description of German Coworking Space Users. *Land* **2021**, *10*, 326. https://doi.org/10.3390/land10030326

Academic Editor: Ruishan Chen

Received: 19 February 2021
Accepted: 16 March 2021
Published: 21 March 2021

1. Introduction

The structure of the article is as follows: we first describe the framework of spatial disparities, the role of work, and the development of remote work in Section 1. In Section 2 we formulate and focus the research question. In Section 3 we describe the methodology through which we address the research question. In Section 4 we display the results of the survey, in Section 5 we discuss the findings from the survey with the literature and finally in Section 6 we formulate conclusions on the findings, limitations and an outlook.

The research on why, where and how urban and rural disparities exist addresses different causes and impacts, such as the differences in health [1,2], education [3] and social services. Most prominently in the urban-rural divide discourse is however the demographic disparity [4,5]. Processes of urbanization in developed and developing countries are showing more or less the same trend, characterized by massive urban growth, rural exodus, urban sprawling—the consumption of land or change of land use [6–9], traffic jams and many other problems, in different stages [10–12]. Such demographic changes result in significant problems in allocations and conversions of land use and land cover, and ultimately land rights and restrictions. Cities and urban agglomerations are growing and the number of their inhabitants are increasing since the first wave of industrialization, faster and faster in the last decades in nearly every part of the world—the so called urbanization process [13]. This leads to a higher demand of land for housing [14] on one hand and a shrinking proportion of living space per person [15,16], due to the rising rental and purchase prices, on the other hand. Meanwhile the living space per person in rural region is higher [17] and growing.

Stimulated by advancements in agricultural science, mechanization and industrialization, the workforce which is needed to feed the population, has been shrinking [18] since decades, especially in developed countries. This reflects the "push" for people to leave rural regions alongside beside the "pull" of urban regions and major cities [19]. Small rural towns and villages are also suffering from automobile traffic, the growth of commercial and residential areas on the outskirts of towns [20], and the simultaneous vacancy and closure of stores and pubs in inner-city areas—the so called Donut-Effect [21–24].

Germany is a country in an advanced stage of development [25,26]. Yet, in several fields of society (economy, democracy, science etc.) it exhibits an advanced stage in rural-urban disparity as well [27]. From the perspective of spatial planning, the problem of demographic disparities between rural and urban region has led to heated debates and contested assessment frameworks. The indicators which are used and the results related to spatial development vary, but the general insight is that spatial inequality is the manifestation of the disparity [28,29]. There is both spatial inequality between German regions [30] and there is a lack of spatial justice [31]. Other countries, such as France or the U.S.A. exhibit similar spatial disparity challenges [32,33], which leads internationally to polarization among EU-member states for example [34]. In Germany, political actors have gradually become aware of these trends and have therefore formulated a policy of "equal living conditions in urban and rural areas", which needs to be implemented at different administrative and institutional levels of the federal states. At a highest administrative level, the target "to preserve coherent living conditions" is formulated in the constitution of the German Federal Republic (GG Art. 72 Abs. 2 Nr. 3) from 1949 [35]. In 1994 the formulation was modified to "production of equivalent living conditions" [36]. In the coalition agreement of the present German administration the target of equivalent living conditions is articulated as well [37]. It led to a report of a government commission with programmatic recommendations for action [38] by the Federal Ministry of the Interior, Building and Community; the Federal Ministry of Food and Agriculture and Family Affairs, Senior Citizens, Women and Youth. At the level of the European Union, support programs for rural development such as LEADER were introduced in the 1990s [39].

Taking into account that local economic strength, the labor market and incomes contribute to the local economy by the generation tax yields, the spending capacity for the people who live there, is the main driver for disparities and divide, the spatial distribution of local economic strength and work is essential for development and equivalent living conditions.

A key to bridge the urban rural divide could be the allocation of work from urban regions to rural regions. The possibility of remote work means that knowledge-based work is no longer necessarily tied to a specific location. A relatively new phenomenon is coworking spaces in which teleworking is performed.

This research aims to detect the habits and demands of users of rural coworking spaces and to estimate and assess the influence on local offers and services for mobility, consumption, shopping, catering, social and cultural. Based on these aims we formulate the following research questions:

1. Which amenities co-workers asked for in rural coworking spaces?
2. How can co-workers in rural-coworking spaces be described according their personal and professional characteristics?
3. How do coworking facilities in rural areas and their users influence local offers, services for mobility, consumption, shopping, catering, social and cultural life?

2. Literature Review

Drivers of spatial disparities in Germany constitute the variations in regional economic strength, availability and spatial distribution of jobs [40], wages and corporate earnings which contribute to the local economy [41]. They entail prosperity, tax yields for the local and regional authorities and spending capacity for the people who live there [42]. Depending on the tax system regional authorities and municipalities receive financial inflows [43,44]. It seems that there is a general link between regional disparities caused by economic strength or weakness, jobs, income and tax yields [45].

Reasons for migration and the decision where to life rely on a complex decision-making process. From a conceptual standpoint, Ravenstein developed the concept of "push and pull" [46] as a trigger for migration. Widely referred are the 4 classes or types structured by Petersen [47], i.e., Primitive, Forced/Impelled, Mass, Free. In this case it can be focused on "Free" migration forced by higher aspirations or a rational choice, because migration due to natural disaster, war or collective behavior could be unconsidered do not take place at the moment in Germany [48].

These base factors play an important role in peoples' decisions where to live. Especially in Germany, rational choices in life and career planning are often made [49,50]. At crucial turning points in life, such as choices for education, in the professional domains and careers, types of jobs or founding a family [51] they make a decision with spatial impact [52]. The conditions in such rational choices are that before a family is founded, an income has to be ensured [53] and a professional career needs to be established [50]. Due to the availability of birth control pills and better education of women, women have become more independent and the decision to found a family has become a more 'rational choice' as opposed to discretionary or random choice or fate (going with the flow) [49]. Many of the complex decisions about the place where to live are directly or indirectly related or influenced by job or educational opportunities [54]. "Work as the basis of human existence" is examined and considered by sociologists and philosopher like Habermas and Marcuse [55]. From an economical perspective, especially classic economists such as Adam Smith and Karl Marx have considered such work decision rationalities [56,57].

In case of better education and job opportunities in cities und agglomerations—Ravenstein would call it a "pull" [46]—people moving there from rural, less dense and less promising—referring to Ravenstein a "push"—regions [51]. This leads to a kind of vicious circle of depopulation, mainly by people who aspire better job opportunities to combat their declining spending capacity, decreasing entrepreneurship opportunities and growing lack of public services in regions and municipalities outside of major and/or university cities [58]. Declining number of inhabitants in this regions could fall below the technical function threshold of facilities of general interest [59] and jeopardize their economic viability [48,60].

The growing rural urban disparities are described is various publications [27,61]. At the same time a majority of people answer the question where they want to live with "villages, small towns" and the "country side" as a survey on behalf of the BHW Bausparkasse (Building and loan association) [62,63] and the German public television and radio broadcaster ARD [64]. Some studies predict a trend that more people want to and will live in small towns and rural regions. This trends are titled e.g., "Progressive Provinzen" [65,66], "KoDorf" [67] or "Urbane Dörfer" [68].

With the contact limitation due to the Covid-19 pandemic the amount of work which is done remotely was increasing rapidly [69–71], in relation to moderate growth in the past years [72,73]. People indicate that they want to continue working remote [74]. That could amplify the trend people moving from major cities to more rural towns as it is indicated in reporting for Germany [75,76] and for North-America too [77,78]. However, some authors are still skeptical if that will baffle the main trend of urbanization [79].

Wouldn't that be an opportunity to combine the wish of more country life and rurality on the one hand and the need of educated, well paid people in rural regions on the other hand to reduce or turn around the trend of rural exodus? So-called "knowledge workers" [80,81], usually an academically educated professional with specific knowledge and skills, combined with a professional social network are currently demanded at the labor market [82]. Richard Florida describes this group of human resources as the 'creative class' [83]. Knowledge, thought or brainpower is mostly non-physical. Without physical demand this kind of work is easy to transfer to another place and transmit the result to another. Thought work or knowledge work often requires access to knowledge repositories, such as libraries, archives, etc. Knowledge management deals with the nature, management, distribution and generation of knowledge, among other things from an entrepreneurial perspective [84,85].

Before the internet was established and before an increasing amount of information was provided by websites, database, virtual libraries etc. accessing information and knowledge required a physical presence or process by entering a library or an archive, finding and reading the desired Information in a book, report or document—or obtain the requested information in some lists, schedules, directories or the like. Today, a huge amount of knowledge and information is accessible via the internet on server, databases, virtual and digital libraries [86] and through social knowledge networks [87].

The idea of remote work was already practiced in the early 1960s, be it however in an analog manner using paper and pencil from home [72]. In the late 1960s the concept of communication substitute transportation was developed [88] and expanded in the 1970s as a substitute for the increasing business travel [89] and eminently by the oil crises [90]. Remote work or Telecommuting [91], which in German is often called "Home Office", could be operated from home. Workers save time by avoiding the daily commute or also just for a single day per week or month by working from home. However, that kind of remote work entail some negative effects or problems. Homes often are missing a separate room equipped with ergonomic office furniture for remote work from home [92]. Beside this, working from home could stress people, due to the double task of work and family, if they live in a family home. If they live alone the can feel lonely and disregarded [93–95]. Coworking spaces as a place between home and work could be a "third place" as described by Oldenburg [96] enlarging social contacts.

An alternative to operate the remote work from home is to perform in a coworking space. The term "coworking" was invented in 1999 by Bernard DeKovan and the first "coworking space" was opened 2005 in San Francisco by Brat Neuberg [97,98]. A coworking space is a kind of office space where people work at the same location on their own project or tasks and have the opportunity to network, socialize or cooperate with their "space mates". The concept of coworking was well described by Clay Spinuzzi in 2012 [99]. Coworking spaces provides options of collaboration and knowledge sharing, brainstorming and creativity [100]. Janet Merkel described coworking as "Coworking is hence not just about working 'alone together' or 'alongside each other' in a flexible and mostly affordable office space. It is also underpinned by a normative cultural model that promotes five values: community, collaboration, openness, diversity, and sustainability." [101]. Coworking spaces were mostly founded by free lancers, but nowadays this form of work is increasingly used by employees as well [102]. While founder of early coworking spaces were often locally engaged people, in recent years large, professional operators such as WeWork and Mindspace have discovered the market for themselves and offer desks in coworking spaces [103].

The social isolation, which comes along with working remote from home, also underlined by research regarding the impact of Covid-19 pandemic [104], could be tackled by working together in a common space, such as a coworking spaces [105]. At this point we should clarify that in Germany working remote from home is called "homeoffice", "home office" or "Tele-Heimarbeit" [90]. Just working not at the employers office is called "Telearbeit" or in recent years "Mobiles Arbeiten" [106]. In recent years the size and number of coworking spaces is increasing [107,108], not only in cities in rural regions in Germany as well [109,110]. The trend of rural coworking spaces is taken up by policies and some promotion initiatives are introduced [111,112].

Distance and occupied time for the daily commute is increasing since years [113,114]. However, generally remote work save time by avoiding the commute which can be used for other purpose, as spending time with family, friends or for sports, being active in a local club. These benefits lead to different preferable results from the perspective of spatial planning. The main benefit could be a declining mobility demand and a reduction of traffic. The commute is increasing since years [115] an reached an average of 10.5 km [116] one way each day, notable the amount of a commute over 50 km increase more strongly [116].

Reducing the demand of mobility for work, the traffic systems (roads, public transport etc.) could be relieves by avoiding the daily commute, reduce it to occasional or periodical commute. If a presence for a meeting is need the employee can travel for this specific to be there in time e.g., 13:00 o'clock and avoid the rush hour in the morning. This could spread the traffic volume over time and diminish the pressure on the traffic systems [117,118].

Further the saved time could be spent with the family or with friends and improve the work-life-balance (WLB), but could also generate stress [106]. "Entgrenzung der Arbeit"—delimitation of work is a broader topic of changes in the field of work to which remote work contributes substantial [94,119].

Not only because of online shopping, but also because of the high commuting share to distant office districts and the paths traveled, most of which do not lead past traditional village or small-town retail locations, town and city centers suffer from a lack of liveliness, a shrinking range of products and services, and consequently from vacancies [120–123]. Due to increasing distances between home and work, mobility in rural areas is dominated by the use of motor vehicles [124–126]. Another effect of the increasing distance between the location of residence and the location of work leads to multi-local lifestyle [127,128].

Recent publications see an opportunity for the revitalization of rural areas through coworking spaces [112], what probably support an allocation of knowledge work to small towns.

Coworking spaces, whether in inner cities or rural areas, usually offer different types of workplaces. The majority, however, are mostly open office structures, with different desks in an open space setting. Just a small amount of desks are located in smaller units as private, single or double offices [129]. The open structure encourage contact, exchange between the users, what is a major opportunity of coworking spaces related to working from home. However, this open structure have difficulties regarding the acoustic [130–132], if people want to work concentrated and undisturbed, make a phone call or participate in a videoconference [132,133].

3. Methodology

The case search and selection process aimed to receive a broad picture of coworking spaces outside of major and university cities. This implied using the criteria "belongs not to the territory of a major city" [134]. This process enabled a broader insight into the concept, excluding major German cities such as Berlin, Hamburg Munich, and Kaiserslautern (population >100.030 [135]), but including smaller cities and the surrounding and hinterland of major cities. We decided not to use the definition of the Federal Institute for Research on Building, Urban Affairs and Spatial Development (BBSR—Bundesinstitut für Bau-, Stadt- und Raumforschung) for rural area [136] as the measurement for our investigation, for two reasons: first we want to exclude in our survey only the typical environment of coworking space which are usually to find in major cities and second we want to include as much as potential respondents as possible. This also considers that users of coworking spaces tend to not respond on surveys, because they feel to be over investigated. Due to that we chose a simple and short online survey to repel as few respondents as possible.

The survey structure was such that it should become possible to reveal the volume and type of demands and preferences of users—and what could be the impact of a growing quantity of remote workers in rural coworking spaces on their vicinity. The survey contained different categories of questions (Table 1). The first section addressed the use of coworking spaces and their services. The second section investigated the spatial relation of the coworking space. The third section regarded the vicinity of the coworking space. The fourth and the fifth section asked for personal information and profession of the coworker. Table 1 provides these details.

Table 1. Categories of the survey.

Category	Aim	Question
use of coworking spaces and their services	Investigate intensity, interval, duration and service	Do you use different coworking spaces? Which coworking spaces are they? (Please enter the name and address of the coworking spaces) How long have you been using the above coworking spaces (yyyy,mm)? How often do you use a work opportunity in a coworking space? On average, how many hours do they use the work opportunity in a coworking space per day of use? What type of workspace do you prefer to use in the coworking space? In which room is the workplace you use usually located? What offers besides a pure workplace do you use in the coworking spaces? What other offers from the coworking space would you like to see?
spatial relation and mobility	Investigate the spatial relation and mobility/transportation	First and second residence (town, postcode, street, house number) Which means of transport do you usually use to cover the distance between your home and the coworking space? How long do you need for the distance (hh:mm)?
vicinity of the coworking space	detect offers and services in the vicinity	Which offers in the vicinity of your coworking spaces do you use? (Multiple answers possible) What offers besides a pure workplace do you use in the coworking spaces? How much more would you be willing to spend on products or services in the vicinity of the coworking space compared to out-of-town offers?
Profession of the coworker	Kind of employment/entrepreneurially, relation to "home office", support by the employer	What kind of (gainful) activity do you do? How much are your regular weekly working hours? In which industry/area are you employed or entrepreneurially active? What share of the costs for your workplace in the coworking space is borne by your employer? How open was your employer to you providing your work leadership in a coworking space? Do you also-or have you previously-worked in a home office? What are the advantages of working in coworking spaces compared to working in a home office? What are the advantages of working in coworking spaces compared to working at a company location? What prompted you to work in a coworking space?

We used different types of questions: closed, multiple-choice questions to make it easy and fast to answer the questions for the respondents. Rating scale questions were used to get a quick and easy to answer overview on e.g., duration, willingness to spend etc. Considering the expected low response rate, we decided to ask mostly open-ended questions to get answers to unusual options or singular types.

An online survey was executed between end of February and April 2020, addressed to respondents who were directly related to rural coworking spaces. The operators of 80 rural coworking spaces were contacted via email and asked to distribute the link to the survey to their user and tenants. The first email contact was followed up by phone calls and follow-up emails. From that group, initially 13 operators confirmed to distribute the survey among their users and tenants.

4. Results

4.1. Responses

The survey was answered by 36 persons, of which four answers were unreliable and therefore excluded. Unfortunately, this number of responses is rather low. We only received answers from tenants or users of 12 coworking spaces. It is not clear why the number of responses was that low as well. Possibly, the spreading of the corona virus and the officially issued lockdown from the middle of March 2020 [137] restrained operators and tenants from distributing and answering the questionnaire. Nevertheless, we consider the responses, which we did obtain still relevant and appropriate, as they still provide a general picture and some first clues regarding the research question. The subsequent sections provide the results. The respondents came from coworking spaces spread over Germany, from the northeast to southwest and southeast. The respondents were located in remote and shrinking cities, which recently dropped below the limit of major city and small villages of a few hundred inhabitants.

We do not know to how many tenants the coworking space operators forwarded our survey, hence we cannot provide a typical response rate.

4.2. Summary of Responses

One quarter of the respondents indicated using more than one coworking space. 7% use three coworking spaces and one is using five coworking spaces. Some of the people who use several coworking space are working in different countries, such as Switzerland and Germany or Austria and Germany. One of the coworking space is frequently in major cities, while the other used coworking space is in more rural regions, but there are exceptions, where people use more than two coworking spaces, although the majority are located in more rural regions (n = 32).

A majority of 41% use their main coworking space for 1 year, 28% for 2 year and only single person are using their primary coworking space for 6 or 9 years. If they are using more than one coworking space, the second most common used coworking space is in use for 1 year (n = 29) (Figure 1).

Less than one third (28%) of the respondents indicate that they are using coworking spaces each day. The same amount of people state that they are using the coworking space 3 days a week. Nearly one third (31%) of people are using coworking spaces 1 or 2 times a week. And 14% are using a coworking space 1 or times in a month (n = 29) (Figure 2).

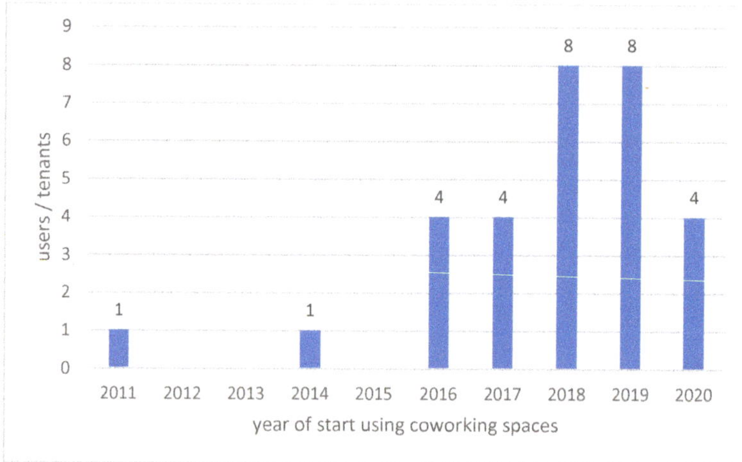

Figure 1. Start year of the work in coworking spaces.

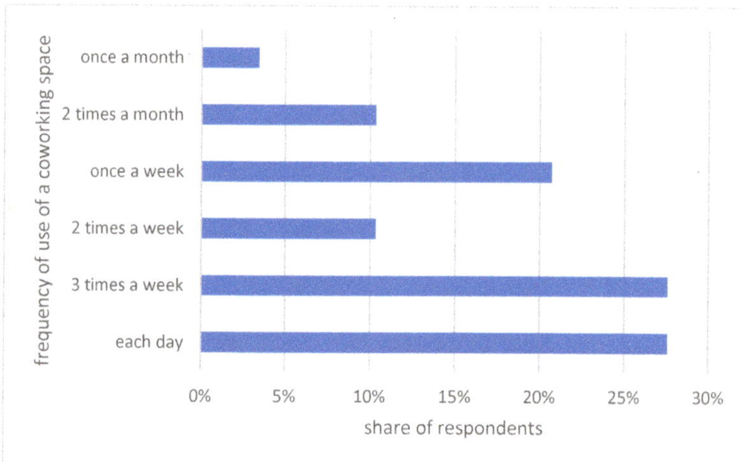

Figure 2. Frequency of use per week or month.

Some 10% of the respondents spend less than 4 h per day spend in a coworking space, whereas 11% spend 4 to 6 h in a coworking space, but most (41%) of the respondents spend 6 to 8 h in a coworking space. 14% of the respondents spend 8 to 10 h in a coworking space and the same amount stay more than 10 h in the coworking space (n = 29).

Most (46%) people work at a fixed or dedicated desk. Additionally, 21% also have a specific own desk, but during their absence, this desk is also usable for others. Nearly a third are working at various desks, so called 'hot desk' (n = 28). A majority of the tenants have a fixed desk, and hot desk users mentioned that they would like to have a fixed desk, if they would not have another desk somewhere else.

A majority of 64% are working in an open space office. 21% of the tenants have their desk in a dual office and 14% are working in a single office (n = 29).

Nearly every user (93%) of a coworking space utilizes the kitchen in the coworking spaces. Printer, WLAN/LAN and social place/meeting point are used by about 80%. Nearly 70% are using conference rooms for 4 to 10 people, 35% are using conference rooms for more than 10 people. A postal address is used by 35%. Two participants mentioned that they are using telephone boxes, to be undisturbed while calling. A single person mentioned lunch service, social events and training are be used (n = 29).

Service as noise protection, carsharing, lockers cubicles are mentioned one time. Multiple named are conference technic, team/social events, café/mess and back office. One person is desiring more colleagues (n = 12).

Nearly half of the respondents (49%) are using a car to get to the coworking space. 7% are going by public transport, 23% are riding a bike and 21% get there by foot. Due to weather conditions or the location of each cowering space, different means of transportation may be used on different days. Taking this into account, multiple responses are possible. Approximately 40% state that they need 10 min commuting time to travel to the coworking space. 11% need only 5 minutes to get there. 14% need 15 min and 18% are 30 min on the way. Another 14% need more than 30 min to get to the coworking space. Two people indicate that their traveling commuting time extends by more than one third, when they go by bicycle instead of by car (Figure 3).

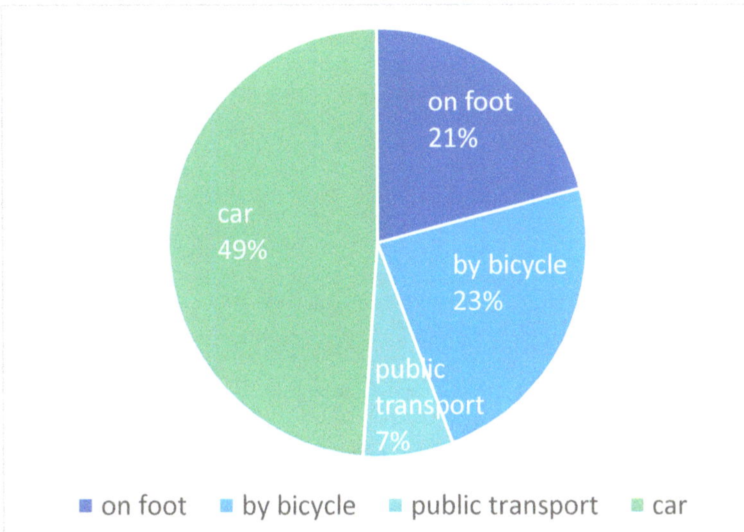

Figure 3. Means of transportation/modal split.

Some tenants of coworking spaces are traveling around and visit different coworking spaces at more than 150 km distance from their residence. Nearly a quarter of the respondents are using two coworking spaces, some are using up to five coworking spaces. Some are switching between commuting by car, bicycle or afoot, even if this doubles or triples their commuting time. It can be assumed that the traveling means depends on weather conditions [138]. Some of the tenants are using two or more coworking spaces, one close to their residence and another one in larger towns (Figure 4). The point shown in Figure 4 is that people may use different coworking spaces at different times, for different reasons and under different circumstances and related to the distance between residence and coworking space the respondents are using different modes of travel.

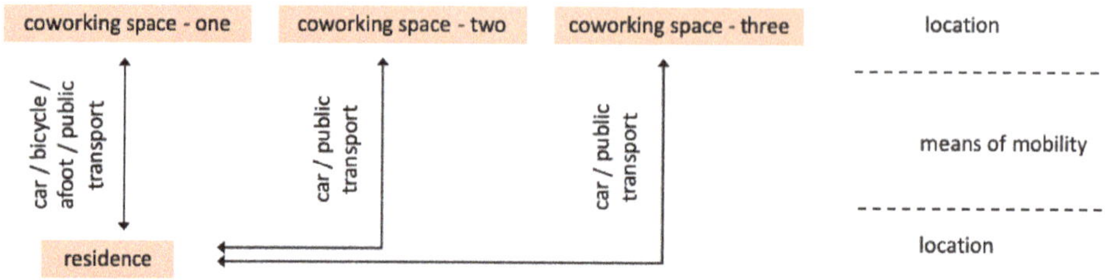

Figure 4. Conjunction of workplace and residence by means of transportation.

The most used services in the vicinity are bakeries, which 89% of the users visit. They spend 5 € per day at most, although some (19%) spend more than 5 € in bakeries. 67% are visiting restaurants or taverns and spend there up to 5 € (22%), from 6 to 10 € (44%) and up to 20 € (33%) per day. 56% of the tenants are going to supermarkets/grocery stores and spend there 20 to 30 € (10%), 10 to 20 € (26%) or up to 10 € (64%) each day. 29.6% of the tenants are going to kiosks, all of them spent less than 5 € each day. Small quantities (4%) are using laundries, childcare, post office and medical practices in the vicinity of the coworking space. Single named are e.g., fishmonger, town hall or drugstore (n = 23).

In response to the question "How much more would you be willing to spend on products or services in the vicinity of the coworking space compared to out-of-town offerings?" 22% of the respondents indicate not being willing to spend more for central offer or service related to an out-of-town offer. However, 55% would spend 5 to 25% more than an out-of-town offer. And another 22% are willing to spent more than 40% up to 50% above the price of out-of-town offers.

Approximately 41% of the tenants of the coworking spaces work for an employer. Others are freelancer or companies using the coworking space as a company headquarter. Some responded being an association/club or volunteer (n = 27) (Figure 5).

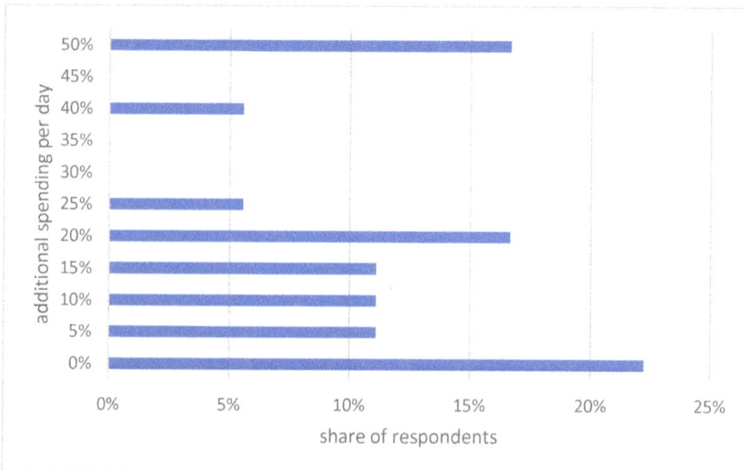

Figure 5. Willingness to spend more on products or services in the vicinity of the coworking space compared to out-of-town offerings.

The largest group (41%) of the coworkers have between 37 up to 40 regular working hours, which corresponds to regular full employment hours in the frame of the German Working Time Act [106]. A group of 20% are regular working between 10 and 30 hours, which indicates a part time employment [107]. Out of this group a majority of more than

60% is working 30 hours per week, some only work 10 hours per week. Indicated by the reduced working hours, it could be assumed that this group of workforce is occupied by other things, like childcare, care of relatives or they have only a part-time position [108]. At the other end of the scale are a group of workers (22%) that state to work regular 50 to 60 hours per week. Considering the Working Time Act [106], this must be a group of freelancers or entrepreneurs (n = 27) (Figure 6).

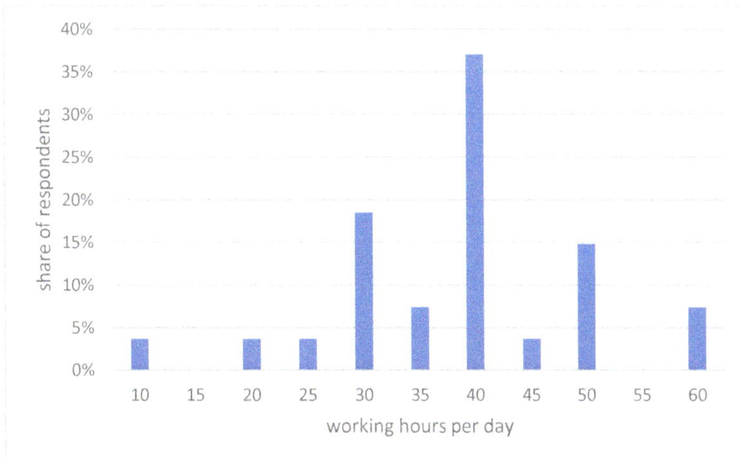

Figure 6. Regular weekly working hours of the tenants of the coworking space.

5. Discussion

The initial findings confirm that location of work and economic activity significantly affects complex spatial relations [41–44] as indicated in the Introduction. As long as wage labor or entrepreneurship is needed to afford life, the location of work is very important and also spatially related to many other aspects of life [113]. The relevancy of the spatial relation of people depends on the life phase, but is dominated by their primary relation to work, education, supplemented by shopping, leisure, sports, social services and child care, amongst others [113]. The work commuting is the main reason to travel [113]. Remote work could reduce the time for commuting [117]. There are also negative effects of remote work, such as social isolation, feeling disregarded or the stress from unseparated private and professional life [94,95]. Considering that, to be able to reach performance levels in a coworking space to a similar extent as in a conventional working place, it is necessary to make it to a kind of a "third place" [96]. If this third place is located close to the place of residence, it could reduce the daily commute [93,116,117], allocated spending capacity [120] and increase the vividness in rural villages and small towns [121,122].

Given these theoretical findings and hypotheses, the core questions remain: (1) which amenities coworkers asked for in rural coworking spaces, (2) how can coworkers in rural coworking spaces be described according their personal and professional characteristics (3) how do coworking facilities in rural areas and their users influence local offers, services for mobility, consumption, shopping, catering, social and cultural life? We elaborate on each of these in the following subsections.

(1) Which amenities coworkers asked for in rural coworking spaces?

As the empirical survey results show, those respondents who work in a coworking space outside of a major city also tend to execute their job in more than one coworking space only (see Figure 4). Some are clearly traveling around in a larger region and are using several coworking spaces. If they are using more than one coworking spaces at least one of them is located in a major city. It seems that a relation to a major city is still relevant for coworkers and that these types of respondents value the need to travel as a crucial part of their job, to meet people in real. In these cases, it seems that the coworking spaces act as a linking point in local networks.

Most coworkers are renting a fix desk in an open space [99], but they are also missing the opportunity to separate in particular to make telephone calls or to work undisturbed. This indicates a demand of variation in the work environment, not only variations from day to day, what several of the participants already experience, also variation during the day. There is a demand for variation in the exchange with others as well as apart from others, such as for concentrated work or undisturbed calls. The demand of noise protection indicates a similar need. Especially in open spaces, tenants seem to be disturbed by noise, presumably from other users. This is classical critique on acoustics in open spaces offices [130]. For certain types of work, these office forms are rather unsuitable [131,132]. On one hand the users in our survey indicates a demand of noise protection and separation—as often in open plan offices [133]—on the other hand they express a desire to meet and network, what also corelate with the literature [129]. It would therefore make sense to offer different types of workstations, one for concentrated, undisturbed work and one for networking. A crucial aspect of using multiple work spaces is apparently not so much related to the facilities offered by the work spaces, but by the specifics of the location where coworkers are going to—such as specific locations or surroundings which make the meetings different than simply by regular communication means.

(2) How can coworkers in rural coworking spaces be described according their personal and professional characteristics?

With 41% the share of employed coworker in this survey is higher than the average numbers in coworking spaces [102]. That could indicate that tenants avoid commuting to headquarters of remote employers, save time for the commuting, relieve traffic infrastructure and with that reduce the CO_2 emission. Considering the increasing traveling distance and time [114,116], it is an advantage to avoid the commuting at least on some days of the week. A majority of the respondents are driving to the coworking space by car, only a small amount could go by bike or on foot (see Figure 3). Reasons for this variation could be a lower population density at certain locations, which increases the catching area for such a coworking space, as long as the number of users of rural coworking spaces is still low. Only a small group of users live so close that they can walk or take the bike. It is typical for rural regions that longer travel distances between home and destination are covered by car [124,126]. Only short distance commuters are saving time saving if it is done on foot or by bicycle [125]. The background of people is also broad. One can find freelancers, entrepreneurs and various types of employees in rural coworking spaces. The age distribution is broader than in coworking spaces in an urban environment [103]. This might be related to the motivation to use rural coworking spaces as a means to spare the traveling time to a distant company office.

(3) How do coworking facilities in rural areas and their users influence local offers, services for mobility, consumption, shopping, catering, social and cultural life?

The usage of the coworking space is temporarily during the day and during the week (see Figure 2). That indicates that work is also operated somewhere else, implying a multi-local lifestyle [127], during travel, at the employers office or at home. Not all regular working hours are used for work. Perhaps the time is also used for childcare, leisure or social engagement. The time which is spent on other activities than working is not only used to recover from and for work, but also to pursue other goals in life. With the saved time, people could get more engaged in social commitments, contribute to associations or similar activities. Their presence in the rural town could thus be higher, resorting in more vitality and vividness in the center of the village and small towns [22,126,127].

Coworking spaces in a rural context are more or less a new form of telecommuting [91]. Most tenants use their main coworking space for 1 or 2 year (see Figure 1). The invention of coworking as a concept started from 1995. The term coworking was introduced in 1999 and the first so designated coworking space opened in 2005 in San Francisco [98]. The number of coworking spaces are increasing rapidly since 2005, mainly in agglomerations. In rural regions, towns or villages are existing some coworking spaces for some time [109]. Since the last 2–3 years the number of coworking spaces outside of major cities is growing rapidly.

The responses reveal that the tenants of coworking spaces use several offers and services, like bakeries, supermarkets, kiosk and taverns or restaurants in the vicinity of the coworking space, nearly each day, where they spend up to 30 € per day. With the rising number of tenants in coworking spaces a considerable spending capacity is available in vicinity of the coworking space. Arriving and leaving the coworking space, as well as the lunch break at the neighboring restaurant or butcher, increase the presence of people—not cars—in public spaces and thus provide more liveliness. In future research it would be interesting to find out, if that potential really reaches local stores, services, groceries and enrich the vitality of their vicinity. Possibly, coworking spaces will bring enough purchasing power to local centers, if they are located there, to enable any retailers that may still be present to generate substantial turnover and continue to exist. Small supermarkets, grocery stores or retailers in the central locations of towns and small cities have been suffering for years from competition with newly established, large-scale retailers on the outskirts of settlements [121]. Vacant, historic buildings can also be converted into coworking spaces and thus find a new use (Figures 7 and 8).

Figure 7. Coworking space in a village, former hayloft, exterior view.

Figure 8. Coworking space in a village, former hayloft, interior view.

In the political program of action "Unser Plan für Deutschland-Gleichwertige Lebensverhältnisse überall" were measures recommended to improve equivalent living conditions, including the support of coworking spaces [38]. Some regional development agencies already provide funding programs, partly direct for coworking spaces [111,112]. However, the eligibility criteria so far do not seem to take into account the impact of each location—in-town or in commercial areas.

6. Conclusions

The research aim was to detect the habit and demands of users of rural coworking spaces and assess the influence on local offers and services for mobility, consumption, shopping, catering, social and cultural. We found that in order to operate remote work in a rural area a coworking space provides several benefits and opportunities for the user/tenant, such as avoiding social isolation, separating private and professional life, avoiding or reducing the commute sometimes. In addition, from a perspective of the municipality a coworking space located in the village or town center support local offers from retailer and services, like grocery stores, bakeries, butcheries, restaurants and cafes—and provide their vicinity with passers-by frequency and vividness. The increasing turnover of local retailers and services could make them more profitable and secure their survival.

These findings are relevant, for several reasons and related to e.g., the increasing distance between residence and workplace [116], the increasing share of remote work, which is growing since years. This brings stress to people and to town centers [21]. When located in the town center, coworking spaces can bring liveliness and spending capacity back to the heart of those towns, which are suffering from the donut-effect. Further research should investigate under which conditions which types of public space, retail and services in the vicinity coworking space are beneficial. So far it seems reasonable that coworking spaces in rural towns should be located at a central location, where daily goods and services are available and where the tenants of the coworking space can contribute to vividness and vitality of the town and to the turnover—and with that maybe to longer persist of the resident suppliers.

With the contact limitation and lockdown due to the Virus Covid-19 the share of remote work got a boost [69–71] and am majority of employees would prefer to continue working remote [70,74]. That will increase remote work [71], people could move to remote rural towns, avoiding high rents and dense populated major cities, work from home or in an rural coworking space. Coming research should find out, if this group of working people tends to work from home, which are usually larger in the countryside and offer more space [17] for domestic study—or would this people go for a desk in an coworking space in the town center?

One of the main influence factors on land management is the kind of land use [6–8]. Depending on the share of work which is done closed to the place of residence and the question where people settle (see Section 1. Introduction) the use of land is strong influenced [9]. Will this trend to remote work lead to a higher demand on settlement area and land consumption at the outskirts of major cities or will this trend bring liveliness, vitality, spending capacity and inhabitants back to rural town centers?

The findings exhibit some limitations, which necessitate further research. As the survey could not trace under which conditions coworking space user opt for increasing or decreasing their economic spending, subsequent research and empirical investigations could investigate the behavior of coworking tenants related to means of transport, their consumer behavior and how the vicinity of coworking spaces get influenced by the presence of coworking spaces and their tenants. Additionally, the survey results could not provide sufficient input for constructing a theoretical cause-effect relational model. Such a construction would require a more extensive data collection, which would investigate the variations in impact of coworking spaces on their surroundings. This includes amongst others how it affects land demand and supply. Practical research into trends of coworking given the experiences due to the crises by the virus Covid-19 and the issued contact restrictions.

There are a number of issues which would require an extension of the investigation to truly understand the causes and effects. First, the conditions under which knowledge workers either live in rural areas, or previously guided their decision to move to rural areas may be a crucial factor for the decision to opt for coworking spaces. Secondly, the workload conducted at a coworking space may differ based on the available facilities. Thirdly, it still remains unclear to which extent supporting conditions, such as access to basic amenities and connections to socio-cultural aspects play a role. Fourthly, what the current research could not find out was whether there exists trend that brings liveliness, vitality, spending capacity and inhabitants back to rural town centers? And will this help to reverse or stop the growth of the "Donut" [22,24] around the towns and villages. Finally, the relations of coworking spaces to the urban-rural divide, urban-rural land use and spatial justice issues needs further empirical research.

Author Contributions: This manuscript is a part of M.H.'s ongoing Ph.D. research. M.H. has written the article under the guidance of W.T.d.V. All authors have read and agreed to the published version of the manuscript.

Data Availability Statement: The data presented in this study are available on request from the corresponding author. The data are not publicly available due to privacy reason.

Acknowledgments: This research received no funds. We thank all participants of the survey and all operators of rural coworking space, who distribute the survey among their tenants.

Conflicts of Interest: The authors declare no conflict of interest.

References

1. Hartley, D. Rural Health Disparities, Population Health, and Rural Culture. *Am J Public Health* **2004**, *94*, 1675–1678. [CrossRef]
2. Spencer, J.C.; Wheeler, S.B.; Rotter, J.S.; Holmes, G.M. Decomposing Mortality Disparities in Urban and Rural U.S. Counties. *Health Serv. Res.* **2018**, *53*, 4310–4331. [CrossRef] [PubMed]
3. Byun, S.; Meece, J.L.; Irvin, M.J. Rural-Nonrural Disparities in Postsecondary Educational Attainment Revisited. *Am. Educ. Res. J.* **2012**, *49*, 412–437. [CrossRef]
4. Bartholomae, F.W.; Popescu, A.M. The Role of Regional Competition for Demography and Regional Disparities in Germany. *Rom. J. Reg. Sci.* **2007**, *1*, 45–70.
5. Kroll, F.; Kabisch, N. The Relation of Diverging Urban Growth Processes and Demographic Change along an Urban–Rural Gradient. *Popul. Space Place* **2012**, *18*, 260–276. [CrossRef]
6. Foley, J.A.; Defries, R.; Asner, G.P.; Barford, C.; Bonan, G.; Carpenter, S.R.; Chapin, F.S.; Coe, M.T.; Daily, G.C.; Gibbs, H.K.; et al. Global Consequences of Land Use. *Science* **2005**, *309*, 570–574. [CrossRef] [PubMed]
7. Larsson, G. *Land Management as Public Policy*; University Press of America: Lanham, MD, USA, 2010; ISBN 978-0-7618-5249-0.
8. De Vries, W.T.; Chigbu, U.E. Responsible Land Management-Concept and Application in a Territorial Rural Context. *Flächenmanagement Bodenordn.* **2017**, *79*, 65–73.
9. Gomes, E. Sustainable Population Growth in Low-Density Areas in a New Technological Era: Prospective Thinking on How to Support Planning Policies Using Complex Spatial Models. *Land* **2020**, *9*, 221. [CrossRef]
10. Bairoch, P.; Goertz, G. Factors of Urbanisation in the Nineteenth Century Developed Countries: A Descriptive and Econometric Analysis. *Urban Stud.* **1986**, *23*, 285–305. [CrossRef]
11. Bhattacharya, P.C. Urbanisation in Developing Countries. *Econ. Political Wkly.* **2002**, *37*, 4219–4228.
12. Ahrens, A.; Lyons, S. Changes in Land Cover and Urban Sprawl in Ireland from a Comparative Perspective Over 1990–2012. *Land* **2019**, *8*, 16. [CrossRef]
13. Martine, G.; Marshall, A.; Brachman, P.; Deligiorgis, D.; Fuersich, C.; Leon, L.; Odelius, A. *Unleashing the Potential of Urban Growth*; Population Fund, Ed.; State of world population; UNFPA: New York, NY, USA, 2007; ISBN 978-0-89714-807-8.
14. Städte-Boom und Baustau: Entwicklungen auf dem Deutschen Wohnungsmarkt 2008–2018. Available online: https://www.destatis.de/DE/Presse/Pressemitteilungen/2019/12/PD19_N012_122.html (accessed on 12 March 2021).
15. Haufe-Lexware GmbH & Co KG Studien: So Viel Wohnung Bekommen Mieter Noch fürs Geld. Available online: https://www.haufe.de/immobilien/entwicklung-vermarktung/marktanalysen/wohnungsmarkt-im-weniger-flaeche-fuer-immer-mehr-geld_84324_469582.html (accessed on 12 March 2021).
16. Sagner, P.; Stockhausen, M.; Voigtländer, M. *Wohnen—Die Neue Soziale Frage?* IW-Analyse; Institut der Deutschen Wirtschaft Köln Medien GmbH: Köln, Germany, 2020; Volume 136, ISBN 978-3-602-45634-5.
17. Kempermann, H.; Sagner, P.; Ewald, J.; Krause, M. *Wohnen in Deutschland 2020—Unterschiede Zwischen Stadt und Land*; IW Consult/Institut der Deutschen Wirtschaft: Köln, Germany, 2020; p. 52.
18. Roser, M. Employment in Agriculture. Available online: https://ourworldindata.org/employment-in-agriculture (accessed on 26 October 2020).
19. Taubenböck, H.; Wurm, M. Globale Urbanisierung—Markenzeichen des 21. Jahrhunderts. *Glob. Urban.* **2015**, 5–10. [CrossRef]
20. Hynes, M. At a Crossroads: Investigating Automobility and Its Implications for Local Urban Transport Policy Design. *Urban Sci.* **2017**, *1*, 14. [CrossRef]

21. Michalski, D.; zur Nedden, M.; Frölich von Bodelschwingh, F.; Pätzold, R.; Stoeckermann, A.S.; Strauss, W.-C. *Stadt und Land*; Nagel, R., Schmedding, A., Eds.; Baukultur Bericht: Potsdam, Germany, 2016; Volume 17, ISBN 978-3-88118-575-2.
22. Nagel, R. *Flächennutzungsmonitoring X: Flächenpolitik-Flächenmanagement-Indikatoren*; Meinel, G., Schumacher, U., Behnisch, M., Krüger, T., Eds.; IÖR Schriften; Rhombos-Verlag: Berlin, Germany, 2018; ISBN 978-3-944101-76-7.
23. Herling, O.; Quaiser, K.; Geier, M. Kommunikation zur Beförderung der Ortsinnenentwicklung—Ein Werkstattbericht aus dem BMBF-Forschungsvorhaben Dorf und Du—Regionalstrategie Ortsinnenentwicklung in der LEADER-Region Wetterau/Oberhessen. *Z. Geodäsie Geoinf. Landmanagement* **2018**, 287–295. [CrossRef]
24. Schröteler-von Brandt, H.; Schmitt, G. Dorferneuerung. In *Stadterneuerung*; Springer Fachmedien Wiesbaden: Wiesbaden, Germany, 2016; pp. 300–337. ISBN 978-3-658-05762-6.
25. Human Development Reports | United Nations Development Programme. Available online: http://hdr.undp.org/ (accessed on 27 October 2020).
26. Pacione, M. *Urban Geography: A Global Perspective*; Routledge: London UK, 2009; ISBN 978-1-134-04308-8.
27. Küpper, P.; Peters, J.C. *Entwicklung Regionaler Disparitäten Hinsichtlich Wirtschaftskraft, Sozialer Lage Sowie Daseinsvorsorge und Infrastruktur in Deutschland und Seinen Ländlichen Räumen*; Johann Heinrich von Thünen-Institut: Braunschweig, Germany, 2019.
28. Fuest, C.; Immel, L. *Ein zunehmend gespaltenes Land?—Regionale Einkommensunterschiede und die Entwicklung des Gefälles Zwischen Stadt und Land Sowie West-und Ostdeutschland*; Ifo-Institit: München, Germany, 2019; pp. 19–28.
29. Hüther, M.; Südekum, J.; Voigtländer, M. (Eds.) *Die Zukunft der Regionen in Deutschland Zwischen Vielfalt und Gleichwertigkeit*; Institut der deutschen Wirtschaft Köln Medien GmbH: Köln, Germany, 2019; ISBN 978-3-602-45621-5.
30. Mießner, M. Die Raumordnungspolitik der Bundesrepublik in der Nachkriegszeit Raumordnungspolitische Konzepte als konservativer Ballast oder Antwort auf drängende Probleme der räumlichen Planung. In *Raumplanung Nach 1945. Kontinuitäten und Neuanfänge in der Bundesrepublik Deutschland*; Campus-Verlag, S.: Frankfurt am Main, Germany, 2015; pp. 197–223, ISBN 978-3-593-50306-6.
31. Hahne, U.; Stielike, J.M. Gleichwertigkeit der Lebensverhältnisse. Zum Wandel der Normierung räumlicher Gerechtigkeit in der Bundesrepublik Deutschland und der Europäischen Union. *Ethik Ges.* **2015**, *2013*, 40. [CrossRef]
32. Combes, P.-P.; Lafourcade, M.; Thisse, J.-F.; Toutain, J.-C. The Rise and Fall of Spatial Inequalities in France: A Long-Run Perspective. *Explor. Econ. Hist.* **2011**, *48*, 243–271. [CrossRef]
33. Hanley, C. A Spatial Perspective on Rising Inequality in the United States. *Int. J. Sociol.* **2010**, *40*, 6–30. [CrossRef]
34. Ezcurra, R.; Pascual, P.; Rapún, M. Spatial Disparities in the European Union: An Analysis of Regional Polarization. *Ann. Reg. Sci.* **2007**, *41*, 401–429. [CrossRef]
35. Parlamentarischer Rat. Grundgesetz für die Bundesrepublik Deutschland. 1949, p. 20. Available online: https://www.bundestag.de/gg (accessed on 19 February 2021).
36. Deutscher Bundestag. Grundgesetz für die Bundesrepublik Deutschland. 1994. Available online: https://www.bundestag.de/gg (accessed on 19 February 2021).
37. CDU/CSU/SPD ein Neuer Aufbruch für Europa, Eine Neue Dynamik für Deutschland, ein Neuer Zusammenhalt für Unser Land, Koalitionsvertrag für die 19. Legislaturperiode 2018. Available online: https://www.bundesregierung.de/resource/blob/975226/847984/5b8bc23590d4cb2892b31c987ad672b7/2018-03-14-koalitionsvertrag-data.pdf?download=1 (accessed on 2 November 2020).
38. Seehofer, H.; Klöckner, J.; Giffey, F. *Unser Plan für Deutschland—Gleichwertige Lebensverhältnisse Überall*; Bundesministerium des Innern, für Bau und Heimat: Berlin, Germany, 2019.
39. Rodríguez Martín, J.A.; Martín Martín, J.M.; Salinas Fernández, J.A.; Zermeño Mejía, K.A.; Añaños Bedriñana, K.G. A Spatial Analysis of the Achievements, in Terms of Regional Development, Accomplished by the Initial EU-Member Cohesion Fund Beneficiaries Using a Synthetic Indicator. *Sustainability* **2019**, *11*, 2343. [CrossRef]
40. Ganesch, F. *Räumliche Mobilität am Arbeitsmarkt: Einfluss-und Erfolgsfaktoren Überbetrieblicher und Überregionaler Mobilitätsprozesse*; Otto-Friedrich-Universität: Bamberg, Germany, 2019.
41. Kriehn, C. *Erwerbstätigkeit in den Ländlichen Landkreisen in Deutschland 1995 bis 2008*; Arbeitsberichte aus der vTI-Agrarökonomie; Institut für Ländliche Räume des Johann Heinrich von Thünen-Instituts: Braunschweig, Germany, 2011; p. 113.
42. Maretzke, S.; Ragnitz, J.; Untiedt, G. *Betrachtung und Analyse von Regionalindikatoren zur Vorbereitung des GRW-Fördergebietes ab 2021 (Raumbetrachtung): Gutachten im Auftrag des Bundesministeriums für Wirtschaft und Energie (BMWi)*; Ifo Institut, Ed.; Dresden Studien; Ifo Institut: München, Germany, 2019; ISBN 978-3-95942-068-6.
43. Boettcher, F.; Freier, R.; Geißler, R.; Schubert, M.; Stollhoff, R. *Kommunaler Finanzreport 2019*; Bertelsmann Stiftung: Gütersloh, Germany, 2019; p. 118.
44. Slack, E. *How Much Local Fiscal Autonomy Do Cities Have? IMFG perspectives*; Institute on Municipal Finance and Governance, Munk School of Global Affairs: Toronto, ON, Canada, 2017; ISBN 978-0-7727-0985-1.
45. Giannakis, E.; Bruggeman, A. Regional Disparities in Economic Resilience in the European Union across the Urban–Rural Divide. *Reg. Stud.* **2020**, *54*, 1200–1213. [CrossRef]
46. Ravenstein, E.G. Die Gesetze der Wanderung I und II. In *Regionale Mobilität: Elf Aufsätze*; György, S., Ed.; Nymphenburger Verlagshandlung: München, Germany, 1972; pp. 41–94. ISBN 3-485-03210-7.
47. Petersen, W. Eine allgemeine Typologie der Wanderung. In *Regionale Mobilität*; Nymphenburger Verlagshandlung: München, Germany, 1972; pp. 95–114.

48. Proske, M. *Demographischer Wandel und Daseinsvorsorge—Auswirkungen Kommunaler Angebote auf die Wohnortwahl*; Technische Universität Kaiserslautern: Kaiserslautern, Germany, 2011.
49. Helfferich, C.; Klindworth, H.; Heine, Y.; Wlosnewski, I. *Familienplanung im Lebenslauf von Frauen: Schwerpunkt: Ungewollte Schwangerschaften*; Bundeszentrale für Gesundheitliche Aufklärung, Ed.; Forschung und Praxis der Sexualaufklärung und Familienplanung; Bundeszentrale für Gesundheitliche Aufklärung (BZgA): Köln, Germany, 2016; ISBN 978-3-942816-77-9.
50. Rupp, M. *Eine Zukunft ohne Kinder*; Verlag Barbara Budrich: Wiesbaden, Germany, 2009; pp. 213–236.
51. Kley, S. *Migration im Lebensverlauf: Der Einfluss von Lebensbedingungen und Lebenslaufereignissen auf den Wohnortwechsel*; VS Verlag für Sozialwissenschaften: Wiesbaden, Germany, 2009; ISBN 978-3-531-16712-1.
52. Müggenburg, H. *Lebensereignisse und Mobilität: Eine generationsübergreifende Untersuchung von Mobilitätsbiographien*; Studien zur Mobilitäts- und Verkehrsforschung; Springer Fachmedien Wiesbaden: Wiesbaden, Germany, 2017; Volume 32, ISBN 978-3-658-16067-8.
53. Schmitt, C. Familiengründung und Erwerbstätigkeit im Lebenslauf. In *Sozialwissenschaftlicher Fachinformationsdienst soFid Familienforschung*; Informationszentrum Sozialwissenschaften der Arbeitsgemeinschaft Sozialwissenschaftlicher Institute e.V.: Bonn, Germany, 2007; Volume 8.
54. Ebertz, A. Die Wohnortwahl privater Haushalte und die Bewertung lokaler Standortfaktoren in den sächsischen Gemeinden. *Ifo Dresd. Ber.* **2008**, *15*, 14–22.
55. Voß, G.G. Kapitel I Arbeit als Grundlage menschlicher Existenz: Was ist Arbeit? Zum Problem eines allgemeinen Arbeitsbegriffs. In *Handbuch Arbeitssoziologie*; Böhle, F., Voß, G.G., Wachtler, G., Eds.; VS Verlag für Sozialwissenschaften: Wiesbaden, Germany, 2010; pp. 23–80. ISBN 978-3-531-15432-9.
56. Recktenwald, H.C. *Adam Smith: Sein Leben Und Sein Werk*; 1. Aufl.; Beck: München, Germany, 1976; ISBN 978-3-406-06298-8.
57. Burkhardt, J.; Priddat, B.P. (Eds.) *Geschichte Der Ökonomie*; Bibliothek der Geschichte und Politik; 1. Aufl.; Deutscher Klassiker Verlag: Frankfurt am Main, Germany, 2000; ISBN 978-3-618-66810-7.
58. Simons, H.; Weiden, L. Schwarmverhalten, Reurbanisierung und Suburbanisierung. *Inf. Raumentwickl.* **2016**, *11*, 263–274.
59. Westphal, C. Dichte als Planungsgröße im Stadtumbau?: Angemessene Dichten zur Gewährleistung der stadttechnischen Daseinsvorsorge in schrumpfenden Städten. *Raumforsch. Raumordn. Spat. Res. Plan.* **2009**, *67*, 7–20. [CrossRef]
60. Schiller, G.; Siedentop, S. Infrastrukturfolgekosten Der Siedlungsentwicklung Unter Schrumpfungsbedingungen. *Disp-Plan. Rev.* **2005**, *41*, 83–93. [CrossRef]
61. Rosenfeld, M.T.W. Demographischer Wandel, unternehmerische Standortentscheidungen und regionale Disparitäten der Standortentwicklung. In *Räumliche Konsequenzen des Demographischen Wandels: T. 6, Demographische Trends in Deutschland-Folgen für Städte und Regionen*; Gans, P., Schmitz-Veltin, A., Eds.; Forschungs- und Sitzungsberichte der ARL; Akademie für Raumforschung und Landesplanung—Leibniz-Forum für Raumwissenschaften: Hannover, Germany, 2006; Volume 226, pp. 65–83, ISBN 978-3-88838-055-6.
62. Ochs, B. Wohnwünsche: Auf dem Dorf Ist es am Schönsten. Available online: https://www.faz.net/1.6622444 (accessed on 9 August 2020).
63. *Kantar GmbH BHW Wohnen 2025*; Kantar GmbH: Bielefeld, Germany, 2020; p. 61.
64. Wohnwunsch: Leben in der Stadt Oder auf dem Land 2018. Available online: https://de.statista.com/statistik/daten/studie/901079/umfrage/bevorzugter-wohnort-in-der-stadt-oder-auf-dem-land/ (accessed on 5 January 2021).
65. Horx, M. Progressive Provinz: Die Neue Heimat der Glokalisten. Available online: https://www.zukunftsinstitut.de/artikel/zukunftsreport/progressive-provinz-die-neue-heimat-der-glokalisten/ (accessed on 5 January 2021).
66. Püschel, J. Prospects of a Rural Renaissance: Will the Smart Economy Compress Regional Disparities? In *Contours of the Illiberal State: Governing Circulation in the Smart Economy*; Campus Verlag: Frankfurt am Main, Germany, 2019; p. 275, ISBN 593510170.
67. Drammeh, N. Dorf der Zukunft. Available online: https://kommunal.de/dorf-ko-dorf-wiesenburg (accessed on 5 January 2021).
68. Dähner, S.; Reibstein, L.; Slupina, M.; Klingholz, R. *Urbane Dörfer: Wie Digitales Arbeiten Städter aufs Land Bringen Kann*; Berlin-Institut für Bevölkerung und Entwicklung und Neuland21 e.V.: Berlin, Germany, 2019; ISBN 978-3-946332-50-3.
69. Hildebrandt, S.; Marschall, J.; Kleinlercher, K.-M.; Nolting, H.-D. *Digitalisierung Und Homeoffie in Der Corona-Krise*; Gesundheitsreport 2020—Beiträge zur Gesundheitsökonomie und Versorgungsforschung; IGES-Institut für DAK Gesundheit: Berlin, Germany, 2020; Volume 33.
70. Alipour, J.-V.; Falck, O.; Schüller, S. *Homeoffice Während der Pandemie und die Implikationen für eine Zeit nach der Krise*; ifo Schnelldienst; Ifo Institut: München, Germany, 2020; p. 7.
71. Alipour, J.-V.; Schüller, S.; Falck, O. Germany's Capacities to Work from Home. *Cesifo Work. Pap.* **2020**, *8227*, 21.
72. Kordey, N.; Korte, W. *Telearbeit Erfolgreich Realisieren: Das Umfassende, Aktuelle Handbuch für Entscheidungsträger und Projektverantwortliche*; Springer: Berlin/Heidelberg, Germany, 2013; ISBN 3-322-86503-7.
73. Brenke, K. Home Office: Möglichkeiten werden bei weitem nicht ausgeschöpft. *Diw Wochenber.* **2016**, *83*, 95–105.
74. Borchers, D. *Mehrheit der Deutschen Angestellten Wünscht Sich Wegen des Corona-Virus Home-Office/Mehrzahl der Arbeitgeber Wäre dazu Technisch in der Lage*; Bundesverband Digitale Wirtschaft (BVDW) e.V.: Berlin, Germany, 2020.
75. Mergel, K. Stadtflucht durch Corona: Wie die Pandemie den Immobilienmarkt Verändert. Available online: https://www.merkur.de/bayern/muenchen-coronavirus-pandemie-immobilienmarkt-arbeitsleben-homeoffice-prognose-zr-90078109.html (accessed on 8 March 2021).

76. Höland, C. Stadtflucht: Wieso die Corona-Pandemie uns aufs Land Treibt. Available online: https://www.rnd.de/wirtschaft/stadtflucht-wieso-die-corona-pandemie-uns-aufs-land-treibt-I264DC3Y7ZHUVMQ5F7Q6CO4M3Y.html (accessed on 8 March 2021).
77. Saad, L. Country Living Enjoys Renewed Appeal in U.S. Available online: https://news.gallup.com/poll/328268/country-living-enjoys-renewed-appeal.aspx (accessed on 8 March 2021).
78. Roper, W. Rural Life Desire Rises in 2020. Available online: https://www.statista.com/chart/23855/rural-urban-living/ (accessed on 8 March 2021).
79. Dinkel, M. Keine Corona-Stadtflucht. Available online: https://www.immobilien-zeitung.de/158586/keine-corona-stadtflucht (accessed on 8 March 2021).
80. Fischbach, K.; Putzke, J. Wissensarbeiter—Enzyklopaedie der Wirtschaftsinformatik. Available online: https://www.enzyklopaedie-der-wirtschaftsinformatik.de/lexikon/daten-wissen/Wissensmanagement/Wissensorganisation--Instrumente-der-/Wissensarbeiter (accessed on 3 August 2020).
81. Fraunhofer-Institut für Arbeitswirtschaft und Organisation IAO Wissensarbeiter—IAO Wiki. Available online: https://wiki.iao.fraunhofer.de/index.php/Wissensarbeiter (accessed on 3 August 2020).
82. Statistik der Bundesagentur für Arbeit. *Blickpunkt Arbeitsmarkt—Akademikerinnen Und Akademiker*; Statistik der Bundesagentur für Arbeit: Nürnberg, Germany, 2019; p. 140.
83. Florida, R.L. *The Rise of the Creative Class*; Basic Book: New York, NY, USA, 2004; ISBN 978-1-5416-1774-2.
84. Krön, E. *Ressource Wissen im Bauprojekt: Ein Wissensmanagement-Prozessmodell für Bauplanungs- und Beratungsleistungen in Kleinen und Mittleren Unternehmen*; Verlag und Datenbank für Geisteswissenschaften: Weimar, Germany, 2009; ISBN 978-3-89739-623-4.
85. Pawlowsky, P. *Wissensmanagement*; Walter de Gruyter GmbH & Co KG: Berlin, Germany, 2019; ISBN 978-3-11-047493-0.
86. Toetzke, K.; Wissen, D. *Die Öffentliche Bibliothek als Nutzer und Anbieter der Virtuellen Bibliothek: Zwanzig Jahre Internet in Deutschen Öffentlichen Bibliotheken*; B.I.T. Online Verlag: Wiesbaden, Germany, 2018; ISBN 978-3-934997-92-9.
87. Körting, C. Netzwerke. In *Wirkungsmodell zu Wissensvermittlungs- und Vernetzungsprozessen in Unternehmen*; Springer: Berlin/Heidelberg, Germany, 2020; pp. 49–89. [CrossRef]
88. Healy, T.J. Transportation or Communications Some Broad Considerations. *IEEE Trans. Commun.* **1968**, *16*, 195–198. [CrossRef]
89. Holz-Rau, C.; Scheiner, J. Raum und Verkehr–ein Feld komplexer Wirkungsbeziehungen. Können Interventionen in die gebaute Umwelt klimawirksame Verkehrsemissionen wirklich senken? *Raumforsch. Und Raumordn. Spat. Res. Plan.* **2016**, *74*, 451–465. [CrossRef]
90. Huber, J. *Telearbeit*; VS Verlag für Sozialwissenschaften: Wiesbaden, Germany, 1987; ISBN 978-3-531-11849-9.
91. Mokhtarian, P.L. Defining Telecommuting. *Transp. Res. Rec.* **1991**, 273–281. Available online: http://onlinepubs.trb.org/Onlinepubs/trr/1991/1305/1305-034.pdf (accessed on 19 February 2021).
92. Gobeli, S.; Krause, A.; Schulze, H. *Ergonomische und gesundheitliche Aspekte eines Home Office Arbeitsplatzes*; Stiftung Produktive Schweiz: Zürich, Switzerland, 2011; pp. 34–37.
93. Vega, G.; Brennan, L. Isolation and Technology: The Human Disconnect. *J. Orgchange Mgmt* **2000**, *13*, 468–481. [CrossRef]
94. Kratzer, N.; Diebig, M.; Funk, M.; Henkel, C.; Kaiser, S.; Klesel, M.; Körner, U.; Kordyaka, B.; Kremer, D.; Monz, A.; et al. *Arbeit der Zukunft Digital, Multilokal, Dynamisch*; Institut für Sozialwissenschaftliche Forschung e.V.: München, Germany, 2019.
95. Tavares, A.I. *Telework and Health Effects Review, and a Research Framework Proposal*; MPRA Paper; Ludwig-Maximilians-Universität München Universitätsbibliothek: München, Germany, 2016.
96. Oldenburg, R. *The Great Good Place: Cafés, Coffee Shops, Community Centers, Beauty Parlors, General Stores, Bars, Hangouts, and How They Get You through the Day*; Paragon House Publishers: New York, NY, USA, 1989; ISBN 1-55778-110-9.
97. Avdikos, V.; Merkel, J. Supporting Open, Shared and Collaborative Workspaces and Hubs: Recent Transformations and Policy Implications. *Urban Res. Pract.* **2019**, 1–10. [CrossRef]
98. Foertsch, C.; Cagnol, R. Es War Einmal … die Geschichte von Coworking in Zahlen. Available online: http://www.deskmag.com/de/die-geschichte-von-coworking-spaces-in-zahlen-zeitleiste-868 (accessed on 15 August 2020).
99. Spinuzzi, C. Working Alone Together: Coworking as Emergent Collaborative Activity. *J. Bus. Tech. Commun.* **2012**, *26*, 399–441. [CrossRef]
100. Castilho, M.; Quandt, C. Collaborative Capability in Coworking Spaces: Convenience Sharing or Community Building? *Technol. Innov. Manag. Rev.* **2017**, *7*, 32–42. [CrossRef]
101. Merkel, J. Coworking in the City. *Ephemer. Theory Politics Organ.* **2015**, *15*, 121–139.
102. Foertsch, C. Die Mitglieder: Wer arbeitet in Coworking Spaces? Available online: https://www.deskmag.com/de/coworkers/die-mitglieder-wer-arbeitet-in-coworking-spaces-coworkers-global-survey-demografie-statistik-977 (accessed on 22 December 2020).
103. Gauger, F.; Pfnür, A.; Skarabi, J. *Arbeitswelten im Wandel: Coworking Spaces. Eine empirische Befragung der Eigenschaften und Nutzerpräferenzen von Coworking Spaces*; Arbeitspapiere zur Immobilienwirtschaftlichen Forschung und Praxis: Darmstadt, Germany, 2020.
104. Lengen, J.C.; Kordsmeyer, A.-C.; Rohwer, E.; Harth, V.; Mache, S. Soziale Isolation im Homeoffice im Kontext der COVID-19-Pandemie. *Zent. Arb. Arb. Ergon.* **2021**, *71*, 63–68. [CrossRef]
105. Robelski, S.; Keller, H.; Harth, V.; Mache, S. Coworking Spaces: The Better Home Office? A Psychosocial and Health-Related Perspective on an Emerging Work Environment. *Int. J. Environ. Res. Public Health* **2019**, *16*, 2379. [CrossRef]

106. Grömling, M.; Schäfer, H. Mobiles Arbeiten in Deutschland Und Europa Eine Auswertung Auf Basis Des European Working Conditions Survey 2015. *Iw-Trends-Vierteljahr. Empir. Wirtsch.* **2017**, *44*, 23.
107. Grevenstein, I. Aktuelle Markterhebung Zeigt Vervierfachung von Coworking-Spaces. Available online: https://www.coworking.jetzt/research/aktuelle-markterhebung-zeigt-vervierfachung-von-coworking-spaces/ (accessed on 16 August 2020).
108. Foertsch, C. 2019 State of Coworking: Over 2 Million Coworking Space Members Expected. Available online: http://www.deskmag.com/en/2019-state-of-coworking-spaces-2-million-members-growth-crisis-market-report-survey-study (accessed on 16 August 2020).
109. Foertsch, C. Der Ländliche Weg des Coworkings. Available online: http://www.deskmag.com/de/laendliche-weg-des-coworking-kleine-staedte-186 (accessed on 16 August 2020).
110. Bähr, U.; Biemann, J.; Hentschel, P.; Lietzau, J. Coworking im ländlichen Raum: Menschen, Modelle, Trends. 2020. Available online: https://coworkinglibrary.com/publication/coworking-im-landlichen-raum/ (accessed on 19 February 2021).
111. Gill, S. Dorfbüros RLP. 2021. Available online: https://dorfbueros-rlp.de/ (accessed on 5 March 2021).
112. Wirtschaftsförderung Land Brandenburg GmbH Co-Working Spaces. Available online: https://gruendung.wfbb.de/en/node/14372 (accessed on 5 March 2021).
113. Nobis, C.; Kuhnimhof, T. *Mobilität in Deutschland—MiD Ergebnisbericht*; Mobilität in Deutschland—MiD; Infas, DLR, IVT infas360, Bundesministerium für Verkehr und Digitale Infrastruktur: Bonn, Germany; Berlin, Germany, 2019; p. 136.
114. Pütz, T. Immer Mehr Beschäftigte Pendeln. Available online: https://www.bbsr.bund.de/BBSR/DE/Home/Topthemen/pendeln.html (accessed on 30 March 2020).
115. Pütz, T. *Verkehrsbild Deutschland—Pendlerströme. Quo Navigant?* Bundesinstitut für Bau-, Stadt- und Raumforschung (BBSR): Bonn, Germany, 2015; p. 20.
116. Dauth, W.; Haller, P. *Berufliches Pendeln Zwischen Wohn- und Arbeitsort: Klarer Trend zu Längeren Pendeldistanzen*; IAB-Kurzbericht; Institut für Arbeitsmarkt- und Berufsforschung (IAB) der Bundesagentur für Arbeit: Nürnberg, Germany, 2018; p. 12.
117. Denzinger, S.; Vogt, W. Datenautobahn statt Autobahn: Löst Telearbeit Verkehrsprobleme? *Neue MedienRaum Verk.* **2000**, 205–224. [CrossRef]
118. Kittler, W. Beeinflussung der Zeitwahl von ÖPNV-Nutzern. Ph.D. Thesis, Technische Universität Darmstadt, Darmstadt, Germany, 2010.
119. Kleemann, F.; Westerheide, J.; Matuschek, I. Arbeit und Leben: Wechselwirkungen und Entgrenzung von Erwerbs- und Privatsphäre. *Arb. Und Subj.* **2019**, 135–164. [CrossRef]
120. Vogt, W.; Lenz, M. Online-Shopping von Lebensmitteln: Ersatz oder Ergänzung physischer Einkäufe? Ersatz oder Ergänzung physischen Verkehrs? *B2c Elektron. Handel Eine Inventur* **2003**, 246–264. [CrossRef]
121. Hilpert, M.; Völkening, N.; Beck, C. Innenstädte im ländlichen Raum. *Standort* **2018**, *42*, 111–116. [CrossRef]
122. Umweltbundesamt Umwelt- und Aufenthaltsqualität in Urbanen Quartieren Empfehlungen Zum Umgang Mit Dichte und Nutzungsmischung 2017. Available online: https://www.umweltbundesamt.de/publikationen/umwelt-aufenthaltsqualitaet-in-urbanen-quartieren (accessed on 17 August 2020).
123. Ulm, H. Tote Hose Innenstadt? *Das Problem der Verödenden Ortskerne in Städten und Dörfern des Ländlichen Raums.* Available online: http://fgg-erlangen.de/fgg/ojs/index.php/mfgg/article/view/53 (accessed on 19 February 2021).
124. Küpper, P. *Auf dem Weg zu Einem Grundangebot von Mobilität in ländlichen Räumen: Probleme, Ursachen und Handlungsoptionen*; Verlag der ARL—Akademie für Raumforschung und Landesplanung: Hannover, Germany, 2011; pp. 152–168, ISBN 978-3-88838-371-7.
125. Lehmphul, K. Wegevergleich: Von Tür zu Tür im Stadtverkehr. Available online: https://www.umweltbundesamt.de/bild/wegevergleich-von-tuer-zu-tuer-im-stadtverkehr (accessed on 21 December 2020).
126. Randelhoff, M.; Reisezeitunterschiede unterschiedlicher Verkehrsarten von Tür zu Tür im Stadtverkehr–Realität und subjektive Wahrnehmungsverzerrung. Zukunft Mobilität. Available online: https://www.zukunft-mobilitaet.net/167997/analyse/tuer-zu-tuer-reisezeit-stadtverkehr-pkw-miv-oepnv-radverkehr-pedelec-gleichheit-subjektive-verzerrung/ (accessed on 21 December 2020).
127. Schier, M. Multilokaler Alltag beruflich mobiler Eltern– (k)ein Handlungsfeld für die betriebliche Gestaltung. In *Mobile Arbeit–Gute Arbeit?* Brandt, C., Ed.; Ver.di: Berlin, Germany, 2010; pp. 101–116.
128. Danielzyk, R.; Dittrich-Wesbuer, A.; Hilti, N.; Tippel, C. *Multilokale Lebensführungen Und Räumliche Entwicklungen: Ein Kompendium*; Forschungsberichte der ARL; ARL–Akademie für Raumentwicklung in der Leibniz-Gemeinschaft: Hannover, Germany, 2020; ISBN 3-88838-097-9.
129. Fuzi, A. Space for Creative and Entrepreneurial Activities? Coworking Spaces in the Entrepreneurial Landscape of an Economically Challenged Region. Ph.D.Thesis, Cardiff Metropolitan University, Cardiff, UK, 2016.
130. Völker, E.J. Akustische Anforderung an Großraumbüros Die Notwendigkeit von Überdeckendem Dauergeräusch. *Frequenz* **1972**, *26*, 78–82. [CrossRef]
131. Stadler, S. *Open Space Büros Eine Studie Über Die Machbarkeit Und Umsetzung von Offenen Bürostrukturen*; Projekt Hans-Böckler-Stiftung: Düsseldorf, Germany, 2011; p. 21.
132. Windlinger, L.; Zäch, N. Wahrnehmungen von Belastungen Und Wohlbefinden Bei Unterschiedlichen Büroformen. *Z. Arb.* **2007**, *61*, 77–85.
133. Kim, J.; de Dear, R. Workspace Satisfaction: The Privacy-Communication Trade-off in Open-Plan Offices. *J. Environ. Psychol.* **2013**, *36*, 18–26. [CrossRef]

134. Bundesamt für Bauwesen und Raumordnung Laufende Stadtbeobachtung-Stadt-und Gemeindetypen in Deutschland. Available online: https://www.bbsr.bund.de/BBSR/DE/forschung/raumbeobachtung/Raumabgrenzungen/deutschland/gemeinden/StadtGemeindetyp/StadtGemeindetyp.html?nn=2544954 (accessed on 31 August 2020).
135. Bevölkerung. Available online: http://www.statistik.rlp.de/de/regional/geowebdienste/bevoelkerung/ (accessed on 10 December 2020).
136. Laufende Raumbeobachtung-Raumabgrenzungen. Available online: https://www.bbsr.bund.de/BBSR/DE/forschung/raumbeobachtung/Raumabgrenzungen/deutschland/kreise/siedlungsstrukturelle-kreistypen/kreistypen.html (accessed on 4 March 2021).
137. Regeln zum Corona-Virus. Available online: https://www.bundesregierung.de/breg-de/leichte-sprache/22-maerz-2020-regeln-zum-corona-virus-1733310 (accessed on 13 December 2020).
138. Jurczok, F. *Fahrrad-Monitor 2017 Ergänzung*; sinus-Markt-und Sozialforschung: Heidelberg, Germany, 2018; p. 67.

land

MDPI

Article

The Integration of New-Type Urbanization and Rural Revitalization Strategies in China: Origin, Reality and Future Trends

Mingxing Chen [1,2,3], Yuan Zhou [1,2,3], Xinrong Huang [1,2,3] and Chao Ye [4,*]

1 Institute of Geographic Sciences and Natural Resources Research, Chinese Academy of Sciences, 11A, Datun Road, Chaoyang District, Beijing 100101, China; chenmx@igsnrr.ac.cn (M.C.); zhouy.19b@igsnrr.ac.cn (Y.Z.); huangsr.19s@igsnrr.ac.cn (X.H.)
2 Key Laboratory of Regional Sustainable Development Modeling, Chinese Academy of Sciences, Beijing 100101, China
3 College of Resource and Environment, University of Chinese Academy of Sciences, Beijing 100049, China
4 College of Geographic Sciences, East China Normal University, Shanghai 200241, China
* Correspondence: cye@geo.ecnu.edu.cn

Abstract: New-type urbanization and rural revitalization have gradually become national strategies, and are an objective requirement for China to be able to enter into a new era of socialism with Chinese characteristics and also an inevitable result of the integration of new-type urbanization and rural development in the new stage. This paper reviews the classic theories and cognition of the research on urban–rural relations at home and abroad, and outlines the stage evolution characteristics of urban–rural relations in China. It is believed that urban-biased urbanization has widened the development gap between urban and rural areas since reform and opening up. Under the guidance of the two strategies of new-type urbanization and rural revitalization, urban and rural areas have transitioned from "one-way flow" to "bilateral interaction", and from "urban bias" to "urban–rural integration". This paper puts forward a research framework and scientific issues regarding the integration of new-type urbanization and rural revitalization from multidisciplinary perspectives. The integration of these two major strategies will contribute to a new situation of the coordinated and high-quality development of urban and rural areas in the new era.

Keywords: new-type urbanization; rural revitalization; urban bias; integration; urban–rural relation; research framework

Citation: Chen, M.; Zhou, Y.; Huang, X.; Ye, C. The Integration of New-Type Urbanization and Rural Revitalization Strategies in China: Origin, Reality and Future Trends. *Land* **2021**, *10*, 207. https://doi.org/10.3390/land10020207

Academic Editor: Luca Salvati

Received: 11 January 2021
Accepted: 10 February 2021
Published: 18 February 2021

1. Introduction

Urbanization and rural development are not only the focus of multidisciplinary research at home and abroad, but also a major strategic issue related to the national economy and people's livelihoods [1]. As a complex process of transition from rural areas to urban cities on a regional scale, urbanization involves rural development. Particularly for China, which is a large developing country based on agriculture for a long time, conforming to the law of urbanization development in the world and simultaneously addressing the practical problems of rural revitalization have become an objective need for China, which has entered a new era and high-quality stage [2–5]. Urbanization and rural revitalization have both become national strategies, which also reflects the inevitability of the integration of these two strategies [6,7]. Unbalanced urban–rural development and inadequate rural development are the key issues in the new era of Chinese society [3,8]. As new-type urbanization and rural revitalization have become national strategies in succession, China is attaching great importance to the coordinated development of urban and rural areas [9]. However, how to combine these two strategies and guide further work awaits to be studied further. A literature review method and comparative analysis are used in this article. With the literature review method, the article summarizes the classic recognition of urbanization and rural relations. With the comparative analysis, the article describes the particularity of urban and rural development in China. This paper reviews the classic recognition of urbanization and rural relations at home and abroad, and outlines the key events related to evolution of urban–rural relations and their impacts on urban–rural relations based on the development history of China, then analyzing the problems of urbanization and rural development since reform and opening up. It is believed that the urban-biased urbanization is an important factor that caused the gap between urban and rural development, and new urbanization will help to reverse the problems existing as a result of the past development. The integration of new-type urbanization and a rural revitalization strategy will further promote the interaction and coupling of urbanization and rural revitalization. From the view of multi-disciplinarity, this paper puts forward the scientific issues of the integration of new-type urbanization and rural revitalization, and constructs a research framework for the integrated development of these two strategies.

2. Historical Evolution of Urbanization and Rural Relations at Home and Abroad

Clarifying the historical evolution and logical relationship between urbanization and rural revitalization strategies at home and abroad, as well as the impact of policy practice, so as to realize the interaction and coupling of the two strategies, and finally achieving collaborative governance, is crucial.

2.1. Overseas Research

Rural development is the inevitable connotation of urbanization. Adam Smith, the originator of classical economics, put forward the "natural order". He believed that there were villages first and then cities; cities came from rural development, and the wealth and poverty of a country mainly depended on the historical geography, institutional culture and other elements of urban and rural development [10]. Influenced by this, the German economic geographer Von Thünen [11], regards urban and rural areas as a whole in his "isolated country" model and explores the law of spatial distribution for different industries between urban and rural areas.

Western urban research can be traced back to the ancient Greek city-state theory where the ancient Greek combined the construction of the city-state with politics and civilization [12,13]. Plato's utopia is the earliest utopia in human history [14]. The proponents of Utopia strongly advocate the integration and further design of urban and rural areas [15]. Subsequently, garden cities, urban agglomeration areas, organic evacuation and other theories also emphasize the integration of urban and rural development [16–19].

From 1940 to 1980, the "dual structure" paradigm dominated the study of urban–rural relations in the West. Due to the one-sided emphasis on urban and industrial development resulting in the development of rural agriculture lagging behind [20], the "dual structure" gave birth to two theoretical schools of urban and rural development. One is the Jorgenson model and Todaro Migration Model developed by reflecting on the Lewis Model [21,22]. The other is the theoretical model of urban and rural spatial polarization development represented by growth poles and a core periphery relationship [23]. Lipton believes that many countries divide urban and rural areas into two classes and implement "urban bias" policies, which lead to serious imbalances in urban and rural development [24,25]. The theory of unbalanced development between urban and rural areas dominated Western academic circles at that time.

After the 1980s, more scholars began to advocate urban–rural connection and coordinated development. Some geographers took the lead in challenging the trend of urban–rural division [26,27], and later, more international studies focused on coordinated development [28–30] and some new concepts were put forward. The pulling force of cities and the pushing force of rural areas in developing countries make cities and villages, and agricultural activities and non-agricultural activities, closely linked, and blur urban–rural boundaries [31]. The regional network development model holds that urban and rural areas generate stronger communication and networks through a series of "flows", thus promoting the integration of urban and rural development [32]. In terms of politics, the economy, society and culture, the interaction and association of urban and rural form an "urban–rural continuum" [33–36]. Based on the five aspects of "food flow, resource flow, people flow, concept flow and capital flow", the interaction between urban and rural areas in developing countries is complex, and the "urban and rural dynamics" can reveal the complexity of urban and rural connection from the point of view of "livelihood strategy" and "resource allocation" [37]. Urbanization is transforming the planet, within and beyond cities, at all spatial scales [38]. The rural area (or countryside or the hinterland) has become key to the process of capitalist urbanization [39,40]. There are also many advanced practices in rural construction in Western countries, such as the "Bavaria experiment" in Germany, "New Village Movement" in South Korea and "city–village merger" in Japan. Taking the Bavaria experiment in Germany as an example, after World War II, the gap between urban and rural areas was further widened, and rural problems were prominent. Under the concept of urban–rural equivalence, district planning, land integration, agricultural mechanization, infrastructure construction, education development and other measures were started in rural areas, to make the rural and urban areas different in type but same in quality. The action is still popular in Germany for making the rural areas better [41].

With the evolution of the urban–rural relation theory, urban–rural governance has roughly gone through the process of the "co-governance of rural and urban, sub-governance of urban and rural, co-governance of urban and rural" [42]. After the industrialization of societies has begun, the gap between urban and rural development has increased, so different management methods have been adopted in urban and rural areas, that is, a separation of the administration of the urban and rural. Western countries have also experienced a process of change from urban–rural dual opposition to urban–rural integrated governance [43]. Since the 1990s, rural construction and governance have gradually become a research hotspot [44–50]. Such issues as rurality, the revitalization of the world's rural areas, the future of rural areas, the relationships between climate change and rural evolution, and food security have become the focuses of research [51–56]. The entrepreneurship of farmers, the development of rural finance and multicommunity cooperation are regarded as the keys to rural revitalization [12,57]. In different countries, governments, other organizations and volunteers play important roles [58–62]. The evolution of urban–rural relation theory reflects the trend of change from attaching extremely great importance to cities to attaching equal importance to both urban and rural areas.

2.2. Domestic Research

Research on urbanization and rural development in China has generally exhibited a shift from an emphasis on rural development to an emphasis on urban development in the modern era and, to date, has developed to achieve the overall planning of urban and rural areas through the integration of the two strategies of new urbanization and rural revitalization (Table 1).

Agriculture was the foundation of society in ancient China. It was a long-term economic policy to emphasize agriculture and suppress commerce. Instead of resulting in a conflict between urban and rural areas, urbanization was based on and even oriented to rural areas, and free communication between urban and rural areas formed the integration of urban and rural areas [63]. However, the Opium War broke this integration in China during that historical period, causing opposition between urban and rural areas gradually. The invasion of capitalism has changed both the city and the countryside. The development of modern industry and commerce has made the city more prosperous than before, and it needed a lot of cheap labor, which led to the continuous migration of the rural population to the city. The urban–rural relations and the dependence of rural areas on cities have been strengthened, and the urban control and exploitation of rural areas have led to the decline of rural areas. A large-scale rural construction movement began to rise, with representatives of Yan Yang-chu, Liang Shu-ming, Lu Zuo-fu, etc. The movement focuses on the rural economy, rural education, the rural environment and rural transportation [21,64–66].

Table 1. Main stages and characteristics of China's urban–rural relations and urbanization evolution.

Historical Period	The Western Zhou Dynasty	From the Eastern Zhou Dynasty to 1840	From 1840 to 1949	From1949 to 1977	From 1978 to 1999	Since 2000
Urban–rural relations	Urban–rural opposition	Urban–rural coordination	Urban–rural opposition	Urban–rural segregation	Reverse of urban–rural segregation	Tendency to urban–rural integration
Urbanization model	Initial stage	Rapid development to stability	Urban development and rural decline	Stagnant urbanization	Rapid and large-scale	New-type urbanization
Urbanization system	Divided land into the residences of dignitaries and the cultivated land of civilians	Taxed and managed by household registration system	Dual structure	Dual structure	Urban bias	Urban–rural integration
Core systems and strategies	Social space of urban and rural; focus on rural social culture	Improving household registration system; joint and several forms of management and punishment	Modern town system: police system	Household registration and welfare system of urban–rural segregation; state monopoly	Efficiency priority; urban-biased construction; differentiated urban–rural public services	Construction of new countryside; new-type urbanization; rural revitalization
Peripheral system and culture	Echelons of administration; foreign policy of both conciliation and control	Prefectures and counties system and administrative province system; policy of promoting agriculture and discouraging commerce	The rise of pro-business culture	Closed economy and planned economy; strategy and ideology of heavy industry	Preliminary socialist market economy system; gradually liberalized urban–rural mobility	Socialist market economy system; gradually liberalizing household registration system; duality system tends to become flexible
Key points	Inequality between urban and rural areas under hierarchy	Taxes and land are separated from household registration system in later period; the strengthening of grassroots social control	Changing the structure of traditional Chinese urbanization; the focus of urbanization changes from regions south of the Yangtze river to coastal areas	Urban–rural dual system decided by regulations on household registration in 1958; unequal rights of urban and rural residents; restricted the migration between urban and rural areas	Urban–rural flow has improved the income level of farmers to some extent, but the absolute income gap between urban and rural areas has been expanding	Gradually establish and improve the system and mechanism as well as policy system of urban–rural integration development; promote the equalization of urban and rural basic public services

Since the 1950s, as a series of policies of industrial priority and urban bias have been adopted, the gap between urban and rural areas has been widening and resulted in the formation of an urban–rural dual structure [67]. The strategy of heavy industry priority under the planned economy promoted industrialization at the expense of agriculture and farmers, which made the urban–rural relations extremely unbalanced [68]. A strict household registration system and a series of economic and social systems derived from it, such as the urban welfare security system, urban–rural dual ownership system, dual citizenship system, and dual exchange and distribution system, led to the deepening of urban–rural isolation and the solidification of the urban–rural dual structure [69]. Since 2000, with the socialist new rural strategy, new-type urbanization strategy and rural revitalization strategy put forward in succession, urban and rural areas are showing a new trend of integrated development.

3. The Reality of Urbanization and Rural Development in China since Reform and Opening up

3.1. The Essential Fact of Urbanization and Rural Development in China

China has witnessed rapid urbanization in the past 40 years or so. The urbanization was 18.92% in 1978 and reached 59.59% in 2018. In 1978, there were only a 17.254 million urban population and 79.014 million rural population. However, there are now an 83.137 million urban population and 56.401 million rural population (Figure 1). In the past 40 years or so, China has made great progress in urbanization, which may have taken about a hundred years for some Western countries.

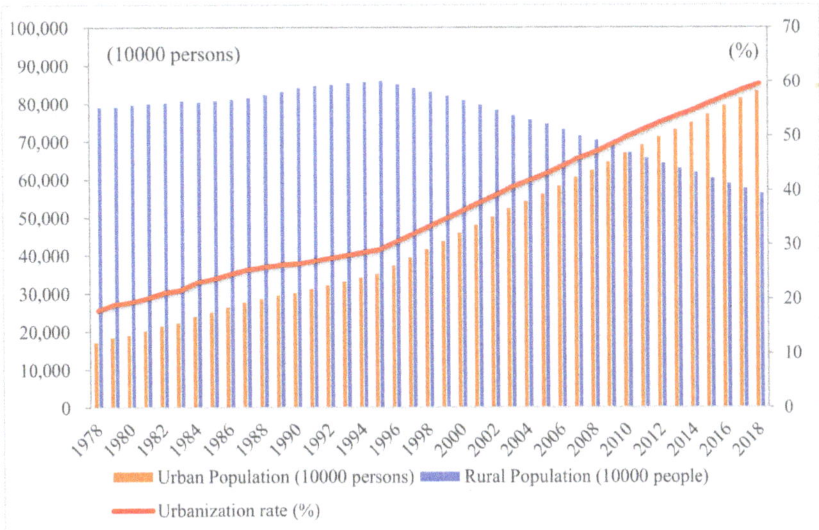

Figure 1. Changes in urban and rural population distributions in China. Source: China Statistical Yearbook 1979–2019.

The various geographical conditions in China result in the differences in population distribution patterns between the East and West, which present as "more in the East and less in the West". There are three steps in the terrain of China, which bring about three natural regions: the Eastern monsoon climate region, northwest arid and semiarid region and Qinghai Tibet alpine region. To a certain extent, the natural conditions determine the basic pattern of China's population distribution and urbanization development. Hu Huanyong, a famous geographer in China, put forward the important dividing line of population geography in 1935. Since Hu Huanyong proposed the line, the macropattern of the population distribution has been basically stable. The population on the western side of the "Hu line" is still less, and that on the eastern side is more (Figure 2).

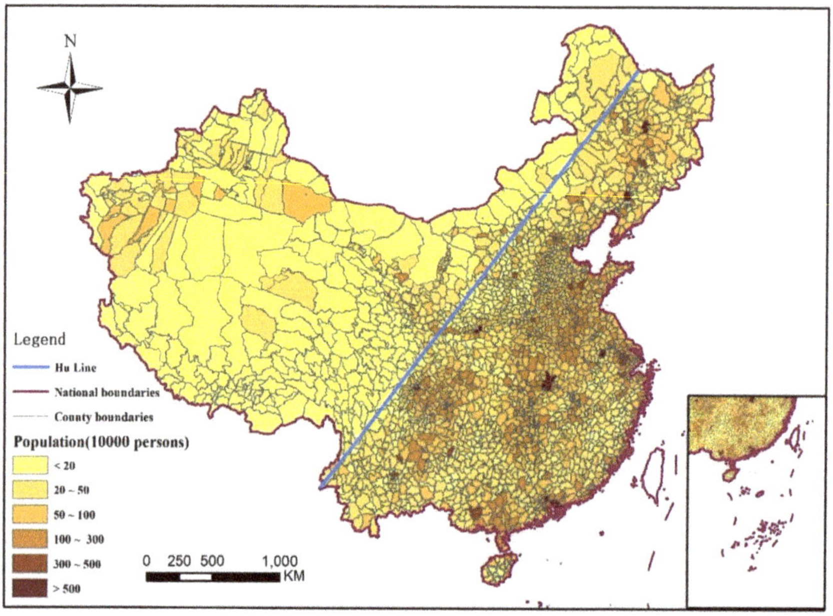

Figure 2. Population distribution of China in 2015. Source: 1% National Population Sample Survey 2015.

China's government has taken various measures to promote urban and rural development recently. Especially in rural construction, targeted poverty alleviation and creating beautiful countryside are two typical measures. Targeted poverty alleviation aims to make all destitute households lifted out of poverty and backwardness. It includes providing jobs, various subsidies, serious illness insurance, etc., to protect the basic lives of them, and small loans, industrial poverty alleviation, and relocation to solve permanent poverty. The construction of beautiful countryside is to allow the exploration of special resources around big cities. Fresh air, leisure life and rural life are known to attract urban tourists, which contributes to farmers' income through the linkage of agriculture and tourism.

3.2. *Urban-Biased Urbanization Widens the Gap Between Urban and Rural Development*

Since reform and opening-up, with the rapid development of urbanization, the gap between urban and rural areas has widened, and the imbalance between rural and urban development has become prominent (Figure 3). The per capita disposable income of urban residents increased from CNY 343.4 in 1978 to CNY 36,396.2 in 2017, while that of rural residents increased from CNY 133.6 to 13,432.4 during the same period. The per capita disposable income of urban and rural residents has increased significantly, but there is still a large gap. The per capita disposable income ratio of urban residents and rural residents fluctuates constantly. The average value of the ratio of disposable income from 1978 to 2017 was 2.81, while the minimum value was 1.86, and the maximum value was 3.14.

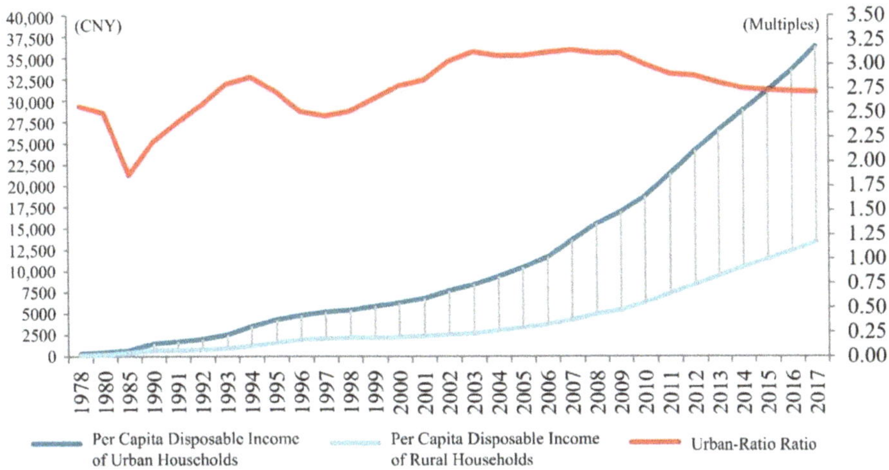

Figure 3. Per capita income gap between urban and rural residents in China (1978–2017). Note: Per capita disposable income is based on current prices. Source: China Statistical Yearbook 1979–2018.

With the increasing inequality between urban and rural areas, the coordinated development of urban and rural areas has become a research hotspot [5,70–73]. The driving forces of population migration in urban and rural areas mainly include the income gap between urban and rural areas, a surplus agricultural labor force, the development of township enterprises, the disintegration of state-owned enterprises and the emergence of private enterprises [74–76]. A series of urban-biased policies in the process of urbanization make cities gather a lot of resources in a short time, but lead to the decline of rural areas [1,5,77] and pose a serious challenge to the overall planning of urban and rural areas. The urban-biased urbanization has led to the expansion of the gap between urban and rural development [78]. For example, the per capita consumption level of urban residents is higher than that of rural residents. In 2017, the per capita consumption expenditure of urban residents was CNY 24,445, while that of rural residents was CNY 10,954.5, with the former being 2.2 times the latter. The total investment in urban fixed assets was CNY 63,168.4 billion, and that in rural households was CNY 955.44 billion, in 2017, accounting for 98.5% and 1.5% of the total investment in social fixed assets, respectively. There is also a gap in the quantity and level of basic public service supply between urban and rural areas; for example, in 2017, the number of health technicians and the number of beds in medical and health institutions per thousand population in urban areas were 10.87 and 8.75, respectively. By comparison, the numbers of health technicians and beds per thousand population in rural areas were 4.28 and 4.19, respectively. The numbers of health technicians and beds in urban areas were 2.54 and 2.01 times those in rural areas, respectively, and that indicates that high-quality medical and educational resources are mostly concentrated in urban areas.

3.3. The Gap between Urban and Rural Development Has Tended to Narrow to Some Extent in Recent Years

In 2017, the per capita GDP of China reached CNY 59,660, and the national urbanization rate reached 58.52%, while the urban–rural income gap reached 2.71 times and was higher than the level at the beginning of reform and opening up [79]. However, in recent years, the gap between urban and rural development has been narrowing [80,81], especially in the past decade; the ratio of the per capita disposable income of urban and rural residents decreased from 3.14 in 2007 to 2.71 in 2017. The new era is a key historic period for the transformation of the social principal contradictions, which are the key issues to be faced and solved in the new era of China [82]. The unbalanced and inadequate development in the new era is reflected in many aspects, but it is mainly reflected in the unbalanced development of urban and rural areas, and the biggest deficiency is the inadequate development of rural areas [83]. The multiples of the urban–rural income gap in China have been in the range of 2–3 for a long time, while those in most countries are below 1.5 [84]. In the new era, "urban and rural China" must change from one-way urbanization to interactive development between urban and rural areas [85–87]. To promote the integrated development of urban and rural areas, we should not only promote the development of rural society and the rural economy, but also strengthen urban development [88]. The continuous development of urbanization and urban innovation are important driving forces for the continuous development of rural revitalization. It is no accident that new-type urbanization and a rural revitalization strategy are being put forward in succession. The coupling of these two strategies is the key for solving the imbalance of urban and rural development in the new era.

4. The Prospect of the Integration of New-Type Urbanization and Rural Revitalization

4.1. Prediction of the Trend of New-Type Urbanization and Rural Revitalization in the New Era

There is often a misunderstanding that urbanization is urban development. In fact, urbanization is a regional process including urban and rural areas and a regional spatial change process in which the labor force, population, land and other elements in rural areas transfer or change to urban areas [89], and it is not a simple urban development problem. There is no practical significance to simply saying that the urbanization level of any urban built-up area is 100%. Therefore, urban–rural relations are essentially one of the important elements of urbanization. New-type urbanization requires the integrated development of urban and rural areas [6,8]. Whether the relations between urban and rural areas are isolated or interactive is related to the long-term development of the country and region. The proposal of the Rural Revitalization Strategy is the best interpretation of the changes in the principal contradictions in the new era. It is a reflection on the past urban bias and rural decline in the process of urbanization and a strategy for dealing with them [1]. It is necessary to promote a collaborative approach involving the government and residents [90], establish an integrated land-use policy framework, formulate and implement effective land-use policies, regulate the process of the conversion of farmland to nonagricultural use, and improve the efficiency of land allocation for urban and rural construction to realize the coordination of land use with the stage of economic development [1,4,91–94]. The weakness of rural development in the process of urbanization should be seized to promote rural revitalization and the high-quality development of new-type urbanization with urban–rural integration [6]. The main trends of new-type urbanization and rural revitalization in the future are predicted to be as follows:

A transition from urban-biased urbanization with the one-way flow of rural–urban to new-type urbanization with a two-way interaction of urban and rural. The "push–pull model" has generally been used to describe the process of rural–urban in the past—that is, the pushing force of rural areas and the pulling force of urban areas, which together lead to a large-scale population migration from rural to urban areas—and it was generally considered as a one-way flow of population from rural to urban areas. With the integration and development of new-type urbanization and rural revitalization, there should be a two-way positive interaction between urban and rural areas, with various resources and elements flowing freely between urban and rural areas.

Urban–rural integration is not only an important trend of the middle–last period of urbanization in China, but also the inherent demand of Chinese traditional culture and the inevitable demand of socialist modernization. The period from 2020 to 2035 is the key period for China to basically realize socialist modernization in, and the gaps between urban and rural areas are one important index of it.

The absolute gap between urban and rural will still exist for a period in the future, but the relative gap between urban and rural is expected to slow-down in growth in

China. Recently, China has made historic achievements in targeted poverty alleviation. China manages to achieve the goal of getting rid of poverty and building an overall well-off society by 2020. It can be expected that with the integrated development of new-type urbanization and a rural revitalization strategy, the gap in basic public services and living standards between urban and rural will continue to narrow.

The urban–rural gap between the east and the west of the "Hu Line" will remain at a high level, but the relative gaps between urban and rural in the east and the west will be narrowed, respectively. It is predicted that the stability of the "Hu Line" will continue to exist, and the distribution pattern of the large populations and the difference in economic development levels in China will not be fundamentally changed in the short term [95]. However, the relative gap in social development and basic public service levels in urban and rural areas between the east and the west of the "Hu Line" will be narrowed [96].

Strengthening the coordinated governance of urban and rural areas is an internal requirement of the integrated development of new-type urbanization and rural revitalization, which is also the key to achieving the integrated development and implementation of these two strategies. Coordinated urban–rural governance mainly includes the free flow of various resources and elements of urban–rural development; the linkage of urban and rural economic industries; the integrated development of primary, secondary and tertiary industries; the integration of urban–rural spatial planning and infrastructure layout [1]; the organic integration of urban and rural development communities; and the renewal of urban and rural communities, with local characteristics as the core.

4.2. The Main Challenges of New-Type Urbanization and Rural Revitalization in the New Era

The strategic coupling and coordinated management of new-type urbanization and rural revitalization is the implementation of "people-oriented" development and the only way to urban–rural integration. The review of overseas research, and the experiences and evolution of urban–rural relations and urbanization processes in Western countries are valuable, and have reference significance to China. However, it is not appropriate for China to adopt Western models completely because there are many differences in urbanization processes between China and developed countries. First, the urbanization in developed countries is gradual, with the evolution of industrialization over a long period of time, and the urbanization of the population, the urbanization of land and industrialization are synchronous. By contrast, before reform and opening up, there was mostly no process of urbanization in China. Then, urbanization experienced a high speed and large scale after reform and opening up. This means that cities in China need to receive large-scale populations from rural areas within a short time and face a rapid increase in the urban public services required by the urban population. The fact that the urbanization of the population, the urbanization of land and industrialization are not synchronous is resulting in the appearance of peri-urbanization. Second, population mobility and settlement in cities face different extents of difficulty in China. The former is related to economic conditions and own needs. The latter is limited because settlement in cities needs to meet the requirements of the household registration systems in different cities, especially in megacities. Except for the roles that government play in the process of urbanization, the national conditions, systems and demand in Western countries and China are different. In such a unique environment, there are many problems brought about by rapid urbanization. The traditional urbanization mode of China shows the important characteristics of "large-scale spatial production", "peri-urbanization", "complexity of multiple factors" and "serious urban–rural isolation", some of which hinder the healthy development of urbanization. Thus, taking Western experiences and models as reference, China should change the urbanization mode considering individual characteristics and needs. The traditional urbanization model, in which space production is taken as the core and population and industrialization are considered as playing leading roles, must be changed. Further urbanization should promote and implement new-type urbanization and a rural revitalization strategy with good human life, local culture, social justice and civil rights, being "people-oriented" at its core and gradually forming a mode of new-type urbanization with Chinese characteristics and rural revitalization.

The principal contradiction facing China in the new era has changed into the contradiction between unbalanced and inadequate development and the people's ever-growing needs for a better life. The contradiction is mainly concentrated in the agricultural and rural areas, so the evolution of urban–rural relations determines the overall change in social contradictions. Due to the existence of historical laws and geographical inertia, there will be the following major challenges and trends in urban–rural relations in China by 2050:

The geographical characteristic of "one side adjacent to the sea" will still be the main factor for the regional imbalance of urban–rural relations in China. Because most of the coastal areas in the world are still in periods of rapid development, the advantages of foreign trade and the urbanization of coastal areas in China will continue, while the inland areas are still facing the constraints of water resources and development paths, so it will be difficult to change the situation of the imbalance in development between the eastern and the western areas in the short term.

The unbalanced and insufficient development in space will continue to embody the characteristics of the "Hu Line" as the geographical boundary. The "Hu Line" not only reflects the long-term demographic geographical pattern of China, but also profoundly reflects the spatial differences in society, the economy, culture and other aspects in China. The east and west of this line have historical inevitability and realistic stability. Therefore, breaking through the current situation of the spatial imbalance of the "Hu Line" will be a long-term process.

The exploration of development paths and transformation driven by innovation. It is crucial to have talent attraction and technological innovation, as well as the transformation of the development mode. In particular, great importance should be attached to the impact of science and technology on the urban–rural relations and regional development. With the continuous promotion of information technology, the Internet has greatly changed the spatial distance of urban–rural areas and different regions, which provides a basic guarantee for the development of emerging industries in rural and inland areas, promoting rural revitalization and inland development into a new era of diversified development.

Deepening the reform of the system and mechanism is the key measure. The proposal of "Rural Revitalization" in the new era has brought important strategic advantages for the integration of urban and rural development, and agricultural and rural areas are even at an unprecedented national strategic height. China will accelerate the reform of the system and mechanism for the integrated development of urban and rural areas to achieve balanced and full development. In July 2014, the State Council agreed to establish an inter-ministerial joint meeting system to promote new-type urbanization with The Letter of State [2014] No. 86, which requires that under the leadership of the State Council, the National Development and Reform Commission should take the lead in comprehensively promoting the implementation of the national new-type urbanization plan and the policy formulation (Figure 4). The State Council held six meetings to summarize and deploy the key tasks of the annual work, which played an important role in effectively promoting the implementation of the national new-type urbanization plan and coordinating and solving major problems. In July 2019, the State Council further agreed to establish an interministerial joint conference system for urbanization and urban–rural integration development, which is the system design and guarantee expected to play an important role in accelerating the high-quality development of urbanization and urban–rural integration. In addition, the evolutionary history of urban–rural relations is basically consistent with the changes in Chinese history and culture, so an urban–rural integration system and cultural system with Chinese characteristics are important parts of the overall development of urban and rural areas.

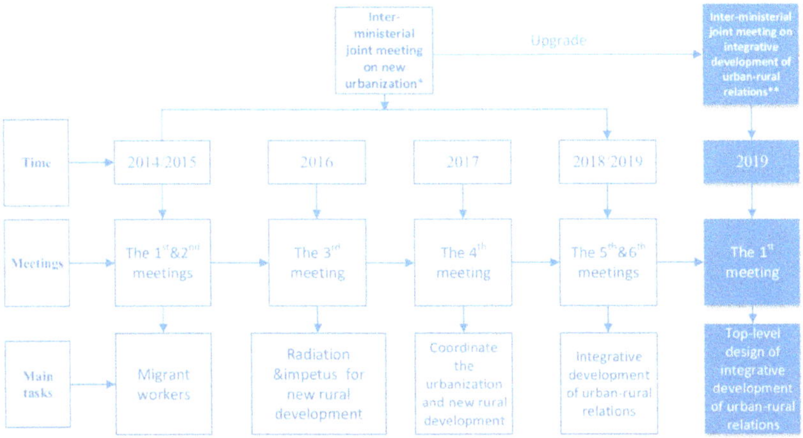

Figure 4. Evolution of inter-ministerial joint meeting on urbanization and integrative development of urban–rural relations (collected from relevant reports).

5. Discussion

The relations between new-type urbanization and rural revitalization are symbiotic. The current research is still insufficient; the discussion hopes to build the following research framework to provide new perspectives and ideas for later researchers. Research on the strategic coupling and collaborative path of these two strategies will help to solve the "three rural" problem and change the development mode of traditional urbanization, thus further realizing the overall development of urban and rural areas. Different disciplines and perspectives have important reference significance and provide inspiration for a comprehensive and profound understanding of the strategic coupling of new-type urbanization and rural revitalization (Figure 5).

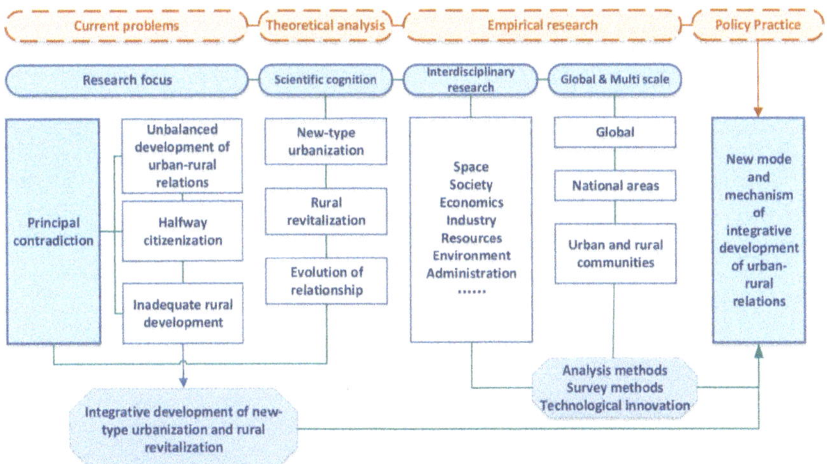

Figure 5. Research framework for integrated development of urbanization and rural revitalization strategy.

Build a research framework for the integration of new-type urbanization and rural revitalization. Research on urbanization and rural research tend to be separated; there is less on the integration of the two strategies. There have been more review studies since the founding of the People's Republic of China and reform and opening up, but fewer prospective studies combined with the background and characteristics of the new era. Therefore, based on the new era, putting forward a strategic coupling research framework for new-type urbanization and rural revitalization, and constructing a new model of collaborative governance, could address the gap in this research.

Establish a comprehensive research paradigm with an interdisciplinary vision. Although many studies focus on urban–rural relations at present, there are often different positions and perspectives between different disciplines, with a lack of dialogue, intersection and integration among the disciplines. Therefore, it is necessary to break the boundaries of disciplines in knowledge or epistemology and conduct interdisciplinary and comprehensive research oriented by problems. The understanding of the requirements of the high-quality development of China in the new era should be deepened, and comprehensive research in multiple dimensions such as space, society, the economy, industry, the resource environment and systems should be carried out.

Focus on the global perspective and multiscale integration of research. It is necessary to carry out solid research systematically comparing and summarizing the theoretical and historical basis of the differences in urban–rural relations between China and foreign countries. It is necessary to establish a global perspective and fully draw on the experience of the development of urban–rural relations in developed countries. New-type urbanization with Chinese characteristics and rural revitalization have been promoted as national strategies, and this institutional design is having an important impact on the evolution of urban–rural relations. Compared with developed countries, it may also form a path of specialization and differentiation. In addition, the urbanization process and rural development in different regions and stages in China are quite different. It is also necessary to analyze the scale differences and regional characteristics from a global perspective at different spatial scales such as countries, urban agglomerations, provinces, big cities, medium- and small-sized cities, counties, towns and villages.

Strengthen technological innovation, and refine and deepen scientific research on urban–rural relations. Collaboration mainly refers to the collaborative operation and governance of policy practice through multiagent, multiple technologies and methods. It requires the comprehensive use of the methods of logical analysis, quantitative analysis, statistical analysis, comparative analysis and case analysis; moreover, attention must be paid to social survey methods such as field surveys, in-depth interviews and questionnaires, and new technology methods such as geographic information technology, big data technology, visualization technology and neural networks should be innovatively explored to solve the complex problems of urban–rural relations. The study of urban–rural relations is a very grounded research field relevant to the real world, which needs to further refine and deepen scientific issues, and carry out in-depth research on the spatial–temporal patterns, integration paths, symbiotic effects and institutional mechanisms of urban–rural relations. In addition, for research on the integration and development of new-type urbanization and rural revitalization, attention must also be paid to facing up to the practical problems so as to draw lessons and avoid risks.

6. Conclusions

With the literature review method and comparative analysis, this paper reviews the historical evolution of urbanization and rural relations at home and abroad, and analyzes the problems of urbanization and rural development since reform and opening up. Based on the above analysis, this article points to the predication and challenges of new-type urbanization and rural revitalization in the new era, and puts forward a research framework for the integrated development of urbanization and a rural revitalization strategy in the discussion.

Both the overseas research and domestic research conclude that urban and rural areas are showing a new trend of integrated development. In the urbanization progress of China, urban-biased urbanization has resulted in a development gap between urban and rural areas since reform and opening up. The gap between urban and rural development has been tending to narrow relatively in recent years, and new-type urbanization and rural revitalization have contributed to this. The main trends in new-type urbanization and rural revitalization in the future are predicted as follows: new-type urbanization with a two-way interaction between urban and rural areas, urban–rural integration, a stable gap between urban and rural areas, a relatively smaller urban–rural gap between the east and the west, and coordinated governance between urban and rural areas. The major challenges in urban–rural relations in China by 2050 will be as follows: The geographical characteristic of "one side adjacent to the sea" will still be the main factor for the regional imbalance of urban–rural relations in China. The unbalanced and insufficient development in space will continue to embody the characteristics of the "Hu Line" as the geographical boundary. The exploration of development paths, transformation driven by innovation, and deepening the reform of the system and mechanism are the key measures and trends.

This study took China as a case, starting from the theory of urban–rural relations and starting from the reality of China's urbanization and rural development, considering the deficits and successes in urbanization practice. Combined with the new-type urbanization strategy and rural revitalization strategy, this article presents powerful actions for the state to use to promote the coordinated development of urban and rural areas in the near future at the policy practice level. It also predicts the trends and challenges for the future urban and rural development in China. This study provides an idea of urban–rural integration for developing countries such as China where the government plays an important role in the context of the global flow of urban–rural elements. In the research of urban and rural development, this study discusses a theoretical framework and emphasizes the importance of multidisciplinary integration, new technology application, and international and domestic vision switching.

Author Contributions: M.C. supervised the conceptualization, designed the research framework, prepared the original draft, and conducted the revision of the manuscript. Y.Z. made the diagrams and wrote a part of the contents. X.H. made the diagrams and wrote a part of the contents. C.Y. reviewed the manuscript, prepared the original draft and wrote a part of the contents. All authors have read and agreed to the published version of the manuscript.

Funding: This research was funded by National Natural Science Foundation of China, grant number 41822104 and grant number41871143, the Chinese Academy of Sciences Basic Frontier Science Research Program from the 0 to 1 Original Innovation Project grant number No. ZDBS-LYDQC005, the Strategic Priority Research Program of the Chinese Academy of Sciences, grant number XDA23100301 and the Youth Innovation Promotion Association of the Chinese Academy of Science, grant number 2017072.

Conflicts of Interest: The authors declare that they have no competing financial interests or other interests influencing this work.

References

1. Liu, Y. Research on the urban-rural integration and rural revitalization in the new era in China. *Acta Geogr. Sin.* **2018**, *73*, 637–650.
2. Liu, Y. Introduction to land use and rural sustainability in China. *Land Use Pol.* **2018**, *74*, 1–4. [CrossRef]
3. Liu, Y. Research on the geography of rural revitalization in the new era. *Geogr. Res.* **2019**, *38*, 461–466.
4. Long, H.L.; Qu, Y. Land use transitions and land management: A mutual feedback perspective. *Land Use Pol.* **2018**, *74*, 111–120. [CrossRef]
5. Chen, M.; Liu, W.; Lu, D. Challenges and the way forward in China's new-type urbanization. *Land Use Pol.* **2016**, *55*, 334–339. [CrossRef]
6. Chen, M.; Ye, C.; Lu, D.; Sui, Y.; Guo, S. Cognition and construction of the theoretical connotations of new urbanization with Chinese characteristics. *J. Geogr. Sci.* **2019**, *29*, 1681–1698. [CrossRef]
7. Ye, C.; Gao, Y. Evolving Relationship between Rural Development and Urbanization in China since 1949. *Econ. Geogr.* **2019**, *39*, 139–145.
8. Chen, M.; Gong, Y.; Lu, D.; Ye, C. Build a people-oriented urbanization: China's new-type urbanization dream and Anhui model. *Land Use Pol.* **2019**, *80*, 1–9. [CrossRef]
9. Ye, C.; Liu, Z. Rural-urban co-governance: Multi-scale practice. *Sci. Bull.* **2020**, *65*, 778–780. [CrossRef]
10. Smith, A. *An Inquiry into the Nature and Causes of the Wealth of Nations*; Random House Press: New York, NY, USA, 1985.

11. Von Thünen, J.H. *Isolated State: A Chinese Edition of Der isolierte Staat*; The Commercial Press: Beijing, China, 1986. (In Chinese)
12. Glacken, C.J. *Traces on the Rhodian Shore: Nature and Culture in Western Thought from Ancient Times to the End of the Eighteenth Century*; University of California Press: Berkeley, CA, USA, 1967.
13. Parker, G. *A Chinese Edition of Sovereign City: The City-State Ancient and Modern*; Shandong Pictorial Publishing House: Jinan, China, 2007. (In Chinese)
14. Plato. *A Chinese Edition of Politeia*; The Commercial Press: Beijing, China, 1986. (In Chinese)
15. More, S.T. *A Chinese Edition of Utopia*; The Commercial Press: Beijing, China, 1982. (In Chinese)
16. Howard, E. *Garden Cities of Tomorrow*; Faber: London, UK, 1946.
17. Geddes, P. *Cities in Evolution: An Introduction to the Town Planning Movement and to the Study of Civics*; Williams: London, UK, 1915.
18. Saarinen, E. *The City. Its Growth. Its Decay. Its Future*; Reinhold Publishing Corporation: New York, NY, USA, 1943.
19. Mumford, L. *The city in history: Its origins, its transformations, and its prospects*; Houghton Mifflin Harcourt: New York, NY, USA, 1961.
20. Lewis, W.A. *Theory of Economic Growth*; Routledge: Abingdon, UK, 2013.
21. Liu, C.L. On Lu Zuo-fu's "rural modernization" construction model. *Chongqing Soc. Sci.* **2004**, *1*, 110–115. (In Chinese)
22. Todaro, M.P. *Economic Development in the Third World: An Introduction to Problems and Policies in a Global Perspective*; Pearson Education: London, UK, 1977.
23. Friedmann, J. *Regional Development Policy: A Case Study of Venezuela*; The MIT Press: Cambridge, MA, USA, 1966.
24. Lipton, M. *Why Poor People Stay Poor: A Study of Urban Bias in World Development*; Australian National University Press: Canberra, Australia, 1977.
25. Krueger, A.O. *Economic Policy Reform in Developing Countries*; Basil Blackwell: Oxford, UK, 1992.
26. Preston, D.A. Rural-urban and inter-settlement interaction: Theory and analytical structure. *Area* **1975**, *7*, 171–174.
27. Potter, R.; Unwin, T. *The Geography of Urban-Rural Interaction in Developing Countries*; Routledge: London, UK, 1989.
28. Venables, A.J. Equilibrium locations of vertically linked industries. *Int. Econ. Rev.* **1996**, *37*, 341–359. [CrossRef]
29. Poncet, S. A fragmented China: Measure and determinants of Chinese domestic market disintegration. *Rev. Int. Econ.* **2005**, *13*, 409–430. [CrossRef]
30. McKinney, M.L. Urbanization, biodiversity, and conservation. *Bioscience* **2002**, *52*, 883–890. [CrossRef]
31. McGee, T.G.; Robinson, I.M. *The Mega-Urban Regions of Southeast Asia*; University of British Colombia: Vancouver, BC, Canada, 1989.
32. Douglass, M. A regional network strategy for reciprocal rural-urban linkages—An agenda for policy research with reference to Indonesia. *Third World Plan. Rev.* **1998**, *20*, 1–33. [CrossRef]
33. Redfield, R. *The Little Community: Viewpoints for the Study of a Human Whole*; University of Chicago Press: Chicago, IL, USA, 1955.
34. Dewey, R. The rural-urban continuum: Real but relatively unimportant. *Am. J. Sociol.* **1960**, *66*, 60–66. [CrossRef]
35. Tacoli, C. Rural-urban interactions: A guide to the literature. *Environ. Urban.* **1998**, *10*, 147–166. [CrossRef]
36. Bell, M.M. The Fruit of Difference: The Rural-Urban Continuum as a System of Identity. *Rural Sociol.* **1992**, *57*, 65–82. [CrossRef]
37. Lynch, K. Rural-urban interaction in the developing world. In Proceedings of the Emerging Issues Along Urban/Rural Interfaces: Linking Science & Society Conference, Atlanta, GA, USA, 13 16 March 2005; 2005.
38. Brenner, N. *Critique of Urbanization: Selected Essays*; Birkhauser: Basel, Germany, 2016.
39. Brenner, N. The Hinterland, Urbanized? *Archit. Des.* **2016**, *86*, 118–127.
40. Brenner, N. *New Urban Spaces: Urban Theory and the Scale Question*; Oxford University Press: New York, NY, USA, 2019.
41. Schlosser, F. *Laendliche Entwicklung im Wandel der Zeit—Zielsetzungen und Wirkungen*; TUM: Munich, Germany, 1999; pp. 184–187.
42. Zheng, G.; Ye, Y.M. The Stage of Chinese Urban-Rural Relation and its Development Model. *J. Renmin Univ. China.* **2009**, *23*, 87–92. (In Chinese)
43. Chen, G.S.; Li, G. From dual opposition of urban-rural to integrated governance: Changes and enlightenment of urban and rural governance model in developed countries. *Southeast Acad. Res.* **2007**, *2007*, 62–68.
44. Marsden, T. Rural geography trend report: The social and political bases of rural restructuring. *Prog. Hum. Geogr.* **1996**, *20*, 246–258. [CrossRef]
45. Marsden, T.; Lowe, P.; Whatmore, S. *Rural Restructuring: Global Processes and their Responses*; David Fulton Publishers Ltd.: London, UK, 1990.
46. Wilson, J. The urbanization of the countryside: Depoliticization and the production of space in Chiapas. *Lat. Am. Perspect.* **2013**, *40*, 218–236. [CrossRef]
47. Long, H.L.; Li, Y.R.; Liu, Y.S.; Woods, M.; Zou, J. Accelerated restructuring in rural China fueled by 'increasing vs. decreasing balance' land-use policy for dealing with hollowed villages. *Land Use Pol.* **2012**, *29*, 11–22. [CrossRef]
48. Nonaka, A.; Ono, H. Revitalization of Rural Economies though the Restructuring the Self-sufficient Realm—Growth in Small-scale Rapeseed Production in Japan. *JARQ Jap. Agric. Res. Q.* **2015**, *49*, 383–390.
49. Zhang, Y.; Li, X.; Song, W.; Thai, L. Land abandonment under rural restructuring in China explained from a cost-benefit perspective. *J. Rural Stud.* **2016**, *47*, 524–532. [CrossRef]
50. Tu, S.; Long, H. Rural restructuring in China: Theory, approaches and research prospect. *J. Geogr. Sci.* **2017**, *27*, 1169–1184. [CrossRef]
51. Whatmore, S. Sustainable rural geographies? *Prog. Hum. Geogr.* **1993**, *17*, 538–547. [CrossRef]

52. Liu, Y.; Li, Y. Revitalize the world's countryside. *Nature* **2017**, *548*, 275–277. [CrossRef] [PubMed]
53. Woods, M. *Rural Geography: Processes, Responses and Experiences in Rural Restructuring*; Sage: New York, NY, USA, 2004.
54. Woods, M. Rural geography: Blurring boundaries and making connections. *Prog. Hum. Geogr.* **2009**, *33*, 849–858. [CrossRef]
55. Woods, M. Performing rurality and practicing rural geography. *Prog. Hum. Geogr.* **2010**, *34*, 835–846. [CrossRef]
56. Chen, M.; Liang, L.; Wang, Z.; Zhang, W.; Yu, J.; Liang, Y. Geographical thoughts on the relationship between 'Beautiful China' and land spatial planning. *J. Geogr. Sci.* **2020**, *30*, 705–723. [CrossRef]
57. Johnson, T.G. Entrepreneurship and development finance—Keys to rural revitalization—Discussion. *Am. J. Agric. Econ.* **1989**, *71*, 1324–1326. [CrossRef]
58. Korsching, P. Multi community collaboration: An evolving rural revitalization strategy. *Rural Dev. News* **1992**, *16*, 1–2.
59. Greene, M.J. *Agriculture Diversification Initiatives: State Government Roles in Rural Revitalization*; Rural Economic Alternatives: Lexington, KY, USA, 1988.
60. Kawate, T. Rural revitalization and reform of rural organizations in contemporary rural Japan. *J. Rural Probl.* **2005**, *40*, 393–402. [CrossRef]
61. Carr, P.J.; Kefalas, M. Hollowing out the middle: The rural brain drain and what it means for America. *J. Rural Soc. Sci.* **2010**, *291*, 30–34.
62. Ayobami, O.; Ismail, H. Host's support for volunteerism: A pragmatic approach to rural revitalization. *Aust. J. Basic Appl. Sci.* **2013**, *7*, 260–272.
63. Ye, C. *The Theory and History on Chinese on China's Rural-Urban Relationship*; Southeast University Press: Nanjing, China, 2014. (In Chinese)
64. Zhang, B.F. Comparison and reference of three rural construction modes in the period of the republic of China. *Modern Econ. Res.* **2006**, *4*, 26–30. (In Chinese)
65. Li, W.S. A comparative study of rural construction thoughts of Yan Yang-chu and Liang Shu-ming. *Acad. Forum* **2004**, *3*, 129–132. (In Chinese)
66. Yang, J.W. Liang Shu-ming's rural construction experiment: Its main purpose and its value in contemporary China. *J. Shandong Univ. Philos. Soc. Sci.* **2006**, *5*, 126–133. (In Chinese)
67. Bai, Y.X. Urban-rural dual structure under the Chinese perspective: Its formation, expansion, and path. *Acad. Mon.* **2012**, *44*, 67–76. (In Chinese)
68. Wu, C.J. The New Development of Rural China. *BEVAS SOBEG.* **1997**, *1*, 101–105.
69. Gao, P.Y. *Comparative Study on Urbanization between China and Foreign Countries*; Nankai University Press: Tianjin, China, 2004. (In Chinese)
70. Zhang, W. Urban-rural coordinated development and planning. *City Plan. Rev.* **2005**, *29*, 79–83. (In Chinese)
71. Wang, J.X. *China Tomorrow: Towards Rural-Urban Integration*; Economic Press China: Beijing, China, 2005. (In Chinese)
72. Yu, A.T.W.; Wu, Y.; Zheng, B.; Zhang, X.; Shen, L. Identifying risk factors of urban-rural conflict in urbanization: A case of China. *Habitat Int.* **2014**, *44*, 177–185. [CrossRef]
73. Chen, J. Rapid urbanization in China: A real challenge to soil protection and food security. *Catena* **2007**, *69*, 1–15. [CrossRef]
74. Iredale, R.; Bilik, N.; Su, W.; Guo, F.; Hoy, C. *Contemporary Minority Migration, Education, and Ethnicity in China*; Edward Elgar: Cheltenham, UK, 2001.
75. Roberts, K.D. China's "tidal wave" of migrant labor: What can we learn from Mexican undocumented migration to the United States? *Int. Migr. Rev.* **1997**, *31*, 249–293. [PubMed]
76. Zhang, K.H.; Song, S.F. Rural-urban migration and urbanization in China: Evidence from time-series and cross-section analyses. *China Econ. Rev.* **2003**, *14*, 386–400. (In Chinese) [CrossRef]
77. Long, H.; Liu, Y.; Li, X.; Chen, Y. Building new countryside in China: A geographical perspective. *Land Use Pol.* **2010**, *27*, 457–470. [CrossRef]
78. Chen, M.; Liu, W.; Tao, X. Evolution and assessment on China's urbanization 1960-2010: Under-urbanization or over-urbanization? *Habitat Int.* **2013**, *38*, 25–33. [CrossRef]
79. National Bureau of Statistics. China Statistical Yearbook 2018. China Statistics Press: Beijing, China, 2018.
80. Chen, M.; Liu, W.; Lu, D.; Chen, H.; Ye, C. Progress of China's new-type urbanization construction since 2014: A preliminary assessment. *Cities* **2018**, *78*, 180–193. [CrossRef]
81. Chen, M.; Sui, Y.; Liu, W.; Liu, H.; Huang, Y. Urbanization patterns and poverty reduction: A new perspective to explore the countries along the Belt and Road. *Habitat Int.* **2019**, *84*, 1–14. [CrossRef]
82. Zhu, M. New era—Principal contradiction—Rural revitalization. *Economist* **2017**, *11*, 1. (In Chinese)
83. Luo, B.L. Formulate development ideas and implement the rural revitalization strategy. *South China J. Econ.* **2017**, *10*, 8–11. (In Chinese)
84. Cai, F. Rural urban income gap and critical point of institutional change. *Soc. Sci. China* **2003**, *5*, 93–111. (In Chinese) [CrossRef]
85. Li, M.; Shao, T.; Liu, S.Y. The international experience of urban-rural integration and its enlightenment to China. *Chin. Rural Econ.* **2014**, *6*, 83–96. (In Chinese)
86. Liu, S.Y. "Rural-urban China" is shifting from one-way urbanization to rural-urban interaction. *Rural Work Bull.* **2017**, *10*, 42. (In Chinese)

87. Fan, J.; Guo, R. Re-recognition of precondition and driving mechanism of new-type urbanization. *Geogr. Res.* **2019**, *38*, 3–12. (In Chinese)
88. Hong, Y.X.; Chen, W. Urbanization and integrating of urban and rural area. *Econ. Theory Bus. Manag.* **2003**, *4*, 28–31. (in Chinese).
89. Friedmann, J. Four theses in the study of China's urbanization. *Int. J. Urban Reg. Res.* **2006**, *30*, 440–451. [CrossRef]
90. Ye, C.; Ma, X.; Cai, Y.; Gao, F. The countryside under multiple high-tension lines: A perspective on the rural construction of Heping Village, Shanghai. *J. Rural Stud.* **2018**, *62*, 53–61. [CrossRef]
91. Liu, Y.; Fang, F.; Li, Y. Key issues of land use in China and implications for policy making. *Land Use Pol.* **2014**, *40*, 6–12. [CrossRef]
92. Liu, Y.; Yang, R.; Long, H.; Gao, J.; Wang, J. Implications of land-use change in rural China: A case study of Yucheng, Shandong province. *Land Use Pol.* **2014**, *40*, 111–118. [CrossRef]
93. Long, H.; Zhang, Y.; Tu, S. Land consolidation and rural vitalization. *Acta Geogr. Sin.* **2018**, *73*, 1837–1849.
94. Long, H.; Ge, D.; Wang, J. Progress and prospects of the coupling research on land use transitions and rural transformation development. *Acta Geogr. Sin.* **2019**, *74*, 2547–2559.
95. Lu, D.D.; Wang, Z.; Feng, Z.M. Academic debates on Hu Huanyong population line. *Geogr. Res.* **2016**, *35*, 805–824. (In Chinese)
96. Chen, M.; Gong, Y.; Li, Y.; Lu, D.; Zhang, H. Population distribution and urbanization on both sides of the Hu Huanyong Line: Answering the Premier's question. *J. Geogr. Sci.* **2016**, *26*, 1593–1610. [CrossRef]

land

MDPI

Article

Building on "Traditional" Land Dispute Resolution Mechanisms in Rural Ghana: Adaptive or Anachronistic?

Festus A. Asaaga

UK Centre for Ecology and Hydrology, Maclean Building, Crowmarsh Gifford, Benson Lane, Wallingford, Oxfordshire OX10 8B, UK; fesasa@ceh.ac.uk or fasaaga@gmail.com

Abstract: Despite the ongoing land administration reforms being implemented across sub-Saharan Africa (SSA), including Ghana, as a viable pathway to achieve tenure security and greater efficiency in land administration, the subject of land dispute resolution has received relatively less attention. Whereas customary tenure institutions play a central role in land administration (controlling ~80% of all land in Ghana), they remain at the fringes of the formal land dispute adjudicatory process. Recognising the pivotal role of traditional institutions as development agents and potential vehicles for promoting good land governance, recent discourses on land tenure have geared toward mainstreaming traditional land dispute institutions into the architecture of the formal judicial process via alternative dispute resolution pathways. Yet, little is known, at least empirically, as to the operations of traditional dispute resolution institutions in the contemporary context. This study therefore explores the importance of traditional dispute resolution institutions in the management of land-related disputes in southcentral and western Ghana, drawing on data collated from 380 farming households operating 746 plots. The results show that contrary to the conventional thinking that traditional institutions are anachronistic and not fit for purpose, they remain strong and a preferred forum for land dispute resolution (proving resilient and adaptable), given the changing socio-economic and tenurial conditions. Yet, these forums have differing implications for different actors within the customary spheres accessing them. The results highlight practical ways for incorporating traditional dispute resolution in the overall land governance setup in Ghana and elsewhere in sub-Saharan Africa. This has implications for redesigning context-specific and appropriate land-use policy interventions that address local land dispute resolution.

Keywords: land dispute; customary land tenure; statutory land tenure; tenure security; Ghana; sub-Saharan Africa

Citation: Asaaga, F.A. Building on "Traditional" Land Dispute Resolution Mechanisms in Rural Ghana: Adaptive or Anachronistic? *Land* **2021**, *10*, 143. https://doi.org/10.3390/land10020143

Academic Editors: Uchendu Eugene Chigbu, Ruishan Chen and Chao Ye

Received: 6 January 2021
Accepted: 28 January 2021
Published: 2 February 2021

1. Introduction

Over the past two decades or so, the good governance agenda has gained currency both in theory and in practice within the context of contemporary land management as a plausible strategy for navigating a sustainable development trajectory in the global South, particularly in sub-Saharan Africa (SSA) [1–3]. The promotion of the good land governance agenda has reinvigorated into the spotlight a raging debate on the land question regarding contextually relevant pathways to engender land tenure security and equitable customary land management in the SSA context [3–6]. Central to this emerging critical discourse is the enduring challenge of widespread customary land disputes exacerbated by the increasing commodification and individualisation of communal lands in most rural parts of SSA [7–9]. Although the wide-ranging socio-economic and political consequences of customary land disputes in affected areas and countries are well-documented in the literature [7,10], it is also acknowledged that the effects of customary land disputes vary across different spatial continuums, from rural to peri-urban to urban areas [7]. There is increasing evidence that rural and peri-urban areas in several SSA countries (including Ghana) have become contested zones in the light of the growing land commodification and large-scale commercial land acquisitions [8,11,12]. The interplay of such local-level land disputes and contestations poses several far-reaching developmental challenges considering the primacy of secure and equitable land access to the achievement of many of the Sustainable Development Goals (SGDs) in SSA countries [11,13].

Against this background, successive SSA governments (with support from the World Bank and other international development partners) have over time pursued Western-styled individualised statutory land tenure as a panacea to the seemingly insurmountable problems of tenure insecurity and contestations over land ownership [4,6,8]. This has been predicated on the underlying logic that customary tenure institutions (although controlling about 80% of SSA's total land area [14]) are anachronistic and not fit for purpose in addressing contemporary tenurial challenges [6,8,15,16]. Advocates of this conceptual view highlight the ambiguity, uncertainty and looseness of customary tenure structures, which render them ineffective or weak in dealing with land disputes and tenure insecurity, necessitating formalisation [17,18].

In spite of the great expectations that supplanting customary land tenure systems with Western-styled statutory tenure systems will afford greater certainty in land rights (tenure security and efficient dispute resolution mechanisms), and by extension economic development [19,20], the implementation outcomes in most parts of SSA (including Ghana) have, at best, been disappointing [21]. The under-performance of statutory systems has prompted many critics to question the suitability of the blanket pursuit of formalisation as a panacea to supposedly insecure customary land rights in the SSA context [15,22,23]. For instance, Bromley's [15] review suggested that formalisation in the SSA context has, in many instances, rather re-created and exacerbated existing land inequalities and contestations over land. While a few recent studies have reported some relative success of land titling intervention programs in SSA countries [24,25], altogether the available statistics still show far less achievement with respect to addressing problems of tenure insecurity and land disputes on a regional scale.

While the reasons for the failure of past land tenure reforms are complex and wide-ranging [16,21], a burgeoning critical scholarship expresses a strong optimism that customary tenure institutions, despite their imperfections, are still relevant and fit for purpose in the contemporary context, especially in under-served rural areas of SSA countries [22,26,27]. In other words, their arguments echo important contextual or place-based differences rather than universalist descriptions that shape and determine tenurial outcomes. Thus, given the right institutional tinkering or tenurial re-engineering of customary tenure institutions (which others have characterised as adaptation [28]), they are better positioned to deliver tenure security and respond to other emergent tenurial challenges [5,9,22]. In any event, an important take-home message from the two opposing conceptual positions rests on how customary tenure institutions fare [ing] on the ground in terms of safeguarding tenure security and, more particularly, effectively addressing land disputes across the sub-region [22,29]. Although a theoretical exposition is useful, a conclusive examination of this hypothesis is certainly an empirical matter.

Yet, to date, in spite of the renewed policy interest in customary tenure institutions, little is known, at least empirically, about the factors that have allowed them to withstand colonial and post-colonial reforms and retain their role in local-level land governance. With the notable exception of a few recent studies [8,9,30,31], there is a relative dearth of empirical focus on the potential role of customary systems in the contemporary context, particularly with respect to land disputes and traditional dispute resolution pathways [32–34]. A more nuanced and detailed contextual understanding of the operations of customary tenure institutions and the limits of their adaptability remains critical to better inform and guide on-going and future interventions towards the integration of customary tenure systems into the formal statutory framework for improved tenure security and effective land dispute resolution [8,9]. As argued by Anyidoho et al. [35], the process of tenurial adaptation cannot happen in isolation from the historical, political and legal context, which change cannot be imposed but must be built upon the institutional structures and practices that have evolved over time. Cleaver [36] also observed that the effectiveness of tenurial interventions is predicated on a socially informed analysis of the content and effects of informal/customary institutional arrangements rather than their form alone.

From the foregoing considerations, three key questions beg answers, namely (1) why have traditional land dispute resolution institutions persisted, (2) are they still fit for purpose or relevant in contemporary land governance and (3) how can customary dispute resolution mechanisms be effectively integrated into the formal statutory framework across socio-spatial contexts? This paper seeks to address these questions by focusing on Ghana's context characterised by a pluralistic tenurial regime, which is currently undergoing a process of harmonisation to enhance tenure security and address land disputes as a point of departure. Synonymous to other SSA countries, Ghana initiated a 25-year land administration reform (LAR) in 2003 as a plausible developmental pathway to enhance tenure security and efficient land administration [37,38]. Central to the LAP agenda is streamlining of the disparate customary and statutory tenure structures for effective local-level land dispute resolution using the customary land secretariats as a key operational vehicle. Although this paper focusses on Ghana's context, the findings of the study are broadly relevant for other SSA countries with a similar tenurial context in providing useful lessons towards the effective implementation of land reforms to achieve beneficial outcomes.

The rest of this paper is structured as follows. The next section provides an overview of contemporary land governance in Ghana, particularly focusing on land disputes and resolution mechanisms to provide a contextual background to situate the subsequent empirical analysis. Section 3 discusses the methods and data used for the paper, followed by a discussion of the results on dynamics of land disputes and resolution pathways in Section 4. The concluding aspect discusses the implications of the findings for integrating customary tenure arrangements into the statutory framework.

2. Contemporary Hybrid Land Governance and Dispute Resolution in Ghana

To sufficiently understand the contemporary debates about the hybridity of land governance in Ghana, it is instructive to consider the historical antecedents of land policies and interventions that have shaped the evolution of neo-customary tenure institutions in the country. Within this context, Ghana is characterised by a pluralistic legal framework consisting of customary and statutory law operationalised within a multi-sectoral governance environment [37,39,40]. Available statistics indicate that 78% of Ghana's total land is classified as customary land, with the remaining 22% falling under the domain of the state (20% exclusively owned by the state and 2% vested lands that are managed by the state but communally owned) [8,41]. The disparate customary and statutory tenure systems have developed over time and undergone several reforms to reach their present state today (for a detailed overview of Ghana's bifurcated tenurial system, see [22]).

While all prominent studies on Ghanaian land tenure [39,40,42] have underscored how customary law is formally recognised and remains an important body of law in all aspects of Ghana's society, the resolution of land disputes has traditionally been in statutory courts. Yet, evidence suggests that the formal court system is clogged with land-related disputes. Over time, increasing contestations regarding land ownership (which is rooted in the legacies of colonialisation; see [43,44]) has meant the introduction of several land policies and interventions by the colonial politico-administrative framework to restructure the supposedly inefficient and insecure customary tenure arrangements [40,45]. A flurry of post-independence legislations also operated to entrench Western-styled statutory tenure, resulting in the dualism of the land governance structures.

Dispute resolution over access to land resources is an important driver of local-level tensions across sub-Saharan Africa, given the marked diversity in the socio-cultural, political and economic spheres [46,47]. Within this context, a critical aspect of the debate in Ghana and SSA generally is the inherent disconnect between customary and statutory tenure systems, which are poorly articulated and seem to be on a collision course [39,40,48]. As the tenurial system defines the conditions of access, use and control of land and its associated resources in a particular socio-political context, it also invariably underpins livelihood security and sustainable land use and management. Nonetheless, issues of inequities in access to and control of land, tenure insecurity and protracted land disputes are characteristic of Ghana's existing tenurial regime [37,49,50]. Within this purview, there seem to be little consensus on the importance of customary tenure arrangements and institutions in promoting equitable land management and sustainable development in general.

Synonymous to other SSA countries, contestations over land in Ghana are acute and insidious, permeating/far-reaching implications for socio-economic development [10]. The increasing commodification and individualisation of land are manifested in the growing land scarcity and disputation over land [8,29]. According to the National Land Policy [37], the causes of land-related disputes have been identified to include multiple sales of land. Whereas the advent of land disputes predates the colonial era, this period was a watershed moment, ushering the indirect rule that served to supplant local customary tenure institutions with statutory tenure. Recent attempts at the harmonisation have witnessed the recognition of customary tenure institutions as central in the effective resolution of land-related disputes, particularly at the local level operating in tandem with the formal state courts [10]. Central to this are debates about how to successfully harmonise the disparate customary and statutory tenure to promote efficiency, enhance the security of tenure and reduce disputes over land. Current debates in the literature revolve around two main issues: (1) whether customary tenure should be supported due to its inherent flexibility, social embeddedness and accessibility and whether it guarantees tenure security [10] and (2) customary tenure is anachronistic and does not adequately safeguard security due to its inherent power imbalances [8,29].

3. Materials and Methods

3.1. Study Sites

This paper is based on a larger DPhil study that examined the dynamics of land tenure and sustainable land management in Ghana [22]. The analysis presented is based on data collated from two study sites, Kakum and Ankasa Conservation Areas situated in the central and western regions of southern Ghana, respectively (see Figure 1). These landscapes are dominated by the permanently protected Kakum National Park and Ankasa National Park and surrounding communities, spanning a total area of 360 km^2 and 509 km^2, respectively. The research was conducted in 19 fringe communities randomly selected in Kakum and Ankasa between December 2013 and September 2015 (Figure 1). Aside from having similar tenurial and social contexts affording comparability, both study sites were selected because of their representativeness in illustrating and exploring the prevailing tenurial situation in Ghana's high-forest zone, which spans the southern third of Ghana. In this context, most of the land in the off-reserve areas is stool land wholly owned and managed by the traditional authorities, with pockets of farmlands owned on a leasehold basis (spanning 2 to 50 years) also present [8,22]. Cocoa, oil palm and food crop farming are the dominant economic activities undertaken by households on relatively smaller plots (<5 ha) in the studied communities, working the land under diverse tenurial arrangements, ranging from customary freehold to customary licenses. The tenurial and ethnic diversity in the studied communities afforded a unique opportunity to explore how differences in socio-cultural dynamics (re-)shape tenurial outcomes and conditions of land rights in the study areas.

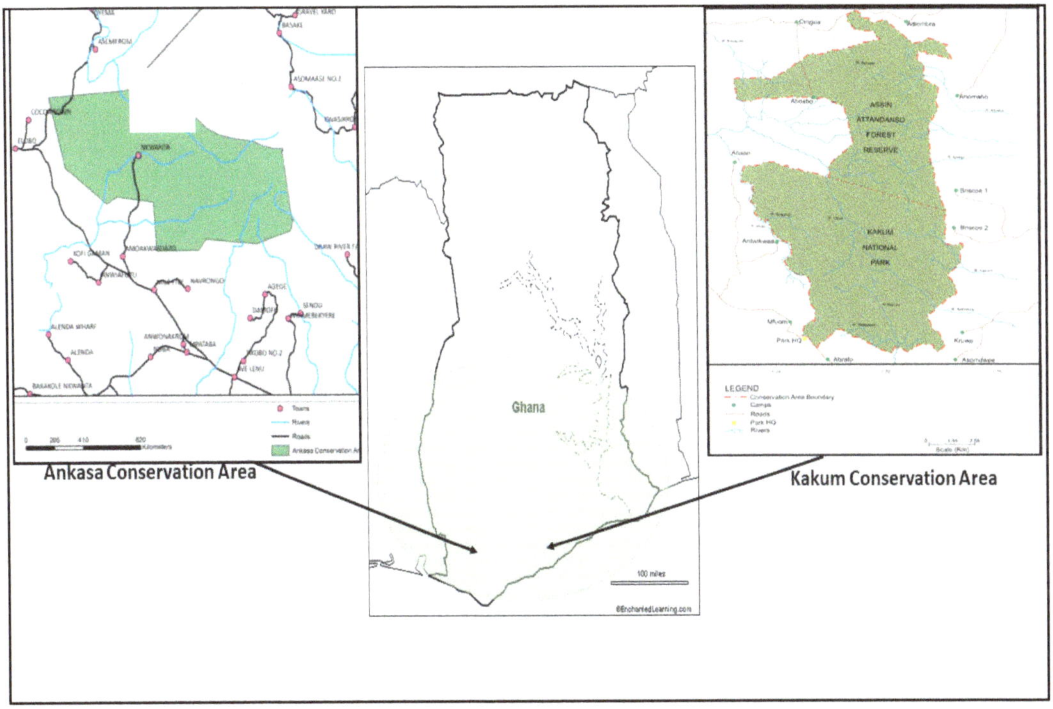

Figure 1. Overview of the two study sites in Ghana. The green area in the map of Ghana illustrates the high-forest zone, with left and right dark arrows pointing to the approximate locations of Ankasa and Kakum conservation areas in the south-western and central regions, respectively. Source: Based on Asaaga and Hirons [8].

3.2. Methods

Within the sampled villages, farming households were selected using stratified random sampling based on the gender and ethnicity of the household head, yielding a total sample size of 380 households[1] (Kakum, n = 232; Ankasa, n = 148). The rationale for sampling the households along gender and ethnicity lines was to ensure that the study sufficiently captures important tenurial differences that exist within and across social groups in the study context [8,22,27,29]. In this regard, the majority of households surveyed were (320, or 84%) male-headed, reflecting the dominance of men in land-use decision-making, even in matrilineal contexts. To the extent that women dominate agriculture in Ghana, this suggests that the seemingly gender-disproportionate sample could be a limitation of this study [8,22]. Nevertheless, the careful constitution of the focus groups vis-à-vis the treatment of gender as an analytical category in this study afforded in-depth qualitative insights into the key differences between men and women regarding perceptions and experiences with local dispute resolution mechanisms.

The administered household surveys were supplemented with focus group interviews (in 8 communities), with farmers purposively selected for their in-depth knowledge of the communities, and key informant interviews with farmers, traditional authorities, farmer cooperative representatives and selected local- and national-level officials of key land sector agencies (including from the Assinman Customary Land Secretariat (CLS), Office

[1] In instances where some household members managed their landholdings independently, they were also interviewed to capture specific variations in tenurial dynamics. This approach afforded an opportunity to capture the views of such household members (particularly female plot managers within male-headed households) who otherwise would have been excluded.

of the Administrator of Stool Lands (OASL) and Ghana National Land Administration Project Secretariat), which afforded an opportunity to sufficiently explain and capture some local-level nuances on tenurial dynamics, which otherwise would have been difficult to capture in a wholly quantitative study. The semi-structured interviews with the key stakeholders in customary land management conducted in Twi[2] focused on collating information on the prevailing tenurial situation in the studied communities, perceptions about land tenure security, land disputes and mechanisms for safeguarding land rights among others. The collated interview data from the semi-structured interviews and Focus Group Discussions (FGDs) were transcribed, followed by content and thematic analysis of the ensuing textual data guided by Miles and Huberman's [51] general strategy for qualitative analysis. The survey data were coded, entered, cross-checked for accuracy and analysed using SPSS (version 20). Cross-tabulation analysis and Pearson's chi-square tests were performed to detect significant differences ($p < 0.05$) between groups based on study area, gender and ethnicity. A dispute perception score was developed to ascertain the severity of land-related disputes in the studied communities. Using a five-point Likert scale, the opinions of respondents were measured by assigning a weight of 5 if 'they felt land disputes are a very serious problem', 4 if 'they felt land disputes are a serious problem,' 3 if 'they were uncertain,' 2 if 'they felt land disputes are somewhat not a problem' and 1 if 'they felt land disputes are not at all a problem'. Therefore, a high score above the median score meant a positive perception to the statement (i.e., severity of land disputes in a said community) and vice-versa. It was assumed that respondents who were uncertain as to whether land-related disputes are problematic in their communities would have a median score, which in this case is 3. The final (mean) score represents the cumulative perception of respondents on a particular statement of the severity of land-related disputes in the surveyed communities.

4. Results and Discussion

This section describes the results based on households' perceptions and experiences about land dispute resolution in the context of the study areas. First, the pattern of land disputes is presented, followed by the different landholders' experiences and perceptions regarding land dispute resolution pathways. Drawing on these findings, the contemporary role of customary land dispute resolution institutions is then discussed, highlighting the challenges and opportunities for integration into the statutory framework towards bolstering local tenure security and equitable land management.

4.1. Patterns of Land Disputes in Ankasa and Kakum

The changing context of customary tenure relations exemplified by widespread monetisation and exclusions in the Ankasa and Kakum regions have been described elsewhere [8]. Within this context of evolutionary changes in customary tenure dynamics, it was pertinent to understand the patterns of emerging land-related disputes and stakeholders' perceptions, as manifested in the study areas[3]. Thus, on the question of whether the surveyed communities had experienced any disputes within the last five years, it appears from the survey data that land-related disputes were more pronounced in the Kakum context. Of the 232 respondents interviewed in Kakum, 60% confirmed the prevalence of land-related disputes in their respective communities relative to just 28% (or 41 respondents) who alluded to same in Ankasa. To further ascertain the extent to which land disputes were problematic, the respondents' views were solicited based on five-point Likert scale measurement (ranging from not a problem to a serious problem), as shown in Table 1.

[2] Twi is the most widely spoken language amongst the Akan tribal groupings of southern Ghana.

[3] Recognising the sensitivity of land issues and the tendency for respondents to either exaggerate or withhold information about the subject, care was taken when soliciting respondents' views on the incidence of land-related disputes in their respective communities. The question was approached from different angles to ascertain the factual situation on the ground.

Table 1. Severity of land disputes in the surveyed communities.

Statement	Severity of Land Disputes		Statement	Severity of Land Disputes	
Kakum	**Responses**	**Mean Score**	**Ankasa**	**Responses**	**Mean Score** [4]
Totoda	30	4.33	Kusasi	10	2.10
Fante	20	3.95	Navrongo	18	3.06
Jerusalem	12	4.83	Nyamebekyere	23	2.74
Nkwantannan	24	3.88	Kanokware	20	3.30
Kwame-Anang	11	4.00	Fawoman	22	3.45
Appiahkrom	26	3.54	Amokwaosuazo	20	3.45
Nkwanta	11	3.00	Ghana-Nungua	20	3.45
Mankata	21	3.81	Old Ankasa	15	3.27
Bunsu	42	3.79	**Total (*N*)**	**148**	**3.16**
Seidukrom	25	2.96			
Kwaku-Mmore	10	3.10			
Total (*N*)	**232**	**3.72**			

It is quite clear from Table 1 that land disputes were prevalent in almost all the surveyed communities in Kakum, except for Seidukrom, Nkwanta and Kwaku-Mmore, which appeared to be the least problematic relative to the other communities. Juxtaposing to Ankasa' s context, a similar pattern is discernible across all the studied communities except for Kusasi and Nyamebekyere, which portrayed a negative perception to the above statement, with mean community scores of 2.10 and 2.74, respectively (less than the median value of 3). It can therefore be suggested that the incidence of land disputes appeared to be very problematic in the communities with high land scarcity, particularly in the Kakum area (e.g., Totoda, Kwame-Anang). This, in a way, highlights the growing tensions and/or disputes with regards to land access, with potentially negative implications for the security of land rights of some local landholders. Paradoxically, research and policy emphasis seem to be devoted to violent and large-scale disputes to the relative neglect of small-scale looming disputes, which have the propensity to escalate into full-blown large-scale disputes [52]. From the FGDs, participants expressed the view that disputes associated with land use created situations of uncertainty and insecurity about land rights. The foregoing analysis begs a typical question as to the nature of such intra-community disputes experienced in the study areas. Figure 2 therefore typifies the nature of land-related disputes, as evidenced in the surveyed communities.

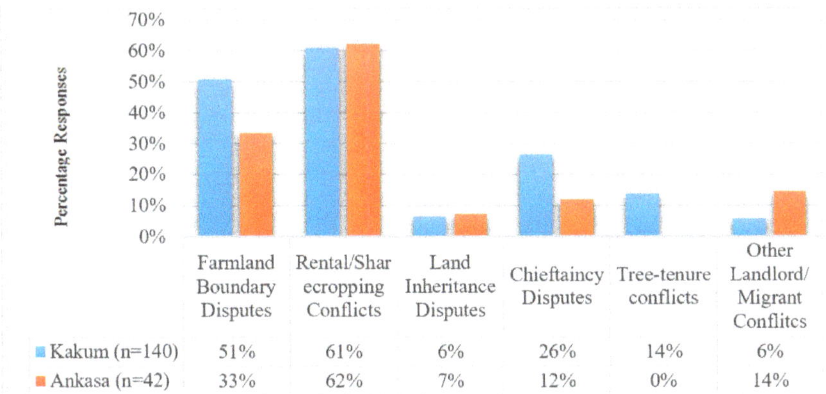

Figure 2. Type of land disputes in the study areas. Figure based on multiple responses on the nature of land disputes.

[4] The mean scores of communities were computed by aggregating individual responses to the statement divided by the total number of respondents surveyed in the respective communities.

As can be seen from Figure 2, sharecropping disputes constituted the commonest form of land-related disputes across the two case study regions. It was gathered from the key informant interviews that the incidence of land rental/sharecropping disputes was attributable to the oral, open-ended and variable nature of terms and conditions of the grant, which renders them highly susceptible to disputes. Generally, while most interviewees downplayed the seriousness of land disputes at the community level, focused and key-informant interviews showed that the impact of land-related disputes at the community level is very far-reaching, with negative consequences for tenure security and access arrangements.

4.2. Household Experiences of Land-Related Disputes: A Plot-Level Analysis

To further explore the dynamics of land-related disputes at the household level, a question was asked as to whether respondents had experienced disputes pertaining to their landholdings. As discernible from Table 2, whereas a breach of sharecropping terms was the common form of land-related disputes in Kakum, farmland boundary disputes were more prevalent in Ankasa. In general, the oral nature of most sharecropping contracts and indeterminate land boundaries have been noted as the common causes of land-related disputes amongst land users in rural Ghana [37,53,54]. Yet, the inherent flexibility in the execution of oral grants in addition to being convenient channels for maintaining social cohesion are some positive attributes, which perhaps explains their pervasive practice across rural communities in Ghana.

Table 2. Experience of previous disputes over household plots (%).

Have You Had a Dispute with Anyone Over This Plot?	Kakum Region		Ankasa Region	
	Household Plots		Household Plots	
	Indigene	Migrant	Indigene	Migrant
Yes	14%	14%	10%	12%
No	86%	86%	90%	88%
Total	**(n = 29)**	**(n = 491)**	**(n = 73)**	**(n = 170)**
Specific nature of the dispute				
Farmland boundary disputes	50%	17%	43%	60%
Multiple claims to land	0%	19%	43%	20%
Breach of terms (sharecropping)	50%	60%	14%	20%
Tree-tenure disputes	0	0	0	0
Other landlord–migrant disputes	0	4%	0	0
Total	**(n = 4)**	**(n = 69)**	**(n = 7)**	**(n = 20)**

The incidence of land-related disputes perhaps gives more impetus to the respondents' quest to protect and secure their property rights. Yet, information from key informants and FGDs indicates that there was the underlying tendency of bias in favour of indigenes, especially in instances of disputes involving migrants and their indigenous landlords. Moreover, groups of people with strong socio-political connections within the community stood a better chance of securing a favourable judgement in the event of disputes over land. This invariably creates room for some level of uncertainty as to the adequate protection and legitimacy of land rights of different social groups. The succeeding section therefore examines in more detail the local mechanisms for land dispute resolution in the study areas. Preceding this is a discussion on the common boundary indicators used in household plot identification in the light of the prevalence of farmland boundary disputes as mentioned above.

Nature of Farmland Boundary Indicators of Household Plots

In view of the relatively high incidence of farmland boundary disputes (especially in Ankasa), respondents were asked about the specific boundary indicators of their respective household landholdings as a way of identifying plausible factors occasioning such disputes (and insecurity) over land, as shown in Figure 3.

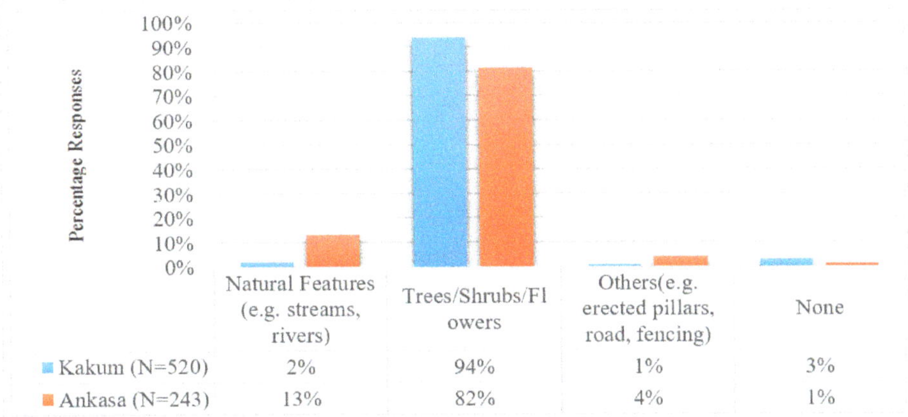

Figure 3. Type of farmland boundary indicators.

It is evident from Figure 3 that non-permanent indicators were commonly used in farmland boundary demarcations as opposed to permanent boundary indicators such as cadastral plans/maps. This finding is unsurprising to the extent that the lack of permanent and accurate boundary indicators has been identified as constituting a major source of land-related disputes associated with the insecurity and uncertainty of rural landholders' property rights [37,53,54]. During the field interviews, the national LAP coordinator emphasised the importance of permanent boundary demarcations pointing to on-going exercises (i.e., Customary Land Demarcation and Rural Parcels Rights Demarcation projects) aimed at mapping the extent of land rights and boundaries at the community and household levels to improve the certainty and security of tenure[5]. While this remains necessary, it is also instructive to note that such exercises could lead to a situation of possible winners and losers, especially in rural communities where land disputes are common. For instance, the permanent boundary demarcation process could be manipulated in favour of powerful and socially connected local actors in terms of further consolidating their land claims (in the case of contested lands) to the detriment of vulnerable actors like women and the poor. It may also result in some natives losing their entitlements (social right) to land, whereas strangers may have an opportunity to bolster the security of their (*de facto*) land rights, which otherwise would have been difficult to accomplish [55]. In any case, the authority of chiefs (as custodians and administrators of customary lands) also comes into play here. Social groups lacking recognition from chiefs could lose their *de facto* land rights (particularly third-party migrant transfers without appropriate legitimation by traditional authorities), whereas those of persons recognised by the chiefs might more likely be protected in the event of any conflicting land claims occasioned by such permanent farmland demarcation exercises [56]. To the extent that the foregoing holds true, this could perhaps result in heightened tensions and social insecurity in said communities (see Section 4.1). This inference derives support from Kasanga and Kotey's [39] and Ayee et al.'s [54] observations that the inefficiency and complicity that plague state land agencies renders them susceptible to manipulations by a few powerful actors (including chiefs and local elites) to the detriment of the less powerful and poor majority. The foregoing also finds expression in Lund's [55] statement that 'the process of securing land rights can often become complex when several

[5] At the time of the fieldwork, the Rural Parcels Rights Demarcation (RPRD) project had taken off on a pilot basis in the western, Ashanti and Brong Ahafo regions, focusing on demarcating 5118 farmlands in the afore-mentioned regions.

competing normative orders may be brought to bear to legitimize specific claims'. In this view, looser demarcations might actually reduce disputes relative to permanent precise measurements when interested parties might start to contest boundaries if they perceive they might permanently lose out particularly in conflict-prone communities. This further underlines the relevance of a sufficient understanding of the local socio-political realities as a precursor and a basis for intervention programmes geared towards enhancing the certainty of rural land rights.

4.3. Dispute Resolution Pathways and Resolution Preference

Despite the different categories of land-related disputes in the study areas, they did not translate into widespread tenure insecurity. The field data suggest that the mechanisms for dispute resolution were largely localised within the socio-political system, with very little variation between the study sites (Table 3). From Table 3, the overall results show that traditional courts (including clan/family heads (Abusuapanyins)) and stool land offices were the main adjudicatory institutions utilised by respondents. On face value, this seems to indicate that several fora were opened to landholders in terms of the enforcement and legitimation of their land rights. Critically, however, this does not reflect the actual situation on the ground, which is seemingly complex and negotiated exemplified by opportunism and the elite capture of the adjudicatory process. Indeed, the quantitative analysis demonstrates statistically significant differences regarding access to dispute resolution institutions across studied groups. For instance, when asked whether they had ever sought help from traditional authorities or statutory courts in the resolution of land-related disputes, a far less proportion of respondents in Kakum (11%) and Ankasa (10%) indicated they had made recourse to the traditional authorities or statutory courts regarding the resolution of land-related disputes (Table 3). This observation was corroborated during the interviews as participants explained that in the event of land disputes, they preferred private resolution through negotiation moderated by the Abusuapanyins and/or Odikro[6] at first instance. In the failure of such private resolution, recourse was made to any of the above-identified institutions.

Table 3. Mechanisms for land dispute adjudication.

Have You Sought Help with Land Dispute Resolution?	Kakum Region			Ankasa Region		
	Ethnicity		Total (N = 232)	Ethnicity		Total (N = 148)
	Indigene (n = 17)	Migrant (n = 215)		Indigene (n = 40)	Migrant (n = 108)	
Yes	12%	11%	11%	10%	9%	10%
No	88%	89%	89%	90%	91%	91%
Institutional preference						
Traditional court	100%	98%	98%	98%	88%	91%
Statutory court	0	1%	1%	2%	2%	2%
Stool land office	0	1%	1%	0	0	0
Don't know	0	0	0	0	10%	7%

Whereas Table 3 demonstrates a limited recourse to adjudicatory mechanisms, there is a sense of trust in the mediation afforded by traditional authorities as opposed to statutory institutions. During the community FGDs, participants cited the inaccessibility, delays in adjudication, cost, public image and fear of being labelled vexatious litigant as the reasons for their preference traditional dispute resolution mechanisms. These customary mechanisms played a pivotal role in maintaining social cohesion in their communities, particularly as most disputing parties are relatives or neighbours in the same village. Traditional authorities (who interpret and administer custom) were well-versed with the

[6] Odikro is the village chief or headman in the Akan traditional areas of southern Ghana.

local customs, norms and rules governing communal landholdings and thus more capable of dealing with contestations over land use and ownership. Two typical remarks by key informants are illustrative of the above observation:

> *'Most people prefer an amicable settlement of their land-related disputes by making recourse to the Odikro and his elders as opposed to the court. This is due to several reasons, including the fact that the courts are expensive, delays in going to and fro and the fear of being labelled a litigant, which could affect or strain social relationships between families and friends in the community ... '. (Interview 1, Kakum)*

> *'In this community we see ourselves as one, and as such, we try to resolve any disputes, especially in relation to land, amongst ourselves, sometimes even without seeking help from Odikro and his elders. Neglect of the customary structures could attract some scorn and/or sanctions from other members of the community ... '. (Interview 1, Kakum)*

The overwhelming stated preference for traditional adjudicatory mechanisms relative to the statutory courts in the study areas also finds expression in the observation that institutional constraints and allegations of corruption impede the effective functioning of the state courts as secure avenues for the protection of land rights irrespective of social status [39,57,58]. Indeed, interviews with some officials of the stool land office and CLS revealed that the local populace were quite sceptical that the statutory mechanisms would best serve their interests and/or protect their land rights as powerful entities could influence decisions in their favour. This is indicative of Platteau's [57] remark that 'in social contexts dominated by differential access to state administration [as in Ghana's case], there is always the fear that the adjudication/registration process will be manipulated by the elite to its advantage'. At the same time, the foregoing also brings to the fore an important question as to the effectiveness of these customary mechanisms in the protection of land rights of different social groups. This is discussed within the context of the respondents' satisfaction with the customary adjudicatory mechanisms.

4.3.1. Satisfaction with Local Dispute Resolution Mechanisms

Given that respondents in both study areas expressed a high preference for traditional over statutory mechanisms of land dispute resolution, it was instructive to further ascertain the extent to which respondents perceived customary mechanisms to be effective in land dispute resolution. Respondents were therefore asked about whether they were satisfied with the local dispute mechanisms in their respective communities, the results of which are shown in Tables 4 and 5.

Table 4. Satisfaction with local dispute resolution mechanisms by ethnicity.

Are You Satisfied with the Local Dispute Resolution Mechanisms?	Kakum Region			Ankasa Region		
	Ethnicity		Total (N = 232)	Ethnicity		Total (N = 148)
	Indigene (n = 17)	Migrant (n = 215)		Indigene (n = 40)	Migrant (n = 108)	
Yes	65%	74%	73%	83% **	57% **	64%
No	35%	25%	26%	17%	14%	15%
Don't know	0	1%	1%	0	29%	21%

** significant at $p \leq 0.05$;

Table 5. Satisfaction with local dispute resolution mechanisms by gender.

Are You Satisfied with the Local Dispute resolution Mechanism?	Kakum Region			Ankasa Region		
	Gender		Total (N = 232)	Gender		Total (N = 148)
	Male (n = 197)	Female (n = 35)		Male (n = 123)	Female (n = 25)	
Yes	75%	66%	73%	63%	68%	64%
No	24%	34%	26%	15%	16%	15%
Don't Know	1%	0	1%	22%	16%	21%

From Table 4, it is quite clear that respondents were generally satisfied with the customary dispute resolution mechanisms in the study areas. In Ankasa, however, a comparatively lower proportion of migrants (57%) as against indigenes (83%) expressed satisfaction with traditional adjudicatory mechanisms ($p < 0.05$; $\chi^2 = 8.01$), which perhaps suggests their waning confidence in traditional authorities as impartial arbiters in land dispute resolution. Corroborating this assertion is a typical remark by a migrant farmer in Ankasa:

> *'If you are a settler farmer here and you have an issue over land with a Nzema [indigene], it is likely that the decision by the traditional authorities would favour your opponent due to his ethnicity. In the eyes of the chiefs, you, the stranger, cannot profess stronger ties to the land than the indigene as you only came to make a living ... '.(Interview 1, Ankasa, June 2015)*

Elsewhere in south-western Ghana, Boone and Duku [59] and Boni [60] reported the incidence of landlord–migrant disputes ostensibly aggravated by local adjudicatory mechanisms operating to favour native landlords at the expense of migrant–tenant farmers.

4.3.2. Gendered Dimensions of Dispute Resolution Preference

Despite the overwhelming preference for customary adjudicatory mechanisms, the results do not necessarily imply that there is complete trust (in terms of fairness and transparency of the procedures employed) in them (Tables 4 and 5). As evidenced in Table 5, the disaggregated results suggest that traditional adjudicatory mechanisms are not entirely immune from gender biases. Indeed, the qualitative interviews highlighted instances of alleged discrimination where traditional authorities used their power to subvert justice, especially in favour of powerful actors within the local political hierarchy and indigenes in the event of contestations with migrant farmers. In the case of women, for example, focus group discussants argued to the effect that customary mechanisms were somewhat discriminatory (gender-biased) to the extent that women even had to rely on spouses and male relatives to enforce their land rights in the event of any contestations. In most of the surveyed communities, it was socially unacceptable (taboo) for women to discuss land matters with strangers without prior approval of their male relatives or husbands. Interestingly, they were equally unenthused about the statutory court as a forum for enforcement of their property rights, citing a myriad of reasons, including fear of possible backlash by the community for not according respect to traditional authorities, marital tensions, leading to instability of marriage, and poverty.

It is inferable from the above-enumerated reasons that fear of community sanctions and poverty perhaps constitute the underlying reasons for respondents' preference for the customary mechanisms and not necessarily because of equity in the adjudicatory processes. The low levels of education could also be a potential barrier to the enforcement of land rights via the formal adjudicatory mechanisms, especially in the case of women (in rural areas) who might be intimidated by the formal procedures that characterise the state courts [61]. This observation is supported by the fact that a far lesser proportion of respondents in Kakum (28%) and Ankasa (7%) who had sought help from either the customary and/or statutory dispute resolution mechanisms were women. The peculiar vulnerability of women is witnessed in the fact they generally constitute the largest segment of the poor, with limited access to requisite financial resources and education, particularly in the rural areas. Yet, the statutory mechanisms, as observed earlier, are touted as expensive and bureaucratic, which invariably implies that seeking enforcement or protection of land rights within such mechanisms could well be outside the reach of the poor and vulnerable. To the extent that this inference is correct, it is not totally surprising that the majority of respondents prefer traditional courts as a forum to assert their land rights notwithstanding issues of discrimination and procedural inequities (both perceived and real) that may characterise customary mechanisms.

In spite of a low preference for the formal mechanisms of dispute resolution, interviews with officials of the Assin Fosu Municipal Court revealed that land-related cases constituted the highest number of cases registered, although official/exact figures were not readily available. Further enquiries at the Assin Fosu and Elubo stool land offices showed that these offices were playing an unconventional role acting as forums for the resolution of land disputes (through mediation) involving farmers and landowners, especially in communities within their jurisdictional scope. It can therefore be concluded that whereas several options are available to landholders as forums for enforcement and protection of their land rights, in practice, they have restricted options, as evidenced by the above-enumerated challenges plaguing the existing mechanisms for dispute resolution. As pointed out by Rao [62], 'Legal pluralism does not imply a normative preference of one legal order or the other, as the choice of the arena for contestation is ultimately a political choice, determining as it does the access to resources.' Thus, the different identities, subject positions (within the local socio-political hierarchy), authority and fairness of procedures used could be instrumental in determining final outcomes [17,62].

5. Conclusions and Policy Implications

This study examined the extent to which customary land dispute resolution mechanisms are fit for purpose in contemporary tenurial relations in rural Ghana. This is against the backdrop of increasing contestations over customary land, exacerbated by land commercialisation and commodification. Yet, the burgeoning scholarship has largely focused on large-scale land-grabbing and inter-community boundary disputes, with relatively little empirical attention to intra-community dynamics on traditional land dispute resolution processes [31–33]. The findings of this study demonstrate that despite concerns about the exclusionary practices in the customary land delivery process [9,49,51], traditional institutions remain the preferred fora for land dispute resolution in the surveyed communities. The overwhelming preference of traditional mechanisms for land dispute resolution rather than state courts (98% and 90% of respondents in Kakum and Ankasa, respectively) is a testament of the strong social legitimacy enjoyed by local customary tenure arrangements regarding land dispute resolution. For instance, reasons adduced by respondents for the high stated preference for traditional dispute resolution mechanisms included accessibility, in-depth knowledge of local tenurial issues by traditional authorities, inexpensive and expeditious settlement of disputes and public image (see Section 4.3). It therefore follows that traditional dispute resolution mechanisms can be supported and strengthened (particularly in terms of procedural equity and enforcement of decisions) as way of facilitating their integration into the statutory dispute resolution system.

In furtherance to this, efforts initiated under the auspices of the recently ended Land Administration Project (LAP) in collaboration with the Ghana Judicial Service to support the creation of Customary Land Secretariat (CLS) internal dispute resolution forums through alternative dispute resolution (ADR) are encouraging and should be up-scaled. In doing so, however, it is important to take cognizance of the myriad of local level challenges including protracted chieftaincy disputes and multiple claims to land that could stifle the harmonisation process (see Section 4). For example, with the existing contestations over land ownership at the paramountcy level in parts of Ankasa and Kakum, there is the underlying tendency that the government's move of ceding greater control over customary land in traditional authorities (through the CLS concept) as part of the broader agenda of strengthening customary tenure institutions could further escalate these disputes (Section 4.3). Besides, a lack of permanent land boundaries and the undocumented nature of land rights also have a propensity to trigger (latent) land disputes and perhaps the loss of land rights of the poor and vulnerable social groups, especially in the wake of growing land scarcity and commodification [8].

This calls for the speedy resolution of chieftaincy disputes and conflicts over allodial ownership, land boundary demarcation and recordation of land rights as critical first steps towards improving certainty of land ownership in the study areas. Moreover, the existing power imbalances vis-à-vis the increasing monetisation of access arrangements have created seeming spaces for the manipulation and abuse of chiefly authority to the detriment of the poor and vulnerable social groups in the study areas [8]. This thus suggests that addressing issues such as transparency, accountability and fairness in customary decisions regarding land use and allocative decision-making is crucial to ensure efficient customary land delivery and safeguard the interests of the poor and vulnerable social groups [63]. Within this context, the legally mandated CLS (under the recently promulgated Lands Act 2020) provides a basis where some personnel and local women leaders may be trained as community volunteer paralegals to provide local support mechanisms for women and other vulnerable groups seeking to enforce their land rights in both customary and statutory spheres.

Furthermore, traditional authorities may be trained in these areas (under the auspices of the traditional councils, regional and national Houses of Chiefs and the Ministry of Chieftaincy and Culture) to ensure fairness in their administration of land and dispute settlement. At the same time, it is recommended that socially disadvantaged groups, including women, migrants and the youth, be allocated seats on the village land management committees to give a voice to these groups and sufficient consideration of their interests in land allocation. An encouraging sign that this could be achieved in Ankasa and Kakum is that a few migrant settlers in some communities have been elevated to the position of village headmen. Besides, adopting such an inclusive stance in the composition of local land management committees could foster institutional trust and possibly dispel any misconceptions of the CLS being a collusion between government and traditional authorities to usurp the land rights of rural landholders. The creation of a permanent gender desk under the CLS would also provide useful support and protection of land rights of women and other marginalised groups. Equally important is the need to review the excessive concentration of power in traditional authorities under the existing regulatory framework on customary land administration to safeguard against potential abuse of authority and discrimination. One way to achieving this end would perhaps be the codification of existing customary rules and norms on land allocation and use in different traditional areas to afford clarity and unbiased interpretation as well as minimise their susceptibility to manipulation by traditional authorities. This is, however, a very complex issue, considering that the codification has the propensity to fossilise fluid customary norms, thereby limiting their flexibility [63–65]. Besides, the fact that land matters are politically sensitive sitting at the cleavage of national politics and tradition in Ghana vis-à-vis the government's policy of non-interference highlights the need for a cautious approach to legal reforms in this direction. Navigating this dilemma thus requires serious political willingness on the part of government and traditional authorities in facilitating the creation of more neutral policy spaces for these sensitive yet important issues to be deliberated amongst politicians, representatives of the regional and national Houses of Chiefs and civil society groups at large. Academia also has an important role to play in continuously undertaking independent evidence-based research to inform policy deliberations and actions, particularly on questions regarding who benefits and who loses from efforts on harmonising traditional and statutory adjudicatory mechanisms in customary land governance.

This study is not without limitations. The cross-sectional nature of the dataset has meant that important differences and tenurial changes over time that might affect land dispute dynamics could not be explored. It is recognised that the use of household panel data will afford an opportunity to capture effects of tenurial changes on local dispute resolution mechanisms and preferences over time rather than cross-sectional variation only. A related issue has to do with the limited interpretability of the findings, as they might be subject to self-reported recall bias. The generalisability of the survey findings to other contexts with varying socio-economic and tenurial characteristics, particularly northern Ghana, should be with caution. Nevertheless, the combination of qualitative and survey data (especially information obtained through key informant interviews and focus group discussions) afforded a unique opportunity to sufficiently explain and capture some local-level nuances in local dispute resolution pathways and preferences, which otherwise would have remained unobserved or hidden in a wholly quantitative study.

Funding: The manuscript writing was supported through the Natural Environment Research Council award number NE/R016429/1 as part of the UK-SCAPE programme delivering National Capability.

Institutional Review Board Statement: The research was carried out as part of larger DPhil study titled "Land Tenure and Sustainable Land Use in Rural Ghana", which was approved by the Central University Research and Ethics Committee (CUREC) of the University of Oxford in 2013.

Informed Consent Statement: Participation in this research was voluntary and all participants gave their full prior-informed verbal and written consent before the conduct of the interviews and household survey. The collated data from the survey and interviews were duly anonymised using de-identifiers or pseudonyms to safeguard the confidentiality of participants.

Data Availability Statement: The datasets generated and/or analysed during the present study are not publicly available in order to protect the privacy of participants but are available from the corresponding author on reasonable request.

Acknowledgments: The author would like to thank Gideon Tuffour and Godfred Bempah for research assistance. We are grateful to the traditional leaders and community members of the Kakum and Ankasa regions and other research participants for their contributions to the research process. We also express our gratitude to the anonymous reviewers and other colleagues at Oxford for their valuable comments and suggestions on earlier drafts of this article. All remaining errors are the author's own.

Conflicts of Interest: The author declares no conflict of interest.

References

1. Union, African. *Land Policy in Africa: A Framework to Strengthen Land Rights, Enhance Productivity and Secure Livelihoods*; African Union and Economic Commission for Africa: Addis Ababa, Ethiopia, 2009. Available online: https://bit.ly/3a1lFkc (accessed on 23 January 2021).
2. FAO. *State of the World's Forest*; FAO: Rome, Italy, 2012. Available online: http://www.fao.org/3/i3010e/i3010e.pdf (accessed on 23 January 2021).
3. Byamugisha, F. Securing Land Tenure and Easing Access to Land. In *Background Paper for African Transformation Report 2016: Transforming Africa's Agriculture*; African Center for Economic Transformation (ACET): Accra, Ghana, 2016; pp. 1–34.
4. Ghebru, H.H.; Edeh, H.; Ali, D.; Deininger, K.; Okumo, A.; Woldeyohannes, S. Tenure Security and Demand for Land Tenure Regularization in Nigeria. 2014. Available online: https://bit.ly/3sRYT8k (accessed on 23 January 2021).
5. Knight, R.S. Statutory Recognition of Customary Land Rights in Africa: An Investigation into Best Practices for Lawmaking and Implementation. FAO Legislative Study 105. 2010. Available online: http://www.fao.org/3/i1945e/i1945e00.pdf (accessed on 23 January 2021).
6. Deininger, K.W. *Land Policies for Growth and Poverty Reduction*; World Bank Publications: Washington, DC, USA, 2003; Available online: https://bit.ly/39d1W2Q (accessed on 23 January 2021).
7. Owusu Ansah, B.; Chigbu, U.E. The Nexus between Peri-Urban Transformation and Customary Land Rights Disputes: Effects on Peri-Urban Development in Trede, Ghana. *Land* **2020**, *9*, 187. [CrossRef]
8. Asaaga, F.A.; Hirons, M.A. Windows of opportunity or windows of exclusion? Changing dynamics of tenurial relations in rural Ghana. *Land Use Policy* **2019**, *87*, 104042. [CrossRef]
9. Akaateba, M.A. The politics of customary land rights transformation in peri-urban Ghana: Powers of exclusion in the era of land commodification. *Land Use Policy* **2019**, *88*, 104197. [CrossRef]
10. Crook, R.C. State Courts and the Regulation of Land Disputes in Ghana: The Litigants' Perspective. 2005. Available online: https://www.ids.ac.uk/download.php?file=files/Wp241.pdf (accessed on 23 January 2021).

11. Baah, K.; Kidido, J.K. Sharecropping arrangement in the contemporary agricultural economy of Ghana: A study of Techiman North District and Sefwi Wiawso Municipality, Ghana. *J. Plan. Land Manag.* **2020**, *1*, 50–62.
12. Yaro, J.A. Customary tenure systems under siege: Contemporary access to land in Northern Ghana. *GeoJournal* **2010**, *75*, 199–214. [CrossRef]
13. UN (United Nations). Global Sustainable Development Report. 2015. Available online: https://bit.ly/3sTQkKi (accessed on 23 January 2021).
14. Augustinus, C. *Comparative Analysis of Land Administration Systems: African Review*; The World Bank: Washington, DC, USA, 2003; Available online: https://bit.ly/3iHoZq0 (accessed on 23 January 2021).
15. Bromley, D.W. Formalising property relations in the developing world: The wrong prescription for the wrong malady. *Land Use Policy* **2009**, *26*, 20–27. [CrossRef]
16. Peters, P.E. Conflicts over land and threats to customary tenure in Africa. *Afr. Aff.* **2013**, *112*, 543–562. [CrossRef]
17. Crook, R.C. Customary Justice Institutions and Local Alternative Dispute Resolution: What Kind of Protection Can They Offer to Customary Landholders. Contesting Land and Custom in Ghana: State, Chief and the Citizen. 2008, pp. 131–154. Available online: https://bit.ly/3iE79nu (accessed on 23 January 2021).
18. Antwi-Agyei, P.; Dougill, A.J.; Stringer, L.C. Impacts of land tenure arrangements on the adaptive capacity of marginalized groups: The case of Ghana's Ejura Sekyedumase and Bongo districts. *Land Use Policy* **2015**, *49*, 203–212. [CrossRef]
19. Robinson, J.A.; Acemoglu, D. *Why Nations Fail: The Origins of Power, Prosperity and Poverty*; Profile: London, UK, 2012.
20. Byamugisha, F.F.K. *Securing Africa's Land for Shared Prosperity: A Program to Scale up Reforms and Investments*; The World Bank: Washington, DC, USA, 2013; Available online: https://bit.ly/2KGcY7s (accessed on 23 January 2021).
21. Onoma, A.K. *The Politics of Property Rights Institutions in Africa*; Cambridge University Press: Cambridge, UK, 2009.
22. Asaaga, F.A. Land Rights, Tenure Security and Sustainable Land Use in Rural Ghana. Ph.D. Thesis, University of Oxford, Oxford, UK, 2017.
23. Bugri, J.T. *Improving Land Sector Governance in Ghana*; World Bank: Washington, DC, USA, 2012; Available online: https://bit.ly/3sRQy4f (accessed on 23 January 2021).
24. Deininger, K.; Ali, D.A.; Alemu, T. Impacts of land certification on tenure security, investment, and land market participation: Evidence from Ethiopia. *Land Econ.* **2011**, *87*, 312–334. [CrossRef]
25. Bezu, S.; Holden, S. Demand for second-stage land certification in Ethiopia: Evidence from household panel data. *Land Use Policy* **2014**, *41*, 193–205. [CrossRef]
26. Lawry, S.; Samii, C.; Hall, R.; Leopold, A.; Hornby, D.; Mtero, F. The impact of land property rights interventions on investment and agricultural productivity in developing countries: A systematic review. *J. Dev. Eff.* **2017**, *9*, 61–81. [CrossRef]
27. Adoko, J.; Akin, J.; Knight, R. Understanding and Strengthening Women's Land Rights under Customary Tenure in Uganda. Land and Equity Movement Uganda. 2011. Available online: https://bit.ly/2NmyC1s (accessed on 23 January 2021).
28. Nkwae, B. Conceptual Framework Modelling and Analysing Periurban Land Problems in Southern Africa. 2006. Available online: http://www2.unb.ca/gge/Pubs/TR235.pdf (accessed on 23 January 2021).
29. Akaateba, M.A.; Huang, H.; Adumpo, E.A. Between co-production and institutional hybridity in land delivery: Insights from local planning practice in peri-urban Tamale, Ghana. *Land Use Policy* **2018**, *72*, 215–226.
30. Obeng-Odoom, F. *Property, Institutions, and Social Stratification in Africa*; Cambridge University Press: Cambridge, UK, 2020.
31. Kansanga, M.M.; Arku, G.; Luginaah, I. Powers of exclusion and counter-exclusion: The political ecology of ethno-territorial customary land boundary conflicts in Ghana. *Land Use Policy* **2019**, *86*, 12–22. [CrossRef]
32. Biitir, S.B.; Nara, B.B. The role of Customary Land Secretariats in promoting good local land governance in Ghana. *Land Use Policy* **2016**, *50*, 528–536. [CrossRef]
33. Kuusaana, E.D.; Kidido, J.K.; Appiah, M.N.; Mireku, K.O. Alternative Dispute Resolution by Chiefs and Tendamba: A Case Study of Kumasi and Wa Traditional Areas. 2013. Available online: https://bit.ly/3qIMzoU (accessed on 23 January 2021).
34. Deininger, K.; Gershon, F. Land Institutions and Land Markets. In *Handbook of Agricultural Economics*; Elsevier: Amsterdam, The Netherlands, 2001; Volume 1, pp. 287–331.
35. Anyidoho, N.A.; Amanquah, S.T.; Clottey, E.A. *Chieftaincy Institutions and Land Tenure Security: Challenges, Responses, and the Potential for Reform*; Institute of Statistical, Social & Economic Research, University of Ghana: Accra, Ghana, 2008.
36. Cleaver, F. Reinventing institutions: Bricolage and the social embeddedness of natural resource management. *Eur. J. Dev. Res.* **2002**, *14*, 11–30. [CrossRef]
37. Ghana National Land Policy. 1999. Available online: https://www.documents.clientearth.org/library/download-info/national-land-policy-1999/ (accessed on 23 January 2021).
38. LAP. *Restructuring of the Ghana Land Administration Project*; Project Paper No. 45705; LAP: Accra, Ghana, 2008.
39. Kasanga, R.K.; Kotey, N.A. Land Management in Ghana: Building on Tradition and Modernity. 2001. Available online: https://bit.ly/369ogbM (accessed on 23 January 2021).
40. Agbosu, L.; Awumbila, M.; Dowuona-Hammond, C.; Tsikata, D. *Customary and Statutory Land Tenure and Land Policy in Ghana*; Institute of Statistical Social and Economic Research (ISSER), University of Ghana: Accra, Ghana, 2007.
41. Government of Ghana. Emerging Land Tenure Issues. 2003. Available online: https://bit.ly/3c4wPbH (accessed on 23 January 2021).

42. Bentsi-Enchill, K. Ghana Land Law, An Exposition, Analysis and Critique. Ghana Land Law, an Exposition, Analysis and Critique. 1964. Available online: https://bit.ly/3pekXaL (accessed on 23 January 2021).
43. Mamdani, M. Historicizing power and responses to power: Indirect rule and its reform. *Soc. Res.* **1999**, *66*, 859–886.
44. Njoh, A.J.; Akiwumi, F. Colonial legacies, land policies and the millennium development goals: Lessons from Cameroon and Sierra Leone. *Habitat Int.* **2012**, *36*, 210–218. [CrossRef]
45. Berry, S.S. *No Condition is Permanent: The Social Dynamics of Agrarian Change in Sub-Saharan Africa*; University of Wisconsin Press: Madison, WI, USA, 1993.
46. De Juan, A. "Traditional" Resolution of Land Conflicts: The Survival of Precolonial Dispute Settlement in Burundi. *Comp. Political Stud.* **2017**, *50*, 1835–1868. [CrossRef]
47. Fred-Mensah, B.K. Capturing ambiguities: Communal conflict management alternative in Ghana. *World Dev.* **1999**, *27*, 951–965. [CrossRef]
48. Obeng-Odoom, F. Urban land policies in Ghana: A case of the emperor's new clothes? *Rev. Black Political Econ.* **2014**, *41*, 119–143. [CrossRef]
49. Whitehead, A.; Tsikata, D. Policy discourses on women's land rights in Sub–Saharan Africa: The implications of the re–turn to the Customary. *J. Agrar. Chang.* **2003**, *3*, 67–112. [CrossRef]
50. Tsikata, D.; Seini, W. Identities, Inequalities and Conflicts in Ghana. 2004. Available online: https://bit.ly/39SO0dG (accessed on 23 January 2021).
51. Miles, M.B.; Huberman, A.M. *Qualitative Data Analysis: An Expanded Sourcebook*; Sage: Thousand Oaks, CA, USA, 1994.
52. Tettey, W.; Gebe, B.Y.; Ansah-Koi, K. *The Politics of Land and Land-Related Conflicts in Ghana*; No. 84; Institute of Statistical, Social & Economic Research, University of Ghana: Accra, Ghana, 2008.
53. Bugri, J.T. The dynamics of tenure security, agricultural production and environmental degradation in Africa: Evidence from stakeholders in north-east Ghana. *Land Use Policy* **2008**, *25*, 271–285. [CrossRef]
54. Ayee, J.R.; Frempong, A.K.; Asante, R.; Boafo-Arthur, K. *Local Power Struggles, Conflicts and Conflict Resolution: The Causes, Dynamics and Policy Implications of Land-Related Conflicts in the Greater Accra and Eastern Regions of Ghana*; CODESRIA: Dakar, Senegal, 2011.
55. Lund, C. Property and citizenship: Conceptually connecting land rights and belonging in Africa. *Afr. Spectr.* **2011**, *46*, 71–75. [CrossRef]
56. Boamah, F. How and why chiefs formalise land use in recent times: The politics of land dispossession through biofuels investments in Ghana. *Rev. Afr. Political Econ.* **2014**, *41*, 406–423. [CrossRef]
57. Platteau, J.-P. The evolutionary theory of land rights as applied to sub-Saharan Africa: A critical assessment. *Dev. Chang.* **1996**, *27*, 29–86. [CrossRef]
58. S Amanor, K.; Ubink, J.M. *Contesting Land and Custom in Ghana. State, Chief and the Citizen*; Leiden University Press: Leiden, The Netherlands, 2008.
59. Boone, C.; Duku, D.K. Ethnic land rights in western Ghana: Landlord–stranger relations in the democratic era. *Dev. Chang.* **2012**, *43*, 671–693. [CrossRef]
60. Boni, S. *Clearing the Ghanaian Forest: Theories and Practices of Acquisition, Transfer and Utilisation of Farming Titles in the Sefwi-Akan Area*; Institute of African Studies, University of Ghana: Accra, Ghana, 2005.
61. Ghana Statistical Service. *Ghana Living Standards Survey Round 6 (GLSS6)*; Ghana Statisical Service: Accra, Ghana, 2014.
62. Rao, N. Custom and the courts: Ensuring women's rights to land, Jharkhand, India. *Dev. Chang.* **2007**, *38*, 299–319. [CrossRef]
63. Arko-Adjei, A. *Adapting Land Administration to the Institutional Framework of Customary Tenure: The Case of Peri-Urban Ghana*; No. 184; IOS Press: Amsterdam, The Netherlands, 2011.
64. Oba, A.A. The future of customary law in Africa. In *The Future of African Customary Law*; Fenrich, J., Galizzi, P., Higgins, T.E., Eds.; Cambridge University Press: Cambridge, UK, 2011.
65. Woodman, G.R. A survey of customary laws in Africa in search of lessons for the future. In *The Future of African Customary Law*; Fenrich, J., Galizzi, P., Higgins, T.E., Eds.; Cambridge University Press: Cambridge, UK, 2011; pp. 9–30.

land

MDPI

Article

Rural Development from a Gender Perspective: The Case of Women Farmers in Southern Spain

Jaime De Pablo Valenciano, Juan Milán-García *, Juan Uribe-Toril and María Angustias Guerrero-Villalba

Area of Applied Economics, Business and Economics Department, University of Almeria, 04120 Almería, Spain; jdepablo@ual.es (J.D.P.V.); juribe@ual.es (J.U.-T.); maguerre@ual.es (M.A.G.-V.)
* Correspondence: jmg483@ual.es

Abstract: This article analyses the contribution to local development by women workers in the fruit- and vegetable-handling sector in Almería (Spain) over the last five years (2015–2019). It is a continuation of research carried out during the period 2000–2014. Using data collected through surveys and focus groups, the aim is to ascertain if the results obtained in this analysis meet the condition of sustainability, i.e., whether the improvement in working women's quality of life has been maintained over time, and whether these beneficial effects have multiplied. The results show that women workers in the fruit- and vegetable-handling sector are satisfied with their jobs and with the company they are working for. The existence of fixed-discontinuous employment contracts facilitates greater flexibility for women in terms of balancing work and family life. This main contribution of this study lies in extrapolating the sustainability of a local development model in regard to other initiatives that aim to increase women's empowerment in the labour market.

Keywords: women empowerment; sustainable local development; food-handling sector; horticultural sector

Citation: De Pablo Valenciano, J.; Milán-García, J.; Uribe-Toril, J.; Guerrero-Villalba, M.A. Rural Development from a Gender Perspective: The Case of Women Farmers in Southern Spain. *Land* **2021**, *10*, 75. https://doi.org/10.3390/land10010075

Received: 20 December 2020
Accepted: 12 January 2021
Published: 15 January 2021

1. Introduction

The role of women in the agricultural sector has been on the rise in recent years. In some countries, such as Nigeria and Pakistan, women are involved in subsistence agricultural production. In addition to subsistence farming, they may also receive income from the formal and informal sectors, depending on their means of living and poverty level. In Africa, some women are willing to engage in collecting forest resources to support their families. However, inequality between men and women remains evident in some parts of Africa. This inequality reduces women's opportunities to improve their family's situation and their household economy. Additionally, women play a crucial role in food security [1].

In some regions, such as Western Europe [2], which also have a high concentration of migrant workers, there is a need for women's participation in the agricultural and industrial sectors. In other regions, such as the African continent, social pressure from ethnic groups makes it difficult for women to engage in economic activities [3].

The analysis presented in this study is based on the province of Almería, located in the south of Spain (Figure 1). This territory was chosen because Almería is one of the leading centres of fruit and vegetable production in Spain, and in Europe in general. In fact, it is known as the orchard of Europe [4]. In this region, agriculture is concentrated in three main areas: La Cañada, Campo de Níjar, and Campo de Dalías.

In terms of freshness, fruits and vegetables in Almería are considered a touchstone in Spain. The area is also considered a global reference in this sector [4,5].

The commercial system in Almería is structured at a business level as follows:

- Establishments that sell at origin (*alhóndigas*). These take the legal form of public limited companies.
- Establishments that sell in destination markets (usually cooperatives or agricultural processing companies (APCs)).

- The San Isidro Agricultural Cooperative (SAIC), a cooperative that sells at origin

For the companies that sell at origin (*alhóndigas* and SAIC), the supply from the agricultural producers and the demand from the commission agents compete with sales being made through reverse (or Dutch) auctions. In turn, these brokers work for third parties, including the cooperatives themselves, APC in Almería and other provinces, and the consignees of central markets and purchasing centres of large supermarkets. While the reverse auction system originated in the Netherlands, it is no longer used in that country. However, it is still used in Belgium (known as *veiling*) and France (known as *cadran*), particularly in the Brittany region.

Companies selling at destination (cooperatives and APC) operate directly with the central purchasing agencies of supermarkets.

Figure 2 presents an overview of the main aspects of the value chain that characterise fruit- and vegetable-handling activities. As can be seen, the activities begin with the entry of products into the warehouse followed by transportation and preparation for wholesale or retail sale.

Figure 1. Almería, polled areas. Compiled by the authors.

This paper analyses the contribution to local development by women workers in the fruit- and vegetable-handling sector in the province of Almería (Spain) over the last five years (2015–2019). It is the continuation of research carried out during the period 2000–2014, which showed that the presence of women in this sector increased their empowerment [6]. Therefore, this paper aims to evaluate whether this process of empowerment has been consolidated in recent years, guaranteeing a sustainable process of local development that permeates all sectors of the population. In addition, it evaluates the safety of women participating in the handling of fruit and vegetables, a fundamental aspect in guaranteeing this sustainability.

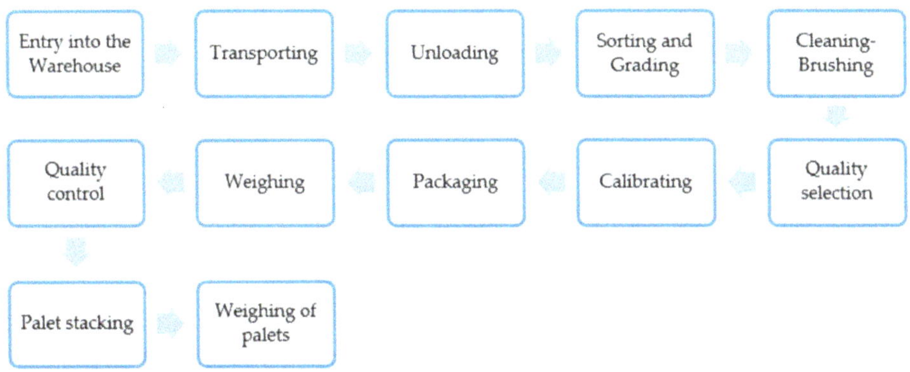

Figure 2. Handling process.

2. Understanding Rural Development

Throughout history, women's employment has been characterised by a low presence in the production chain and restrictive working conditions that have hindered their ability to balance work and family life [7]. This situation is further complicated if women's backgrounds are taken into account, as prejudices further raise barriers to women's access to the labour market [8] and the sector in which they work [9].

In recent years, general awareness of women's participation in the labour market has grown [10], highlighting the need to integrate a gender perspective into any business, as well as into social and cultural projects [11]. However, the data show that women still occupy a secondary position in the work environment, suffering from phenomena such as the well-known glass ceiling, which refers to the difficulty women have accessing high-level positions within companies [12].

At the international level, differences between countries regarding women's participation in the labour market are significant in both quantity and quality [13]. There are also differences between urban and rural environments, with the latter being the most difficult for women [14]. However, some research papers state that when men move to urban areas, it leads to a gradual feminisation of agriculture [15]. In an African context, the difference between women in rural and urban areas is significant [16]. Resources for subsistence farming in rural areas are both insufficient and lack quality. According to [17], life is neither satisfying nor decent for millions of women in rural Nigeria.

In comparison, women's participation in the fruit- and vegetable-handling sector is higher than in other primary-sector activities (fishing, livestock) or other industries (construction, technology). This is due to the characteristics of the food-handling activities, which require greater flexibility and agility [18].

The empowerment of women in the workplace is manifested in their satisfaction with the tasks they perform, with the environment they experience during the workday, as well as with the company that hires them [19]. Scientific literature has addressed the satisfaction of women workers in the fruit- and vegetable-handling sector. Researchers highlight aspects that affect satisfaction, such as the devaluation of women's work [20] or the need to improve working conditions in order to increase the satisfaction of working women, both in terms of physical and mental health [21–23].

The concept of rural development refers to the improvement of the quality of life of rural citizens through initiatives based on endogenous resources that are economically and environmentally sustainable in the long term [24]. The fruit and vegetable sector has been key to the rural development of the Almería area [4], as well as other areas of Spain, such as Huelva [25].

In rural areas, the role of women is key to developing innovative rural entrepreneurship initiatives [26]. Therefore, in Spain, a series of strategies have been articulated to increase the empowerment of women in rural environments, such as the Strategy for Rural Modernisation and Diversification or the Strategic Plan for Equal Opportunities [27].

3. Materials and Methods

Through a systematisation of experiences, the papers referred to in the previous section were analysed. This methodology was chosen because it is considered an innovative and creative approach to learning from a multitude of different experiences [28]. Furthermore, a survey was developed based on expert interviews.

3.1. Phases of the Study

- First phase: Based on the following goals, we tested our initial hypothesis:
  - Detect whether certain situations of economic growth and business dynamics (organisational and technological changes) influence the greater participation of women in the workplace.
  - Understand how the processes of economic growth and business dynamics in the sector have affected working women, particularly women in positions of responsibility.

To carry out this research, we focussed on data collection and interpretation of information involving trade unions and workers in a number of companies.

- Second phase: Based on the information collected, a questionnaire was developed for the companies, combining personal questions and specific details related to the sector. The research was completed following the subsequent steps:
  - A series of surveys and interviews were used to cross-reference the information from the handling companies.
  - In-depth interviews with men and women working in the horticultural sector were conducted, including interviews with social agents and institutions.
  - Technological changes in the local production system were analysed.
  - The effects of innovation on working women in the horticultural sector were evaluated.
- Third phase: Diagnosing study.

The qualitative part of the study was based on in-depth interviews carried out within the sector and was accompanied by hypothesis testing and analyses of the data obtained from the survey conducted in the second phase of the study.

3.2. Companies Studied

Most companies in the sector were established before 1985, a year which is considered the starting point of the development of horticultural companies in Almería. The early 1980s saw the establishment of cooperatives and APCs. In recent years, there has been an increase in the number of *alhóndigas* (places used as both warehouses and marketplaces), while the same has not happened in social economy enterprises. It is important to consider the age of the companies to understand the evolution of working conditions and to verify if feminisation has been observed in certain positions. The 25 companies in the fruit-and-vegetable handling sector in Almería with the highest turnover during the period 2014–2019 were selected since they employ a significant part of the female workforce. All companies had a turnover of more than 75 million euros per year, with two of them exceeding 230 million euros per year.

4. Results

4.1. Typology of Workers and Shareholders

The presence of women in positions of responsibility in fruit and vegetable companies has been debated over the past few years. Compared to 2014, there are now significantly more social economy companies and *alhóndigas* (Table 1), particularly in sectors such as field technicians, marketing, occupational-risk prevention, and quality-control departments. However, in high-responsibility positions, such as management, the number is insignificant.

Table 1. Shareholders.

	Social Economy Enterprises		Alhóndigas (Auctions)	
	2014	**2019**	**2014**	**2019**
Total number of shareholders	17.051	18.067	425	502
% Women	6.82%	9.65%	8.2%	10.02%

Source: Compiled by the authors.

4.2. *Age and Nationality of Women in the Food-Handling Sector*

The data analyses reveal that, in the food-handling sector, over 90% of workers are women. Most of them are Spanish and married (Table 2). The results obtained from the partial surveys do not evince significant differences among the data by area.

Table 2. Horticultural areas studied.

Questions	Campo Dalias	La Cañada	Campo de Níjar
Age (%)			
<20	4	5	0
20–30	27	32	39
30–40	28	37	27
>40	41	26	34
Nationality (%)			
Spanish	76	83	78
Moroccan	7		
Romanian	4		
Bulgarian		9	
Polish		3	
Lithuanian		3	
Ecuadorian	3		5.5
Peruvian	3		5.5
Colombian			5.5
Chilean			5.5
Others	7	2	
Marital Status (%Married)	55	52	39

Source: Compiled by the authors.

The distribution is even across age groups, with the exception of women under 20. This is evidence that most women have been working in the sector for years.

4.3. *Education Level*

Concerning the level of education of women workers in the fruit- and vegetable-handling sector, there has been an increase in the percentage of women with secondary education in recent years, while the percentage of women with no education has decreased (Figure 3).

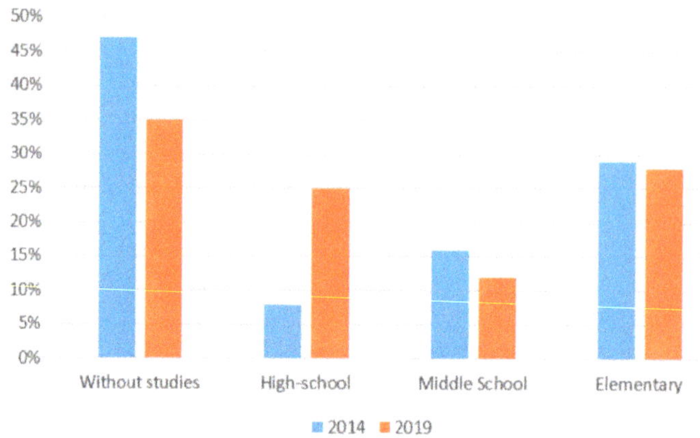

Figure 3. Education level of workers in fruit- and vegetable-processing companies. Source: Compiled by the authors.

4.4. *Types of Contract*

The taxonomy of professionals can be found in the collective employment agreement (ECA). There are different professional subcategories charged with handling and packaging in the fruit, vegetable, and flower sectors. The taxonomy has been designed according to legally-established criteria and based on contractual relation, i.e., according to the type of working contract chosen and formalised by the parties.

- Fixed/Permanent: Those that work for the business on a regular and continual basis.
- Fixed-discontinuous: Those that carry out regular, discontinuous work, which is seasonal or intermittent, for horticultural campaigns.
- Casual/Temporary workers: Those with an indefinite relationship with the business, whose employment is conditioned by market circumstances or workload, including during regular campaigns. The maximum length of these contracts is about nine months over a twelve-month period, calculated from the moment the relationship is initiated.
- Interims: Those that are contracted as substitutions of fixed-term or fixed-discontinuous staff.
- Single job or service: These positions are aimed at not overfilling the staff with fixed-discontinuous workers. Workers are not guaranteed employment during the whole season, so the parties agree that contracts for a single job or service can be completed for typical campaign tasks. However, this kind of contract should not constitute more than 50 percent of the staff.

The survey shows that in the last five years, the number of fixed-term and fixed-discontinuous contracts has vastly increased, giving women the security that comes with stable income streams (Figure 4).

As discussed in the previous study, the number of working hours per day ranges from six to ten, depending on the seasonal activity (Table 3).

Table 3. Hours of work per day.

Campaign/Hours	<5	6 to 9	10 to 15	<15
High Activity	4%	42%	52%	2%
Low Activity	71%	29%	0%	0%

Source: [4].

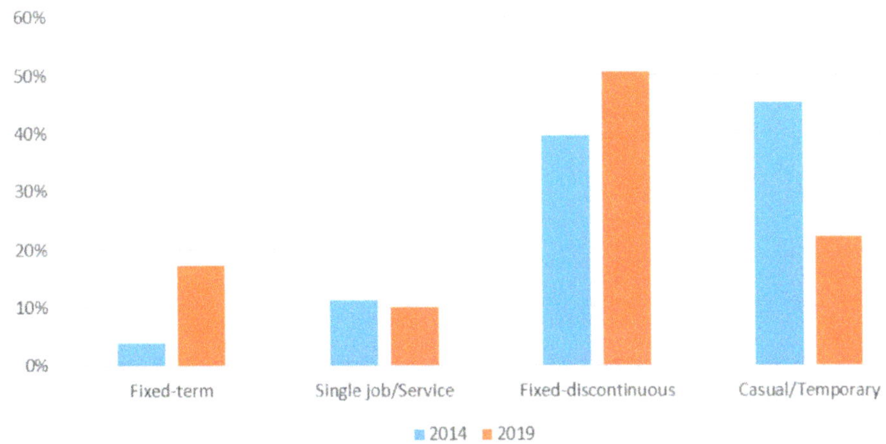

Figure 4. Types of contracts. Source: Compiled by the authors.

4.5. *Impact of Workplace Accidents and Occupational Hazards in the Food-Handling Industry*

At this juncture, it is essential to understand how company size can influence the number of accidents. Thus, information was collected from a sample set of 25 companies in the sector (calculating the average over the years 2019 and 2018), where NA = number of accidents, TC = tons sold, and NW = number of workers. First, asimple correlation analysis of the variables (first panel of Table 4) was conducted. As shown, all variables are closely related. It is evident that the number of accidents is correlated with the trade total (r = 0.78) and the size of the business. This is to be expected, although it also indicates that a bigger company demands a higher level of organisation and greater control in order to reduce the NA. However, this has not been verified, as can be confirmed by looking at the correlation coefficient between NW and NA (0.94). Using ratios helps deepen the analysis. The *productivity index* (PI = TC/NW) and *incidence rate* (IR = NA/NW) are two variables used in the analysis. One might think that the company with the highest productivity (PI) would use the resources more intensively, thus affecting a larger NA, a hypothesis that is not supported by the data.

Table 4. Correlation between variables.

	NW	NA	TC	IR	PI
NW		0.94	0.63	0.14	−0.25
NA	0.94		0.78	0.36	−0.22
TC	0.63	0.78		0.41	0.06
II	0.14	0.36	0.41		−0.02
IP	−0.25	−0.22	0.06	−0.02	

Sample of 25 companies. Source: Compiled by the authors.

The companies studied employ approximately 40% of the total number of women who work in the food-handling sector of the three main areas: La Cañada, Campo de Níjar, and Campo de Dalías.

According to data from the Andalusian Regional Government's occupational-risk prevention service, there has been a considerable decrease in the incidence rate over the last five years (Table 5).

Table 5. Evolution of the magnitudes of incidence rates.

Year	Number of Firms [1]	Average Number of Employees per Firm (1)	Average Number of Accidents per Firm (2)	Incidence Index [(2)/(1)] × 100	Lost Days per Company
2014	56	128	23	17.9	364
2015	60	141	25	17.7	353
2016	59	138	24	17.4	335
2017	58	135	22	16.3	330
2018	57	130	21	16.2	325
2019	60	142	23	16.2	325
Average (2015–2019)	58.8	137.2	23	16.76	333.6

[1] The firms represented are all social enterprises and fall into two categories: cooperatives and agricultural transformation societies. Accidents without sick leave are included. Source: [29].

The average accident rate for the period 2000–2014 was 26 accidents per company, and the incidence rate was almost 20%. In contrast, the period 2015–2019 was characterised by a decrease in both rates, with the average number of accidents per company falling to 23 and the average rate of incidents decreasing to 16.76%. Therefore, the period 2015–2019 evinced the efforts made to increase the occupational safety of workers.

An analysis of women's empowerment and job satisfaction should include an understanding of the degree of risk involved in their work. The workplace incidents with the highest index of danger are overexertion and falling objects. (Table 6).

Table 6. Main position and risk analysis.

Position	Main Risks
Junior Warehouse Worker	Using work equipment: - Falling boxes - Removal of boxes - Manual pallet removal.
Forklift Operator	High workload: - Forklift - Truck (falls, overexertion, and crashes).
Handling and Packaging Workers	Quantity and monotony - Overexertion - Falls - Bumping into mobile objects - Machine blockages.

Most of the injuries suffered by women are specific to the lower and upper limbs (Figure 5).

To understand the recent improvements in the workplace, it is important to consider that the fruit- and vegetable-handling companies are applying the Hazard Analysis and Critical Control Points (HACCP), as well as occupational-risk prevention. This widespread use is due, among other reasons, to the market (customers), which demands their application. It is an excellent example of private initiatives demanding stricter limits that have surpassed public regulatory capacities.

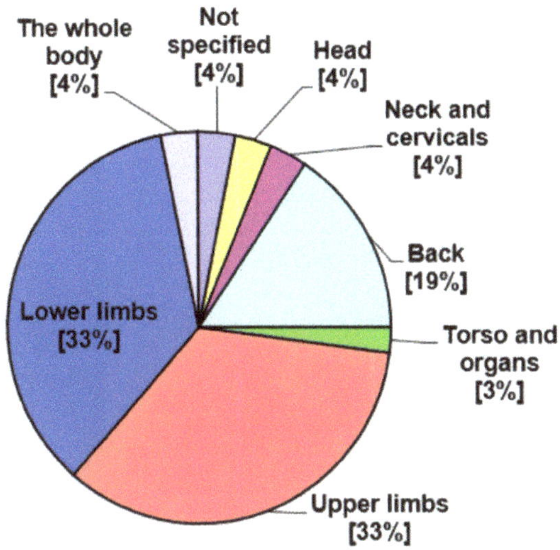

Figure 5. Accidents according to the anatomical region of the body affected, 2019. Source: Compiled by the authors.

4.6. Level of Job Satisfaction

A total of 86% of respondents replied that they did not suffer any discrimination in their working life, compared to 14% who replied that they experienced discrimination during the period.

According to the respondents, the leading causes of discrimination were the following:

- machismo (women consider that men work less but earn more than them)
- pregnancy
- between colleagues (temporary vs. indefinite term workers and between workers in different family situations).

In the final year studied, most women rated their satisfaction level with their supervisors highly (Figures 6 and 7).

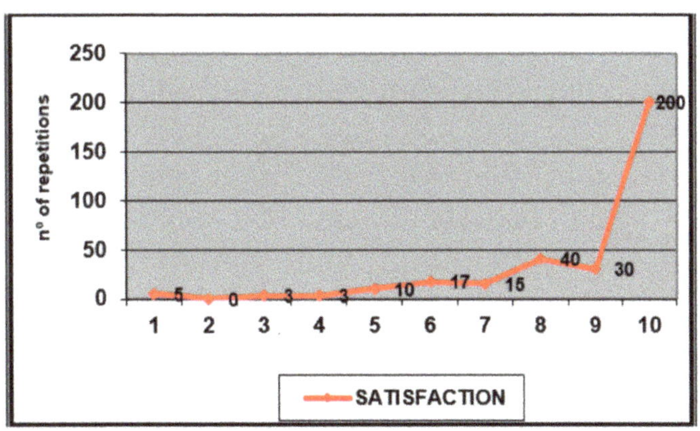

Figure 6. Satisfaction with women supervisors, 2019. Source: Compiled by the authors.

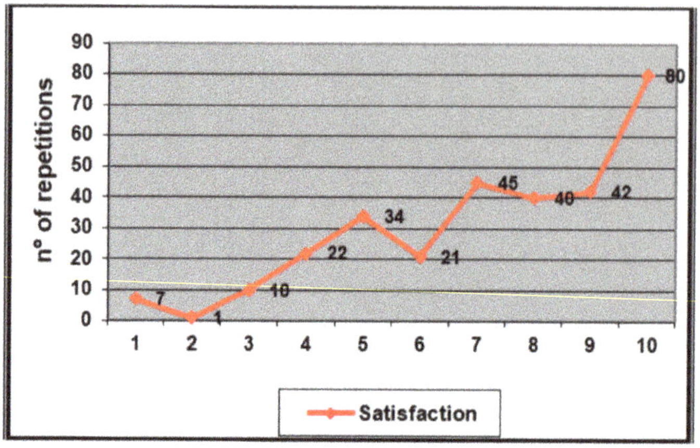

Figure 7. Satisfaction with men supervisors, 2019. Source: Compiled by the authors.

The women were asked about their satisfaction in the workplace using a rating scale of one to ten, where one was not at all satisfied and ten was completely satisfied. Less than 10% of women had a satisfaction level of less than five points, while more than 30% rated both male and female supervisors at ten points. However, the overall satisfaction level was higher when the supervisor was a woman. This dramatically contradicts the results obtained in the period 2000–2014, where participants reported higher levels of satisfaction with male supervisors.

5. Implications of Results

The results show progress in women's empowerment in the food-handling sector in the province of Almería. This has resulted in an increasing number of women in positions of responsibility within fruit and vegetable companies in Almería. However, the level of empowerment is lower when involving the women who occupy positions of responsibility [30]. Historically, the fruit and vegetable sector of Almería has been governed and managed by men, in the greenhouse-farming subsector as well as in other types of business. For example, although women work within greenhouses, the majority of these businesses are owned by men. A similar situation exists within marketing companies, where, traditionally, men have held the majority of technical, executive, and management positions, as well as serving on the board of directors. Despite this, compared to the previous period studied, this subsector can be considered to have initiated a process of feminisation [6].

Within this feminisation process, the improvement in the level of satisfaction experienced by women is relevant. It is linked to the decrease in the number of workplace accidents due to the strict application of the law regarding the prevention of occupational hazards, which has resulted in an improvement in the conditions of the workplace, the use of machinery and vehicles, and the physical and psychological characteristics of the workers (ergonomics) [31–33].

These results are in line with the situation in other countries within the European Union, which brings to light the feminine supremacy in the processing sector. Back in 2013, according to research by the European Parliament [34], in countries such as Bulgaria, France, Poland, and Lithuania, there was a greater percentage of women than men in the processing sector.

The fruit and vegetable sector has not been affected by the economic-financial crisis of 2007 [35,36] nor by the Covid-19 pandemic [37,38]. Fruit and vegetables are considered necessities and are in demand despite adversities. According to the OECD and FAO, "because food is a basic necessity, the agriculture sector is showing more resilience to the global economic crisis than other industries." Moreover, this has meant that women working within the sector have not lost their jobs, unlike other feminised sectors such as the hotel-chambermaid sector [39,40]. Furthermore, the role of trade unions as representatives advocating for working women is fundamental [41–43].

Despite this, there is still much progress to be made to consolidate women's empowerment in all sectors and socio-economic areas [44]. First, this study has considered the role of women in the workplace from an employee perspective. However, the degree of empowerment is lower when women's entrepreneurial activities are taken into account [45]. Furthermore, as mentioned in the introduction, this low level of empowerment is accentuated in the rural environment.

The situation for women in rural areas is characterised by aspects such as the feminisation of the rural exodus, social pressure, and the sexual division of labour, which speaks of a double-discrimination phenomenon: discrimination for being a woman and for living in a rural environment. Issues related to women's empowerment, gender, family, and work-life balance are incorporated into public policies [46]. In the case of Spain, action plans such as the Rural Modernisation and Diversification Strategy or the Strategic Plan for Equal Opportunities have been implemented [27]. All of this is in line with the European Union's guidelines in relation to women in the rural world [47].

6. Conclusions

This study has analysed the evolution of the role of women workers in the fruit- and vegetable-handling sector to ascertain whether the progress in women's empowerment has been consolidated in recent years.

The vegetable sector in Almería is critical to both Spain and the rest of Europe and is fundamentally based on two types of companies, *alhóndigas* and social enterprises. Furthermore, most of the workers in the food-handling departments are women, which means it can be considered a feminised sector. Most of them are Spanish nationals, and there are many fixed-discontinuous employment contracts. During the agricultural season, they work, while, in the summer months, they are unemployed and receive unemployment benefits, which helps them balance their work and family life during the summer months when their children are on vacation. Thus, they receive an income every month of the year.

Concerning women workers' satisfaction in the workplace, differences can be seen before and after 2014. While in the period 2000–2014 women expressed greater satisfaction with male supervisors, there is a change of opinion in the period 2015–2019. This is evidence of greater empowerment of women workers who are no longer afraid to express more controversial opinions as they have job security in the form of a greater number of fixed and fixed-discontinuous employment contracts. In their own words: "It is not the best job I have ever had, but it's my job." The main recommendation for maintaining this positive effect on women's satisfaction and empowerment is to increase the number of women in managerial and technical positions in the handling sector.

To improve the work-life situation of these women, the most powerful tool in the medium and long term is the education of younger generations. Teaching and learning processes must include the gender perspective at all stages of education and in all areas of knowledge, from nursery school to university, and in subjects like mathematics, natural sciences, history, and foreign languages. There is ample proof that gender equality promotes sustainable local development for all sectors of society. Likewise, public policies should encourage creating mechanisms that improve the degree of family and work-life balance for women, such as the establishment of child-care centres within companies of a significant size.

Regarding future lines of research, extrapolations of this analysis to other sectors such as, for example, at the national level, the strawberry and red-fruit sector of Huelva, Spain or, at the international level, the vegetable sector in Morocco are possible. However, this sector's main limitation is the pending negotiations for the collective agreement for the handling sector in the province of Almería, a negotiation process that has been stalled since 2018.

Author Contributions: Conceptualization, J.D.P.V., J.M.-G., J.U.-T. and M.A.G.-V.; methodology, J.D.P.V., J.M.-G., J.U.-T. and M.A.G.-V.; writing—original draft preparation, J.D.P.V., J.M.-G., J.U.-T. and M.A.G.-V.; writing—review and editing, J.D.P.V., J.M.-G., J.U.-T. and M.A.G.-V. All authors have read and agreed to the published version of the manuscript.

Funding: This research received no external funding.

Data Availability Statement: The data presented in this study are available on request from the corresponding author.

Conflicts of Interest: The authors declare no conflict of interest.

References

1. Dekens, J.; Voora, V. *Rural Women, Partner in the Development Process*; IISD Commentary; The International Institute for Sustainable Development: Winnipeg, MB, Canada, 2014; pp. 1–6.
2. Morgan, K.J. Path shifting of the welfare state: Electoral competition and the expansion of work-family policies in Western Europe. *World Polit.* **2013**, *65*, 73–115. [CrossRef]
3. Duflo, E. Women empowerment and economic development. *J. Econ. Liter.* **2012**, *50*, 1051–1079. [CrossRef]
4. Wolosin, R.T. El Milagro de Almería, España: A Political Ecology of Landscape Change and Greenhouse Agriculture. Master's Thesis, Texas State University, San Marcos, TX, USA, 2008.
5. Cantliffe, D.J.; Vansickle, J.J. Competitiveness of the Spanish and Dutch Greenhouse Industries with the Florida Fresh Vegetable Industry. University of Florida, 2012. Available online: http://edis.ifas.ufl.edu/pdffiles/CV/CV28400.pdf (accessed on 3 May 2013).
6. De Pablo-Valenciano, J.; Bernal-Jodar, L.M. The role of women in the horticultural sector in the mediterranean region. In Proceedings of the 20th APDR Congress, Evora, Portugal, 10–11 July 2014; pp. 1319–1329.
7. Dommati, D.; Chittedi, K.R. Socio-economic conditions of agricultural women labour in Andhra Pradesh: A case study of Karimnagar district. *Int. J. Busin. Econ. Manag. Res.* **2011**, *2*, 115–135.
8. Gutierrez, L.M. Working with women of color: An empowerment perspective. *Soc. Work* **1990**, *35*, 149–153.
9. Mittal, M.; Bhakar, S.S. Examining the impact of role overload on job stress, job satisfaction and job performance-a study among married working women in banking sector. *Int. J. Manag. Stud.* **2018**, *2*, 1. [CrossRef]
10. Varghese, T. Women empowerment in Oman: A study based on Women Empowerment Index. *Far East J. Psychol. Busin.* **2011**, *2*, 37–53.
11. Duflo, E. Gender equality in development. *BREAD Policy Pap.* **2005**, *11*.
12. Kiaye, R.E.; Singh, A.M. The glass ceiling: A perspective of women working in Durban. *Gender Manag.* **2013**, *28*, 28. [CrossRef]
13. Nordenmark, M. Does gender ideology explain differences between countries regarding the involvement of women and of men in paid and unpaid work? *Int. J. Soc. Welfare* **2004**, *13*, 233–243. [CrossRef]
14. Ferdoos, A. Social Status of Rural and Urban Working Women in Pakistan—A Comparative Study. Ph.D. Thesis, Osnabrück University, Osnabrück, Germany, 2007.
15. Srivastava, N. Feminisation of agriculture: What do survey data tell us? *J. Rural Dev.* **2011**, *30*, 341–359.
16. Bello-Bravo, J. Rural-urban migration: A path for empowering women through entrepreneurial activities in West Africa. *J. Glob. Entrepreneur. Res.* **2015**, *5*, 9. [CrossRef]
17. Adedayo, A.G.; Oyun, M.B.; Kadeba, O. Access of rural women to forest resources and its impact on rural household welfare in North Central Nigeria. *Forest Policy Econ.* **2010**, *12*, 439–450. [CrossRef]
18. Mingo, E.; Bober, G. Inserciones laborales de trabajadoras agrícolas: Nociones culturales y articulaciones domésticas en los casos del Valle de Uco (Mendoza) y Exaltación de la Cruz (Buenos Aires). *Revista Margen* **2009**, *54*, 1–16.
19. Shikdar, A.A.; Das, B. The relationship between worker satisfaction and productivity in a repetitive industrial task. *Appl. Ergonom.* **2003**, *34*, 603–610. [CrossRef]
20. De Castro, C.; Reigada, A.; Gadea, E. The devaluation of female labour in fruit and vegetable packaging plants in Spanish Mediterranean agriculture. *Organization* **2020**, *27*, 232–250. [CrossRef]
21. Bhavya, B.; Nithyapriyadharshini, V.; Rani, K.J. Occupational hazard and its impact on life satisfaction and wellbeing of women in unorganised sector with special context to fisheries. *Int. J. Adv. Res. Manag. Soc. Sci.* **2016**, *5*, 267–284.
22. Pattussi, M.P.; Olinto, M.T.A.; Canuto, R.; da Silva Garcez, A.; Paniz, V.M.V.; Kawachi, I. Workplace social capital, mental health and health behaviors among Brazilian female workers. *Soc. Psychiatry Psychiat. Epidemiol.* **2016**, *51*, 1321–1330. [CrossRef]
23. Ehlert, C.R.; Mithöfer, D.; Waibel, H. Worker welfare on Kenyan export vegetable farms. *Food Pol.* **2014**, *46*, 66–73. [CrossRef]
24. Milán-García, J.; Uribe-Toril, J.; Ruiz-Real, J.L.; de Pablo Valenciano, J. Sustainable local development: An overview of the state of knowledge. *Resources* **2019**, *8*, 31. [CrossRef]

25. Silva Pérez, R. Territorio, redes e innovaciónen el sistema agrocomercial de la fresa de Huelva. *Cuadernos de Estudios Agroalimentarios (CEA)* **2012**, *2*, 109–131.
26. Ogunlela, Y.I.; Mukhtar, A.A. Gender issues in agriculture and rural development in Nigeria: The role of women. *Hum. Soc. Sci. J.* **2009**, *4*, 19–30.
27. Cobano-Delgado, V.; Llorent-Bedmar, V. Women's well-being and rural development in depopulated spain. *Int. J. Environ. Res. Public Health* **2020**, *17*, 1966. [CrossRef] [PubMed]
28. Herout, P.; Schmid, E. Doing, knowing, learning: Systematization of experiences based on the knowledge management of HORIZONT3000. *Know. Manag. Dev. J.* **2015**, *11*, 64–76.
29. Andalusian Regional Government. *Occupational Risk Prevention Plan*; Andalusian Regional Government: Seville, Spain, 2020.
30. Sharma, E. Women and politics: A case study of political empowerment of Indian women. *Int. J. Sociol. Soc. Pol.* **2020**, *40*, 607–626. [CrossRef]
31. Bergqvist, U.; Wolgast, E.; Nilsson, B.; Voss, M. Musculoskeletal disorders among visual display terminal workers: Individual, ergonomic, and work organizational factors. *Ergonomics* **1995**, *38*, 763–776. [CrossRef]
32. Salerno, S.; Giliberti, C. Non-vehicle commuting in Italy: Need for ergonomic action for women's lower limbs? *Appl. Ergonom.* **2020**, *83*, 102982. [CrossRef]
33. Mistarihi, M.Z. A data set on anthropometric measurements and degree of discomfort of physically disabled workers for ergonomic requirements in work space design. *Data Brief* **2020**, *30*, 105420. [CrossRef]
34. European Parliament. *Women in Fisheries: A European Perspective*; European Parliament: Brussels, Belgium, 2013.
35. Shah, D.K. Global financial and economic crisis: Implications for agricultural sector in India. *Indian J. Agric. Econ.* **2010**, *65*, 477–486.
36. Von Braun, J. *Food and Financial Crises: Implications for Agriculture and the Poor*; International Food Policy Research Institute: Washington, DC, USA, 2008; Volume 20.
37. Poudel, P.B.; Poudel, M.R.; Gautam, A.; Phuyal, S.; Tiwari, C.K.; Bashyal, N.; Bashyal, S. COVID-19 and its global impact on food and agriculture. *J. Biol. Today's World* **2020**, *9*, 221.
38. Singh, A.K.; Singh, L.; Kumar, S. Impact of COVID-19 on agriculture and allied sectors. *J. Commun. Mobilizat. Sustain. Dev.* **2020**, *15*, 8–16.
39. Cañada, E. Intensificación del trabajo en los hoteles: La percepción de las Kellys. *Iglesia Viva* **2018**, *275*, 117–126.
40. Moral Martín, D. ¿Por qué se han organizado las camareras de piso? Algunas claves e interpretaciones desde la revitalización sindical. *Revista Española Sociología* **2020**, *26*, 97–115. [CrossRef]
41. Hamann, K.; Martínez-Lucio, M. Strategies of union revitalization in Spain: Negotiating change and fragmentation. *Eur. J. Ind. Relat.* **2003**, *9*, 61–78. [CrossRef]
42. Béroud, S. Une campagne de syndicalisation au féminin. Une expérience militante dans le secteur de l'aide à domicile. *Travail, Genre Sociétés* **2013**, *2*, 111–128. [CrossRef]
43. Dufresne, A.; Vandewattyne, J. Le syndicalisme en quête d'autonomie et de renouvellement en Europe: Études de cas: Grèce, Espagne, Portugal et France. *Relat. Ind.* **2015**, *70*, 201–209. [CrossRef]
44. Bhuyan, P.M. Women empowerment: Issues and challenges. *PalArch's J. Archaeol. Egypt/Egyptol.* **2020**, *17*, 2393–2398.
45. Pachorkar, S.; Kawishwar, S.; Sharda, P. Women entrepreneurship and women empowerment in India: A case study of Jwala Mahila Samiti. *Prest. Int. J. Manag. Res.* **2020**, *12*, 254–264.
46. Golkar Fard, M. Relationship between entrepreneurship and empowerment dimensions of rural women in Fars province. *J. Entreneur. Agricult.* **2019**, *6*, 1–10. [CrossRef]
47. Vida, M.N.M. Crónica Legislativa de Seguridad Social y materias conexas-nº 16. *Revista Derecho Seguridad Social Laborum* **2018**, *16*, 243–258.

land

MDPI

Article

Performance Evaluation of the Urban Cadastral System in Addis Ababa, Ethiopia

Solomon Dargie Chekole [1], Walter Timo de Vries [2], Pamela Durán-Díaz [2,*] and Gebeyehu Belay Shibeshi [1]

[1] Institute of Land Administration, Bahir Dar University, 5001 Bahir Dar, Ethiopia; solomon.dargie@bdu.edu.et (S.D.C.); gebeyehu.belays@bdu.edu.et (G.B.S.)

[2] Chair of Land Management, Department of Aerospace and Geodesy, Technical University of Munich, 80333 Munich, Germany; wt.de-vries@tum.de

* Correspondence: pamela.duran@tum.de; Tel.: +49-89-289-257-89

Received: 17 November 2020; Accepted: 7 December 2020; Published: 9 December 2020

Abstract: The cadastral system is a land management and land administration tool to provide a safe and reliable real property registration system. In Ethiopia, however, the attempts to implement a reliable urban cadastral system have not been successful, which translates into a deficient land administration system. This paper is an evaluation of the performance of the urban cadastral system of Addis Ababa, based on the European Foundation for Quality Management (EFQM) excellence model. The nine criteria of the model were used as independent and dependent variables. Data were collected through interviews, Likert-type questionnaires, and focus group discussions, and validated with method-to-method technique. Qualitative and quantitative data analysis techniques (ordinal logistics regression model) were employed. In order to ascertain reliability of the data, Cronbach's alpha reliability test was performed in SPSS, and a coefficient of 0.883 was calculated, confirming that the items (questions) have relatively high internal consistency. According to the statistical result from the independent variables, the people result criteria estimated the achievement of cadastral organization at most (1.724). The societal result predicted with a coefficient of 0.281 less. This indicates that the people criterion determines more importantly than other variables. Overall, the independent variables scored the performance of the cadastral organization 24.92 out of 40 points. Findings from interviews and group discussion also confirmed that the most bottlenecks for the organizational achievement are the strategic plan, quality of leadership, bureaucratic processes, and supply of resources. Therefore, we suggest that the responsible authorities need to pay more attention to the enabler criteria (especially, the design of policy and strategy, quality of leadership, provision of resource and partnership, and the process), in order to improve the achievements of the urban cadastral system organization.

Keywords: urban land; urban cadaster; performance evaluation; land management; Ethiopia

1. Introduction

Land is the ultimate resource, without which, life on earth cannot be sustained [1]. The 2030 Agenda for Sustainable Development puts land at the center of accelerating and achieving the Sustainable Development Goals (SDGs) worldwide. The reason is that land plays a significant role in sustainable development due to its multiple economic, social, political, and cultural dimensions. In terms of the economic aspect, it serves as a basis for livelihood; in terms of the social aspect, land is a space for interaction; in terms of politics, land is a source of power; in terms of culture, land is a symbol of collective identity [2]. Hence, significant efforts have been invested all over the world for the correct management of land, including the development of reliable cadastral systems towards a secure land recordation [3,4].

The human dimension of land relates to the concepts of land governance, land management, and land administration. Although these concepts are interconnected, the land administration guideline [1] defines them separately and specifically. According to the land administration guideline, land governance is the process by which decisions are made regarding the access to and use of land, the manner in which those decisions are implemented, and the way that conflicting interests in land are reconciled. Land management is the process by which the resources of land are put to good effect; it covers all activities concerned with the management of land as a resource, both from an environmental and from an economic perspective. Land administration is the processes of recording and disseminating information about the ownership, value, use, and development of land. The land management paradigm [5] turns the cadastral system into the engine of land administration, such that cadastral information assists the functions of land tenure, land value, land use, and land development. In this way, the cadastral system becomes the core technical engine delivering the capacity to control and manage land through the four land administration functions [6]. Soto [7] and Larsson [8], Yildiz [9], and Milka [10] also recognized that accurate and reliable cadastral systems are fundamental to the economic development of any nation.

The main issue of the cadastral system is documenting land information in support of land management, and its definition varies depending on each country's circumstance and context [11,12]. In addition to this, the level of understanding and operation of cadastral systems in different countries are different due to the fact that there are different interpretations of the concept as a consequence of cultural, legal, social, and institutional differences [13]. According to Williamson [14,15], the cadastral system is the foundation and an integral component of parcel-based land information systems (LIS) that contain a record of interests in land. These systems are the central component of the land administration and land management in a state or jurisdiction [16]. Bogaerts [17] defined cadastral systems on the basis of their constituents, in which the cadastral system is a blend of a land registration and a cadaster. In the same way, Zevenbergen et al. [18] stated that a cadastral system consisted of the land registration and the cadaster. For Silva et al. [13], it is the combination of a cadaster with a spatial focus, and a land register with a legal focus including all aspects of the juridical, fiscal, and regulatory cadaster, and developed and assessed considering its political, legislative, economic, technological, and social aspects and relationships. Other scholars defined the term as a subsystem of LIS, which incorporates other subsystems; juridical, regulatory, and fiscal cadastral systems [19,20]. The cadastral template [21] defined it as the system that includes the cadaster, title registry, and the associated processes of land transfer, subdivision, and adjudication, often termed land administration. To Enemark [5,22,23], a mature cadastral system could be considered as a land administration system.

Although there is no universal definition of a cadastral system [15,24], for the purposes of this research paper, it is defined as a system that refers to the operations that a cadastral organization is conducting [25].

Ethiopia, as one of the fastest developing countries in Africa, is in the process of implementing a modern urban cadastral system at the country level. In response to this, new legal and institutional frameworks have been introduced. In this regard, the Constitution of the Federal Democratic Republic of Ethiopia (FDRE) [26], under Article 40 §3, states that the mandate of administering both urban and rural land is given exclusively to the regional states. As an integral part of the land administration, the Constitution promotes the implementation of cadastral systems. Given this empowerment, the urban land is administered and managed by the legal frameworks of Proc. Nos. 721/211 [27] and 818/2014 [28]. These proclamations dictate the modality of urban land acquisition and registration, respectively. The institution undertaking these mandates is the Ministry of Urban Development, Housing and Construction [29].

In addition to these legal frameworks, a five-year strategy called the Growth and Transformation Plan (GTP) was introduced to implement the cadastral system policy. According to this strategy, adjudication and registration of 1.6 million and 1.2 million landholdings, respectively, across 91 cities, are planned with 200,000 adjudicated and 150,000 registered in the first year across the prioritized

23 cities [30]. Ethiopia's urban cadastral system is carried out with the goal of providing a safe and reliable real property registration system in order to foster land management which, in turn, achieves sustainable development goals (SDGs) [31]. Literature studies [16,25,32–35] have documented that Ethiopia's urban cadastral system has not been successful. In view of this, Daniel [20] argued that Ethiopia has experienced a poor urban land registration system due to the past land registration laws and also because strategic directions were not comprehensive. With regard to operational cadastral registration, Deininger [36] revealed that the early 1990s attempts of land titling in Ethiopia were unsuccessful. According to Tigistu [33], the problems and challenges faced in implementing cadastral systems basically fall within the realm of policy and legislative gaps, technical deficiencies, and inadequate institutional arrangements. Likewise, Chekole [25] reported that, though there have been many projects developed to implement the urban cadastral system, none of them could be successful. Each of these projects contained trials for implementing cadastral systems, yet these were often not complementary to earlier projects. This has resulted in overlaps, redundancies, and ill-functioning and inconsistent cadastral systems throughout the country. The aforementioned issues are results of the absence of a progress performance evaluation of the project in each project phase. In other words, there is no systematic assessment and evaluation of the strengths and weaknesses of earlier projects, and there is no systematic set of guidelines used at the start of projects.

Therefore, the purpose of this research is to investigate whether the quality of leadership, strategic planning, excellence of professional expertise, level of partnership, and mode of process affect the organizational performance of the urban cadastral system significantly or not. The underlying research question is: does quality of leadership, strategic planning, excellence of professional expertise, level of partnership, and mode of process affect the organizational performance of the urban cadastral system significantly?

2. Methods and Materials

2.1. Description of the Study Area

According to the Ethiopian constitution, Ethiopia is a federal state, administratively structured into nine regional states and two city administration councils (FDRE, 1995). Based on the nature of the research question, Addis Ababa, the capital city of Ethiopia and headquarters of the African Union, was selected as a case study for the following reasons: the city is giving urgent attention to the cadastral system; it is also the place where the urban cadastral system is being undertaken extensively compared to other regional cities; in addition, the city was selected to serve as a pilot area for the rest of the regional cities.

2.2. Data Obtained and Used

To achieve the intended objective of the study, both primary and secondary data sources were used. Primary data were collected through physical observation and semi-structured interviews. Secondary data were collected from published journals, books, and grey literature (reports, proclamations, regulations, directives, standards, and legislations).

2.3. Experimental Design

The study applied a true experimental research design that relies on statistical analysis to determine the extent to which cadastral system organizational achievements are met. To do so, data were collected in all (ten) sub-cities of Addis Ababa. Likert-type open and close-ended questionnaires were administered to a sample of 150 land administration and management professionals, who were selected purposively. Additionally, ten cadastral system directors from all sub-cities provided their responses in focus group discussions consisting of 6–8 experts in each sub-city.

2.4. *Questionnaire Design*

Both close-ended and open-ended questions (see Appendix A) were designed and administered to the sampled participants. Respondents were requested to rate their level of satisfaction for 33 closed questions raised for 9 categories (5 enablers and 4 results in the EFQM excellence model) related to their organizational circumstances. Benchmarking this model, a number of scholars [37–43] have evaluated the perceptions of respondents based on Likert scale data. The questionnaire contained both positively and negatively worded items to identify random responses. Close-ended responses are bounded between 1–5 scores and represent "Not at all satisfied", "Slightly satisfied", "Moderately satisfied", "Very satisfied", and "Completely satisfied".

2.5. *Methods and Theoretical Framework*

The research applied desk review and case study research methods. The desk review focuses on exploring and looking into existing literature on cadastral systems performance evaluation. Case study was used to get real-life situations and better understandings into the detailed actions of cadastral system.

The research has benchmarked Connel's [44] theory of change to guide the overall theoretical aspect of cadastral systems. Given that the origin of the theory of change lies in the field of monitoring and evaluation, it is an adequate framework for an urban cadastral system performance evaluation in order to determine how much of the intended result of a given intervention is achieved as a result. The EFQM Excellence model is used as an analytical tool to comprehensively measure and evaluate the performance of the urban cadastral system (see Figure 1). The intervention criterion evaluates what and how an organization does in response to the achievement that the organization aimed for.

The enabling interventions are leadership, strategy, people, resource and partnership, and process, which, when effectively implemented, aids the organization to achieve the intended result. For the sake of analysis, these enablers and results are independent and dependent variables, respectively, which can be quantified within the concept of cadastral systems. Each criterion has its own weight, which needs to be considered. The sum of all criteria accounts an overall weight of 100%.

Thus, the overall processes of the theory of change follow a cause–effect relationship as in the Logical model (a hypothesized description of the chain of causes and effects leading to an outcome of interest). It depicts the relationship between organizational activities and its intended effects. When evaluating an urban cadastral system, enablers such as policies and strategies, leadership, excellence of professional expertise, partnerships, and processes are benchmarked as enabling requirements, while organizational results (business results) are benchmarked as organizational achievements.

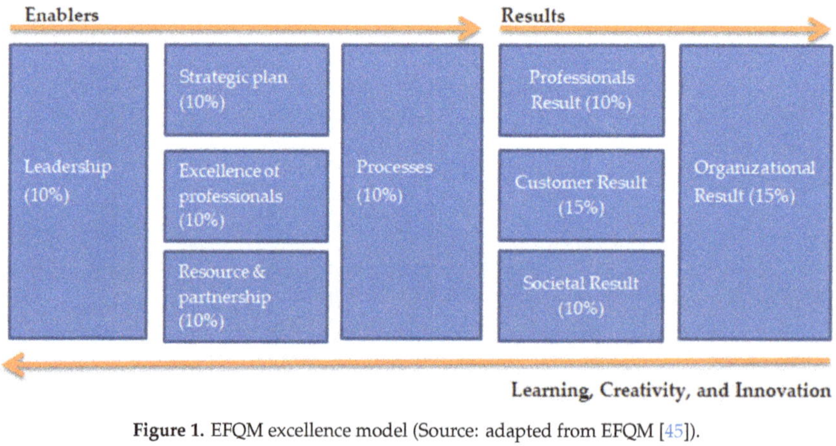

Figure 1. EFQM excellence model (Source: adapted from EFQM [45]).

2.6. Statistical Methods

After collecting the required data, Cronbach's alpha (a measure of internal consistency) was used to check how closely related a set of items (questions) are. Cronbach's alpha, which ensures that either the response given from each respondent was consistent or not, was used to validate and ensure that the data is reliable. Based on the collected data, qualitative and quantitative data analysis techniques were employed. In undertaking the quantitative analysis, a regression model was used to investigate the extent to which independent variables affect the dependent variable. The correlations among the independent variables were also investigated to see which variable has more effects. Data collected from interviews, open-ended questionnaires, observations, and focus group discussions were described qualitatively.

In order to make sure that the collected data are correct, consistent, and useful for its accuracy and reliability against respondents, the data were verified through Cronbach's alpha in SPSS. The alpha coefficient for this study is 0.883, suggesting that the items have relatively high internal consistency. A reliability coefficient of 0.70 or higher is considered "acceptable" in most research studies. After ensuring that the data reliability was correct and certain, Equation (1) was used to calculate the overall organizational achievement of the cadastral system. The equation represents the influence of independent variables X on the dependent variable Y, which is a fundamental concept in ordinal regression.

$$Y = \beta_0 + \beta_1 X_1 W_1 + \beta_2 X_2 W_2 + \ldots + \beta_n X_n W_n \quad (1)$$

where Y is the dependent variable (overall organizational performance), β_0 is the intercept, β_n is coefficients (estimates), X_n is the mean value of independent variables, and n refers to the number of independent variables, which in this case is 8. The intercept for ordinal regression model is zero since it starts from the origin.

Table 1 shows basic descriptive statistical information about the response statistics. The minimum and maximum numbers are bounded between 1 and 5 to represent level of satisfaction. The Mean describes the average of the responses; median explains the value separating the higher half from the lower half of a response, while the mode explains the number that appears most frequently.

Table 1. Response statistics of the questionnaire (n = 150).

		Strategy	Leadership	People	Resource & Partnership	Process	People Result	Customer Result	Societal Result	Organizational Result
N	Valid	150	150	150	150	150	150	150	150	150
	Missing	0	0	0	0	0	0	0	0	0
Mean		3.12	3.25	2.97	3.95	3.97	3.37	2.65	3.41	3.11
Median		3.00	3.20	3.00	4.00	4.00	3.25	3.00	3.00	3.10
Mode		2.75	3.40	3.00	4.00	4.00	3.25	3.00	3.00	3.00

Assumptions of Ordinal logistic regression model: ordinal logistic regression (often just called "ordinal regression") is used to predict an ordinal dependent variable given one or more independent variables. There are four assumptions to validate this model. (1) Dependent variables should be measured at the ordinal level; (2) independent variables must be treated as either continuous or categorical, they cannot be treated as ordinal variables; (3) two or more independent variables that should not be highly correlated with each other (no multicollinearity); (4) each independent variable has an identical effect at each cumulative split of the ordinal dependent variable.

Regression model fitting information: Since the collected data from questionnaire are in the form of order (rank) through Likert scale, ordinal regression model was performed to extract meaningful information. This type of regression model has five conditions to be fulfilled.

Model fitting: this is the measure of how well the model fits the data. The significance level of alpha is 0.05, which limits the level of significance value. The result from this model is 0.00 (See Table 2), which is less than the common alpha level of 0.05, which indicates that it is statistically significant, telling that the model gives better predictions. The statistical significance indicates that changes in the independent variables (enablers) correlate with shifts in the dependent variable (organizational performance).

Goodness of fit: The goodness of fit of a statistical model describes how well it fits a set of observations. Measures of goodness of fit typically summarize the discrepancy between observed values and the values expected under the model in question. The significance value for the goodness of fit (Pearson) is 1 (see Table 2), which is greater than the common alpha level of 0.05, which indicates that it is statistically significant and suggests that the model fits the data very well.

Table 2. Regression model fitting information that justifies appropriateness of the model.

Model Fitting Information				
Model	−2 Log Likelihood	Chi-Square	df	Sig.
Intercept Only	485.218			
Final	309.316	175.902	43	0.00
Link function: Logit.				
Goodness-of-Fit				
	Chi-Square	df	Sig.	
Pearson	558.504	851	1.00	
Deviance	309.316	851	1.00	
Link function: Logit.				
Pseudo R-Square				
Cox and Snell		0.690		
Nagelkerke		0.719		
McFadden		0.363		
Link function: Logit.				
Test of Parallel Lines [a]				
Model	−2 Log Likelihood	Chi-Square	df	Sig.
Null Hypothesis	363.724			
General	338.524 [b]	25.200 [c]	40	0.967

The null hypothesis states that the location parameters (slope coefficients) are the same across response categories. [a] Link function: Logit. [b] The log-likelihood value cannot be further increased after maximum number of step-halving. [c] The Chi-Square statistic is computed based on the log-likelihood value of the last iteration of the general model. Validity of the test is uncertain.

Pseudo R^2: This statistic indicates the percentage of the variance in the dependent variable (organizational performance) that the independent variables (enablers) explain collectively. If R^2 (Nagelkerke) is greater than 0.7, it indicates that 70% of the independent variables explain the dependent variable, which in this case is 0.719.

Test of parallel line: this is the test according to the assumption of proportional odds. This is a key assumption in ordinal regression. The assumption is that the effects of any explanatory variables are consistent (proportional) across the different thresholds (by thresholds we mean the splits between each pair of categories of your ordinal outcome variable). In other words, that the explanatory variables have the same effect on the odds regardless of the threshold.

2.7. Uncertainties and Shortcomings of the Study

Since the study was conducted in the capital city of Ethiopia, the result may not be certain to make generalization to other cities of the country due to geographical differences. Due to the COVID-19 situation, some of the interviews with key informants were made through virtual communication, which may limit the study in getting detailed information compared with physical presence with the interviewee. In addition to the EFQM excellence model, it would have been better if the study had benchmarked other evaluation frameworks.

3. Results

3.1. Socio-Economic Profile of the Respondents

Age and Marital Status: The age of respondents affects their views about the particular problems; usually age indicates level of maturity of individuals, in that sense age becomes more important to examine the response. The perceptions and attitudes of the person can also differ by the marital status due to the acquired responsibilities through marriage. Table 3 shows that 64.7% of the respondents were within the range of 20–30 years; 30% were within the range of 31–40 years; and 5.3% were within the range of 41–50. This indicates that the majority (64.7%) of the respondents were professionals within the age of 31–40 years. An overwhelming number of the respondents (73.3%) were married and the remaining (26.7%) were unmarried.

Education and educational background: education is one of the most important characteristics that might affect a person's attitudes and way of looking and understanding any particular social phenomena. The educational background of respondents also affects the credibility of the responses. A considerable number of respondents (80.7%) were BSc graduates and 19.3% were MSc graduates who are working in the cadastral system organization. In relation to this, most of the experts (70%) are land administration graduates, while lawyers, geomatics and civil engineers, IT, geographers, and related fields share the remaining 30% (see Table 3). It is believed that the organization has recruited appropriate professionals who can accomplish the intended missions of the cadastral system.

Experience and level of income: experience gives firsthand knowledge of what is actually working in the industry. In Table 3, the majority of the respondents (78.7%) have about 4 years of experience, while the remaining 21.3% fall within the range of 8–11 years. It was ensured that the respondents were all very experienced staff members who are (or have been) involved in all aspects of land administration activities. The collective responses can thus be considered significant. Respondents' level of income plays an important role in shaping the economic conditions of an individual, which, in turn, is likely to have a bearing on the responses to a posed problem. Most of the respondents (79.3%) were earning a monthly income of between 12,001 and 15,000 birr, 12.7% of the respondents earn between 15,001 and 18,000 birr, while the rest (8%) earn above 18,000 birr.

Table 3. Profile of the respondents.

Characteristics	Frequency	Percent	Characteristics	Frequency	Percent
Age			Educational Background		
20–30	97	64.7	Land Administration	105	70
31–40	45	30	Geomatics	7	4.7
41–50	8	5.3	Law	14	9.3
Marital Status			Geography	6	4
Single	40	26.7	Civil Engineering	7	4.7
Married	110	73.3	IT	8	5.3
Education			Others	3	2
BSc	121	80.7	Income (ETB)		
MSc	29	19.3	12,001–15,000	119	79.3
Experience (years)			15,001–18,000	19	12.7
4–7	118	78.7	>18,000	12	8
8–11	32	21.3			

3.2. Correlation Results

Correlation (see Table 4) explains the level of association between involved variables. According to the assumption of the ordinal regression model, the level of association between or among independent variables should not be greater than 0.7.

Table 4. Correlation results among variables.

	Policy & Strategy	Leadership	People	Resource & Partnership	Process	People Result	Customer Result	Societal Result	Original Result
Policy & Strategy	1								
Leadership	0.371	1							
People	0.140	0.202	1						
Resource & Partnership	0.1	0.423	0.206	1					
Process	0.062	0.253	0.146	0.662	1				
People Result	−0.081	0.013	0.037	0.036	0.04	1			
Customer Result	0.001	0.054	0.027	−0.003	−0.004	0.042	1		
Societal Result	−0.006	−0.037	−0.045	0.092	0.046	−0.031	−0.077	1	
Organizational Result	−0.034	−0.060	0.12	−0.035	−0.040	0.508	0.252	0.455	1

3.3. *Ordinal Logistic Regression Model Estimation Result*

Ordinal logistic regression model was used to predict the relationship between the ordinal outcome and independent variables towards urban cadastral system level of excellence. From Table 5, it can be noted that policy and strategy, leadership, resource and partnership, process, and customer result are variables which are not statistically significant (p values > 0.05).

Table 5. Coefficients that estimates the influence of independent variables on the dependent.

Variables	Estimate (β)	Exp (β)	Std. Error	df	p Value
Policy & Strategy	0.354	1.425	0.476	1	0.457
Leadership	0.489	1.631	0.653	1	0.454
People	1.120	3.065	0.510	1	0.037
Resource & Partnership	0.508	1.662	0.735	1	0.489
Process	0.540	1.716	0.681	1	0.875
People Result	1.724	5.607	0.618	1	0.000
Customer Result	1.657	5.244	0.377	1	0.085
Societal Result	0.281	1.324	0.312	1	0.000

Based on this evidence, we retain the null hypothesis and reject the alternative hypothesis. To interpret the result, cadastral system policy and strategy, quality of leadership, provided resource and partnership, existing process to deliver services, and the satisfaction result of the customer have no significant effect on the organizational achievements. This does not mean that those variables do not affect, rather they affect the performance of the cadastral organization with less significance. For instance, the quality of cadastral policy and strategy affects the organization with 0.354 amounts. When the independent variable (policy and strategy) increases with 1 unit, the dependent variable (organizational result) will increase with 0.354 amounts.

On the other hand, People, People Result, and Societal Result are statistically significant (p values < 0.05), which in this case reject the null hypothesis and accept the alternative hypothesis. Based on significance values (p values), People, People Result, and Societal Result have a significant effect on the success of the organizational achievements.

Table 6 provides the results of the ordinal logistic regression model. According to the results, all thresholds are statistically significant at the significance level of 0.05.

Table 6. Estimated coefficients, assigned weights, and mean response rate.

Independent Variables	Calculated Coefficients or Estimates (β)	Assigned Weight (W)	Mean ($\bar{X}$) Values	Dependent Variable (Calculated)
Policy & Strategy	0.354	1.0	3.12	
Leadership	0.489	0.8	3.25	
People	1.120	0.9	2.97	
Resource & Partnership	0.508	0.9	3.95	Y = 24.916
Process	0.540	1.4	3.97	
People Result	1.724	0.9	3.37	
Customer Result	1.657	2.0	2.65	
Societal Result	0.281	0.6	3.41	

The column of Calculated Coefficients or Estimates (β) provides the values for β_1 to β_n for this equation; the column Assigned Weight (taken from the EFQM excellence model) presents the weights for the respective independent variables, and the column Mean (X) Values presents the average values for all respondents in each variable. Expressed in terms of the variables used in this table, the regression equation for the overall performance of the organization is calculated based on Equation (1).

$$\begin{aligned} &\textit{Cadastral system achievement } (Y) \\ &= (0.354 * 1 * 3.12) + (0.489 * 0.8 * 3.25) + (1.120\& * 0.9 * 2.97) + (0.508 * 0.9\& * 3.95) \\ &+ (0.540 * 1.4 * 3.97) + (1.724 * 0.9 * 3.37) + (1.657 * 2.0 * 2.65) \\ &+ (0.281 * 0.6 * 3.41) = 24.916 \end{aligned} \tag{2}$$

3.4. Interview and Focus Group Discussion Results

The interview was made (see Table 7) for directors of the urban cadastral system agency in each of the 10 sub-cities. Apart from this, focus group discussions among land administration experts were conducted to crosscheck and validate the responses from the directors.

Table 7. Interview score from the cadastral system directors of ten sub-cities.

Statements	Very Poor	Poor	Fair	Good	Excellent	Total
Quality of policy and strategic plan		8	2			10
Quality of organizational leadership		8	2			10
Excellence of expertise (People)			4	4	2	10
Provision of resource		8	2			10
Quality of organizational (process)		8	2			10
People Result			5	4	1	10
Customer Result		4	4	2		10
Societal Result		1	5	4		10

4. Discussion

The EFQM Excellence Model is based on the logical assumption that excellence in enablers will lead to superior results, and thus leadership drives policy and strategy, people management, and partnerships and resources, and these three elements influence the results through suitable processes [45]. As a quality model, the EFQM Excellence Model explains, through its enabler criteria, the areas that the organization should consider as input to improve its results, as well as the result indicators that must be achieved. In this regard, the EFQM Excellence model provides a pattern of relationships both between enablers and results, and between the criteria. Empirical evidence shows that significant relationships exist between the result elements, where results on one level contribute to outcomes on others [46,47]. The excellence model assumes that customer results, people results, and society results will, together, ultimately infer organizational performance. Research on the relationship between enablers and results indicates that weaknesses in leadership can affect people, customer, societal, and key organizational results [48,49].

Based on the findings of this research, reliability of the empirical data was tested through Cronbach's alpha (0.883) and passed the required level of significance, which is 0.7. This result can be compared with the result by Carlos [50], Cronbach's alpha of 0.71. Hence, the result confirmed that the items (questions) have relatively high internal consistency. The selected regression model (ordinal logistics regression model) was proved to be a good fit with our data and passed the required level of significance (see Table 2). In addition to this, there was an assumption for the ordinal regression model to be met in relation to correlation, which states that correlation among independent variables should not be highly correlated. On the basis of this assumption, it can be observed from the Table 4 that there is a positive and negative correlation between independent variables. If the correlation is greater than 0.7, it is said to be highly correlated (multicollinearity). Accordingly, the assumption is met on the basis of this benchmark. The maximum correlation among independent variables is 0.662, for partnership and process variables, while the lowest correlation is −0.003 for partnership and customer results. These statistics indicate that the ordinal regression is an appropriate model to analyze and interpret the data.

We also considered the results of model fitting, goodness of fit, Pseudo R^2 and test of parallel line. Nagelkerke's (R^2) statistics showed that the independent variables explain about 71.9% of the variations in the outcomes. All those statistics confirm that the model is a good fit to explain the outcome. Based on these test results, estimates (coefficients) presented in Table 5 are calculated. Those estimates or coefficients of the independent variables (Strategy, Leadership, People, Partnership, Process, People Result, Customer Result, and Societal Result) determine the dependent variable (organizational performance). Coefficients are the change in the response associated with a one-unit change of the independent, all other independents being held constant. In short, these parameters are values for the regression equation to predict the dependent variable.

The performance of an urban cadastral system is measured through 8 independent variables, each with a satisfaction score between 1 (not at all satisfied) and 5 (completely satisfied), thus the sum of the overall performance of the cadastral organization could achieve a minimum of 8 and a maximum of 40. Since the performance of an urban cadastral system is measured only through the eight independent variables, there is no intercept (β_0) that can be added with the rest of the estimated coefficients. Hence, based on the results obtained in Table 6, each independent variable predicted the dependent variable in different weights. In the case of the urban cadastral system of Addis Ababa, people result and customer result were estimated with the highest scores among all the independent variables, which are 1.724 and 1.657, respectively, while strategy and leadership were estimated with the lowest values, which are 0.354 and 0.489, respectively. Accordingly, people result was affected the most compared to the rest of the variables, and on the other hand, policy and strategy has less effect on the overall organizational result. Thus, the high value of the regression estimation parameter implies that there is strong causal relationship between the independent and dependent variables. Therefore, the overall performance of the organization is evaluated out of 40. Based on the findings of this research, the urban cadastral system organization scored 24.196 out of 40. In percentage, the Addis Ababa urban cadastral system organization has an overall performance of 62.3%.

With regard to interviews, ten cadastral sub-city directors were interviewed to respond to questions related to the performance of their organization in relation to EFQM criteria. They responded that the most bottlenecks for the achievement of their organization are strategic planning, quality of leadership, bureaucratic processes, and supply of resources. In this regard, most of the respondents (80%) agreed with this idea, in that it was not designed in a way that responds to the existing circumstances of the city.

In addition to this, results from the focus group discussion confirm that most problems emanate from the strategic plan, supply of resources, and leadership skill. Comparing these responses with the interview and questionnaire, it measures the reality of the organization.

5. Conclusions

The objective of this paper was to evaluate the performance (achievement) of the urban cadastral system of Ethiopia, using the case of Addis Ababa, based on the EFQM excellence model. The main idea behind the research is about cadastral information, which is a basis for property valuation, land-use planning, land tenure, and land development. Cadastral information offers accurate inventories of land parcels, provides a true and exact description of the legal situation of rights in land, provides a standardized database for management of public lands, serves as a basis for valuation and taxation, and serves as evidence of ownership for legal cases. Thus, the research benefits land administration stakeholders with these and other related land administration and management functions. This study was conducted based on an international performance evaluation model called EFQM. In this regard, the research is unique in that there is no study conducted in Addis Ababa using this type of evaluation framework/model. It is an empirical study on the urban cadastral system of Addis Ababa using primary data based on direct observation. Hence, it can serve as literature to the scientific community. The research will benefit the land administration and management stakeholders (governmental organizations, NGOs, private sector, etc.) to be aware of the experiences of the urban cadastral system of Addis Ababa. This, in turn, makes the stakeholders well informed in their decision making. Apart from that, some of the interviews with key informants were made through webinar, which may limit the study in getting in-depth information compared with physical presence with the interviewee. The findings showed that the urban cadastral system organization achieves an overall performance of 62.3% with major problems identified in the strategic plan, supply of resources, leadership skill, and the processes bureaucracy.

Policy implications: land administration and management proclamations, regulations, directives, and standards are issued by policy makers. Designing these laws needs well-informed decision makers. Decision makers within the domain of the land governance theme should be aware of what the reality on the ground looks like and about the achievements of the cadastral system organization for their financial planning. Hence, this research provides insights for policy makers in making well-informed decisions. In addition to this, we recommend that the responsible cadastral authorities need to pay serious attention to the enabler criteria (especially, the design of policy and strategy, quality of leadership, provision of resource and partnership, and the process), in order to improve the performance of urban cadastral system achievements.

Future works: this research has conducted the performance of the urban cadastral system organization in Addis Ababa using the EFQM Excellence model. Other scholars in the field may undertake studies using other evaluation frameworks apart from EFQM.

Author Contributions: Conceptualization, S.D.C., W.T.d.V., P.D.-D., and G.B.S.; methodology S.D.C.; validation, S.D.C., W.T.d.V., P.D.-D., and G.B.S.; formal analysis, S.D.C.; investigation, S.D.C.; resources, S.D.C.; data curation, S.D.C.; writing—original draft preparation, S.D.C., W.T.d.V., P.D.-D., and G.B.S.; writing—review and editing S.D.C., W.T.d.V., P.D.-D., and G.B.S.; visualization, S.D.C.; supervision, W.T.d.V., P.D.-D., and G.B.S.; funding acquisition, P.D.-D. All authors have read and agreed to the published version of the manuscript.

Funding: This research received external funding from DAAD, in the form of a Short Term Research Scholarship for In-Country/In region Scholarship holders, grant numer 57520399. The Article Processing Charges of this research were funded by the Technical University of Munich (TUM).

Acknowledgments: The authors would like to acknowledge the editor in chief and anonymous reviewers for their critical review and constructive comments in improving the paper. The authors would also like to thank the Technical University of Munich (TUM), DAAD, and the Institute of Land Administration (ILA) of Bahir Dar University (BDU) for providing supporting materials. Special thanks to Walter Dachaga for proofreading this paper.

Conflicts of Interest: The authors declare no conflict of interest.

Appendix A. Evaluation Criteria of the Cadastral System Organization

Within the framework of the EFQM excellence model, nine criteria were developed and distributed to the respondents in the form of Likert scale questions. The concept behind this research is that people results, customer results, and societal results are achieved through leadership-driving cadastral policy and strategy, people, partnership and resources and processes, which leads ultimately to excellence in cadastral organizational performance achievements. There are included sub-criteria, though broad questions are only presented here.

1. Leadership: how do leaders develop mission and vision, and implement them via appropriate plans for urban cadastral system organization?
2. Strategy: does the urban cadastral system organization implement their mission and vision by developing a stakeholder focused strategy?
3. People: does the urban cadastral system organization value its people and create a culture that allows the mutually beneficial achievement of organizational goals?
4. Partnership and resource: how does the urban cadastral system organization plan and manage external partnerships, suppliers, and internal resources in order to support their strategy, policies and the effective operation of processes?
5. Process: does the urban cadastral system organization design, manage, and improve processes to generate quality services for customers and stakeholders?
6. People results: does the urban cadastral system organization achieve and sustain outstanding results that meet or exceed the needs and expectations of their people?
7. Customer results: does the urban cadastral system organization achieve and sustain outstanding results that meet or exceed the needs and expectations of their customers?
8. Societal results: does the urban cadastral system organization achieve and sustain outstanding results that meet or exceed the needs and expectations of relevant stakeholders within society?
9. Organizational results: does the urban cadastral system organization achieve and sustain outstanding results that meet or exceed the needs and expectations of their business stakeholders?

References

1. UNECE. *Land Administration Guidelines. With Special Reference to Countries in Transition*; United Nations: New York, NY, USA, 1996; ISBN 92-1-116644-6.
2. Dale, P.; McLaren, R. GIS in land administration. In *Geographical Information Systems: Principles, Techniques, Management and Applications*, Abridged Edition; Longey, P., Ed.; John Wiley & Sons, Inc.: New York, NY, USA, 2005; pp. 859–875.
3. Krigsholm, P.; Zavialova, S.; Riekkinen, K.; Ståhle, P.; Viitanen, K. Understanding the future of the Finnish cadastral system—A Delphi study. *Land Use Policy* **2017**, *68*, 133–140. [CrossRef]
4. Dawidowicz, A.; Źróbek, R. A methodological evaluation of the Polish cadastral system based on the global cadastral model. *Land Use Policy* **2018**, *73*, 59–72. [CrossRef]
5. Enemark, S. Understanding the Land Management Paradigm. In Proceedings of the Fig COM 7 Symposium on Innovative Technologies for Land Administration, Madison, WI, USA, 19–25 June 2005.
6. Plimmer, F.; Williamson, I.; Enemark, S.; Wallace, J.; Rajabifard, A. *Land Administration for Sustainable Development*; ESRI Press Academic: Redlands, CA, USA, 2010; ISBN 978-1-58948-041-4.
7. De Soto, H. *The Mystery of Capital. Why Capitalism Triumphs in the West and Fails Everywhere Else*, 1st ed.; Basic Books: New York, NY, USA, 2006; ISBN 978-0465016150.
8. Larsson, G. *Land Registration and Cadastral Systems. Tools for Land Information and Management*; Longman Scientific & Technical: Harlow, UK, 1991; ISBN 0582089522.
9. Yìldiz, O.; Coruhlu, Y.E.; Biyik, C. Registration of agricultural areas towards the development of a future Turkish cadastral system. *Land Use Policy* **2018**, *78*, 207–218. [CrossRef]
10. Mika, M. Modernisation of the Cadastre in Poland as a tool to improve the land management and administration process. *Surv. Rev.* **2020**, *52*, 224–234. [CrossRef]
11. Kocur-Bera, K. Data compatibility between the Land and Building Cadaster (LBC) and the Land Parcel Identification System (LPIS) in the context of area-based payments: A case study in the Polish Region of Warmia and Mazury. *Land Use Policy* **2019**, *80*, 370–379. [CrossRef]

12. Polat, Z.A.; Alkan, M.; Sürmeneli, H.G. Determining strategies for the cadastre 2034 vision using an AHP-Based SWOT analysis: A case study for the Turkish cadastral and land administration system. *Land Use Policy* **2017**, *67*, 151–166. [CrossRef]
13. Silva, M.A.; Stubkjær, E. A review of methodologies used in research on cadastral development. *Comput. Environ. Urban Syst.* **2002**, *26*, 403–423. [CrossRef]
14. Williamson, I.P. Cadasters and Land Information Systems in Common Law Jurisdictions. *Surv. Rev.* **1985**, *28*, 186–195. [CrossRef]
15. Williamson, I.P. Cadastral and land information systems in developing countries. *Aust. Surv.* **1986**, *33*, 27–43. [CrossRef]
16. Williamson, I.; Henssen, J. Land Registration, Cadaster and its Interaction: A World Perspective. In Proceedings of the XIXth International Congress of Surveyors, Helsinki, Finland, 10–19 June 1990.
17. Bogaerts, T.; Zevenbergen, J. Cadastral systems—Alternatives. *Comput. Environ. Urban Syst.* **2001**, *25*, 325–337. [CrossRef]
18. Zevenbergen, J.; Augustinus, C.; Antonio, D.; Bennett, R. Pro-poor land administration: Principles for recording the land rights of the underrepresented. *Land Use Policy* **2013**, *31*, 595–604. [CrossRef]
19. Hull, S.; Whittal, J. *Towards a Framework for Assessing the Impact of Cadastral Development on Land Rights-Holders*; International Federation of Surveyors: Copenhagen, Denmark, 2016.
20. Tadesse, D. Reflections on the situation of urban cadaster in Ethiopia. In Proceedings of the Africa Local Government Action Forum (ALGAF) Phase VI, Addis Ababa, Ethiopia, 7 April 2006.
21. Rajabifard, A.; Steudler, D.; Aien, A.; Kalantari, M. The Cadastral Template 2.0, from Design to Implementation. In Proceedings of the FIG Congress 2014 Engaging the Challenges, Enhancing the Relevance, Kuala Lumpur, Malaysia, 16–21 June 2014.
22. Enemark, S. Land Administration and Cadastral Systems in Support of Sustainable Land Governance-A Global Approach. In Proceedings of the 3rd Land Administration Forum for the Asia and Pacific Region, Tehran, Iran, 24–26 May 2009.
23. Enemark, S.; Williamson, I.; Wallace, J. Building modern land administration systems in developed economies. *J. Spat. Sci.* **2005**, *50*, 51–68. [CrossRef]
24. Steudler, D. (Ed.) *Cadastre 2014 and Beyond*; FIG: Copenhagen, Denmark, 2014; ISBN 978-87-92853-12-7.
25. Chekole, S.D.; de Vries, W.T.; Shibeshi, G.B. An Evaluation Framework for Urban Cadastral System Policy in Ethiopia. *Land* **2020**, *9*, 60. [CrossRef]
26. *Constitution of the Federal Democratic Republic of Ethiopia*; FDRE: Addis Ababa, Ethiopia, 1995.
27. *Urban Land Lease Holding Proclamation*; Proclamation No. 721/2011; Negarit Gazetta; Federal Democratic Republic of Ethiopia: Addis Ababa, Ethiopia, 2011.
28. *Urban Landholding Adjudication and Registration Proclamation 818/2014*; House of People Representatives: Addis Ababa, Ethiopia, 2014.
29. *Strategic Plan of 2020–2025 for Ministry of Urban Development and Construction*; Ministry of Urban Development and Construction: Addis Ababa, Ethiopia, 2020.
30. *Growth and Transformation Plan II (GTP II): Federal Urban Land and Land Related Registration and Information Agency*; FDRE: Addis Ababa, Ethiopia, 2015.
31. Enemark, S. *Fit-For-Purpose Land Administration. Joint FIG/World Bank Publication*; FIG: Copenhagen, Denmark, 2014; ISBN 978-87-92853-10-3.
32. Deininger, K.; Selod, H.; Burns, A. *The Land Governance Assessment Framework: Identifying and Monitoring Good Practice in the Land Sector*; World Bank: Washington, DC, USA, 2012.
33. Tigistu, G.A. Experience and future direction in Ethiopian rural land administration. In Proceedings of the Conference on Land and Poverty, Washington, DC, USA, 18–20 April 2011.
34. Fairlie, K.; Burns, T.; Zhang, Y.; Adlington, G.; Tamrat, I.; Shibeshi, G.; McDowell, A.; Kebede, S.; Zelul, A. Establishing a Legal Cadaster for Good Governance in Ethiopia: Identifying Bottlenecks and Steps towards Scale-Up 2017. In Proceedings of the Annual World Bank Land and Poverty Conference, Washington, DC, USA, 20–24 March 2017.
35. Rajabifard, A. *Sustainable Development Goals Connectivity Dilemma*; CRC Press: Boca Raton, FL, USA, 2019; ISBN 9780429290626.
36. Deininger, K.; Ali, D.A.; Holden, S.; Zevenbergen, J. Rural Land Certification in Ethiopia: Process, Initial Impact, and Implications for Other African Countries. *World Dev.* **2008**, *36*, 1786–1812. [CrossRef]

37. Laarakker, P.; DeVries, W.; Wouters, R. Land Registration and Cadaster, One or Two Agencies?: Stage 2 of the research. In Proceedings of the World Bank Conference on Land and Poverty, Washington, DC, USA, 14–18 March 2016.
38. Kambiz, S.; Hamidreza, A. Mathematical model to rank companies provider EFQM with context-dependent data envelopment analysis (case study: Iran auto industry). *Aust. J. Basic Appl. Sci.* **2011**, *7*, 295–302.
39. Nabitz, U.; Klazinga, N.; Walburg, J. The EFQM excellence model: European and Dutch experiences with the EFQM approach in health care. European Foundation for Quality Management. *Int. J. Qual. Health Care* **2000**, *12*, 191–201. [CrossRef]
40. Turisova, R.; Sinay, J.; Pacaiova, H.; Kotianova, Z.; Glatz, J. Application of the EFQM Model to Assess the Readiness and Sustainability of the Implementation of I4.0 in Slovakian Companies. *Sustainability* **2020**, *12*, 5591. [CrossRef]
41. Vykydal, D.; Folta, M.; Nenadál, J. A Study of Quality Assessment in Higher Education within the Context of Sustainable Development: A Case Study from Czech Republic. *Sustainability* **2020**, *12*, 4769. [CrossRef]
42. Pohle, A.; Blind, K.; Neustroev, D. The Impact of International Management Standards on Academic Research. *Sustainability* **2018**, *10*, 4656. [CrossRef]
43. Wang, G.; Liu, H.; Li, H.; Luo, X.; Liu, J. A Building Project-Based Industrialized Construction Maturity Model Involving Organizational Enablers: A Multi-Case Study in China. *Sustainability* **2020**, *12*, 4029. [CrossRef]
44. Connell, J.P. *New Approaches to Evaluating Community Initiatives. Concepts, Methods, and Contexts*; Aspen Institute: Queenstown, MD, USA, 1995; ISBN 0-89843-167-0.
45. EFQM. European Foundation Quality Management: European Foundation Quality Management (EFQM) Excellence Model. Available online: https://www.efqm.org/ (accessed on 30 October 2019).
46. Oakland, J.S.; Oakland, S. The links between people management, customer satisfaction and business results. *Total Qual. Manag.* **1998**, *9*, 185–190. [CrossRef]
47. Sadikoglu, E.; Olcay, H. The Effects of Total Quality Management Practices on Performance and the Reasons of and the Barriers to TQM Practices in Turkey. *Adv. Decis. Sci.* **2014**, *2014*, 1–17. [CrossRef]
48. Eskildsen, J.K.; Dahlgaard, J.J. A causal model for employee satisfaction. *Total Qual. Manag.* **2000**, *11*, 1081–1094. [CrossRef]
49. Eskildsen, J.K. Identifying the vital few using the European Foundation for Quality Management model. *Total Qual. Manag.* **1998**, *9*, 92–94. [CrossRef]
50. Llusar, J.C.B.; Tena, A.B.E.; Roca-Puig, V.; Beltrán-Martín, I. An empirical assessment of the EFQM Excellence Model: Evaluation as a TQM framework relative to the MBNQA Model. *J. Oper. Manag.* **2009**, *27*, 1–22. [CrossRef]

land

MDPI

Review

Land Tenure Security and Health Nexus: A Conceptual Framework for Navigating the Connections between Land Tenure Security and Health

Walter Dachaga * and Walter Timo de Vries

Chair of Land Management, Department of Aerospace and Geodesy, Technical University of Munich (TUM), 80333 Munich, Germany; wt.de-vries@tum.de
* Correspondence: walter.dachaga@tum.de

Abstract: The rise of urban populations has rendered cities in both developed and developing countries vulnerable to poor health and diseases that are associated with urban living conditions and environments. Therefore, there is a growing consensus that while personal factors are critical in determining health, the urban environment exacerbates or mitigates health outcomes, and as such the solution for improving health outcomes in urban settings can be found in addressing socio-environmental factors that shape urban environments. Land tenure security is a social environmental factor of health that has been understudied by urban geographers despite its obvious role in shaping urban environments, housing conditions, and health. We interpret literature and infer possible pathways through which land tenure security connects to health and propose a land tenure security and health nexus conceptual framework for modeling and investigating the extent of this connection. Based on a narrative review of literature, this inter-disciplinary paper shows that land tenure security can influence health outcomes via four pathways—infrastructure access, environmental justice, psycho-ontological security, and social cohesion. Going forward, a subsequent investigation can focus on developing an index of land tenure security health insults, based on which an empirical investigation of the relationship between land tenure security and health disease is possible.

Keywords: tenure security; health; urban health; land tenure; urban planning; social determinants; land tenure security and health nexus; environmental justice; land use; land management

Citation: Dachaga, W.; de Vries, W.T. Land Tenure Security and Health Nexus: A Conceptual Framework for Navigating the Connections between Land Tenure Security and Health. *Land* **2021**, *10*, 257. https://doi.org/10.3390/land10030257

Academic Editor: Liz Alden Wily

Received: 11 February 2021
Accepted: 28 February 2021
Published: 3 March 2021

1. Introduction

City life is a reality for many and is rapidly becoming so for most of the world's population [1–3]. The world today is experiencing a fast pace of urbanization with nearly 70% of the global population projected to live in urban areas by 2050 [4]. Conventionally, urbanization has been linked to development, and development with health, but in the face of development is the growth of slums, which are characterized by poor health [5]. Cities are therefore characterized by three processes, namely the movement of people into urban centers, development of informal settlements due to lack of tenure, and emergence of environmental and health problems. Thus, while urbanism is idealized to entail social improvements with consequential better quality-of-life for urban residents, on the contrary, in many low-income and some middle-income countries, urbanization conveys inequality and exclusion. This creates cities and dwellings characterized by poverty, overcrowding, poor housing, severe pollution, absence of basic services, and poor health which is defined by the unequal social context that surrounds the daily life of the disadvantaged, and often, socially excluded groups [6]. The rise of urban populations has rendered cities in both developed and developing countries vulnerable to poor health and diseases that are associated with urban living conditions and environments [3]. In response, there is a wealth of literature that recognizes the need to understand the impacts of urban environments and living conditions on the health of urban populations [5,7–13]. There is now a growing consensus that while personal factors are critical in determining health, the urban environment exacerbates or mitigates health and well-being outcomes [14,15]. This is because cities are so ubiquitous and their impact so pervasive that it is difficult to consider any aspect of health ignoring urban settings [16]. To emphasize this, Corburn [17] posits that if the global community is serious about the Sustainable Development Goals (SDGs), greater attention must be paid to understanding and acting to improve urban places, living conditions, and the social and economic conditions that can promote health equity. Hence, the solution for improving health outcomes in urban settings lies in addressing urban environments and associated land–people relations than narrowly focusing on healthcare systems—a stance aligned with socio-environmental determinants of global health [18].

The literature on addressing urban environments to promote health connections is diverse but distributed in professional silos. Most of the literature has subscribed to a planning approach and either praised or blamed urban planning approaches or their absence for the health conditions in urban areas or advocated urban planning as a solution to urban health conditions [13,19–21]. Many have attributed urban health conditions to urban green spaces [22–26], urban density [27,28], transport planning [29,30], environmental pollution including air, water, and noise pollution [31–33], urban land use changes [34–36], and lack of basic infrastructure and water and sanitation facilities [12]. For many, the picture that comes to mind at the mention of urban health is the striking informal settlement that comes with urbanization in most developing countries. The mention of informal settlements creates a picture of poor health and high disease burden and informality which is identified with lack of space and tenure insecurity [13,37,38].

People to land relations and land tenure security are central to the informality, high disease burden, environmental pollution, density, blue-green spaces, land use change, sanitation, and basic infrastructure which are identified with health in urban areas. Land tenure security provides rights that enable access to blue-green spaces, urban infrastructure, and services, while also providing the incentive for environmental stewardship and investment in household infrastructure such as water and sanitation facilities to improve living conditions. Yet, urban planning literature [13,19,21] has made implicit the complex role of land tenure and tenure security in understanding urban environments and health. Some scholars [13,20], in advancing an urban planning course, have argued in the wake of recent pandemics that urban planning, a profession that emerged in the late nineteenth century with a goal of improving the health of the least well-off urban residents, has lost its focus throughout the twentieth century. However, we think that urban planning has not necessarily lost focus, rather it has failed to explicitly leverage the role of land tenure security in urban planning to unleash the full health benefits. While the urban planning and health nexus articulated in literature cannot be underestimated [13], we argue that underlying planning is land tenure which, when combined with planning, leads to improvement in livelihoods [39], quality of life [40] and health. Therefore, focusing on urban planning as a solution to urban health conditions without tenure is a recipe for gentrification [41], which in itself has implications on health [42–44]. Planning and land tenure have land use in common, and while planning organizes space according to uses, tenure connects people to these organized spaces. Therefore focusing on planning alone leads to "urban planning dilemma", where those with legitimate rights to land are different from those deciding how and what to use the land for [45]—a situation which has implications for urban environmental outcomes and health.

Land tenure security is a socio-environmental factor of health that is hardly referenced in the debates about urban built environments and health connections. Nevertheless, it plays an obvious role in shaping urban environments, housing conditions, and for that matter health. Made explicit, Watson [46] emphasizes the importance of land tenure and health in the following excerpt:

"Our identity as human beings remains tied to our land, to our cultural practices, our systems of authority and social control, our intellectual traditions, our concepts of spirituality, and to our systems of resource ownership and exchange. Destroy this relationship and you damage—sometimes irrevocably—individual human beings and their health."

Although this quote clearly recognizes that land tenure is a determinant of health and well-being, the pathway for such association is multi-directional [46] and less understood. In fact, it still sits at the heart of much academic and policy debates which promote tenure choices to deliver certain outcomes including sustainable cities and housing, green cities, clean cities, land degradation neutrality, environmental conservation, and various SDGs [4,38,47–49]. A growing land tenure insecurity correlates with a growing poverty level and volume. A growing poverty level and volume correlates with an increase in compromising of health issues. An increase of compromising of health issues increases the amount of disadvantaged urban population. Yet, it remains unclear how, where, when, through which mechanisms, and under which conditions land tenure insecurity drives and affects health issues. A fundamental issue in land management literature is that there is little reference to health. Likewise, health literature hardly refers to land tenure or land tenure security. In effect, there are insufficient provisions for health in the formulation of land policies and insufficient provisions of land tenure in health policy formulation. Hence, health issues are never a comprehensive part of land policy formulation or are the immediate objectives of land tenure security interventions. Consequently, there are no explicit formal connections between land tenure security and health, which is an omission in both science, policy, and practice. There is no comprehensive study that connects land tenure security with health despite several studies which suggest potential linkages between the two variables [49–54]. The ones that have attempted this connection [47,50,51,55,56] have either ended up alleging a link between land tenure security and health without showing the mechanisms and pathways of this link [51], are not grounded on theoretical and conceptual frameworks [50], or based on a limited scope of what constitutes tenure [47].

This study fills this gap by a process of inferring from existing documentation what the connections between land tenure security and health could be and how this could translate into a land tenure security and health nexus conceptual framework. Such a framework would make the pathways through which variations in land tenure security on the one hand and health issues on the other hand influence or affect each other. The objective of this inference review is to derive the scope, extent, nature, and gaps of existing research on urban environments and health from a land tenure lens/perspective, to identify the pathways through which land tenure security and health are linked, and to develop a land tenure security and health nexus conceptual framework for understanding these linkages. The study sets the stage for further investigation for measuring and empirically investigating the relationship between land tenure security and health outcomes and diseases.

In the next section, we demystify the concepts of health, tenure security, and theoretical foundations of the land tenure security and health nexus. Section 3 describes the methods used in this review. In Section 4, we synthesize findings on the land tenure security and health nexus highlighting the existing evidence of the potential linkages between land tenure security and health. We then present, in Section 5, the land tenure security and health nexus conceptual framework and conclude with implications and recommendations for further research.

2. Land Tenure Security and Social Determinants of Health: Theoretical Framing

Ansari et al. [57] argue that there is a need for (inter-disciplinary) theoretical frameworks that encompass the role of social and environmental determinants of health while acknowledging the crucial inter-connectedness of social context, behavior, and biology. Subsequently, theoretical models such as that by Barton [15] emphasize this need by showing the determinants of health in urban settlements which go beyond the individual to encompass social context in which people live. Although theory by itself is not a panacea, it can help to select key goals and measurable indicators needed to formulate health policies [58]. The strong correlation between proximate determinants of health often obscures the impact of distal social factors, including land tenure security, that can influence the causal pathways for health outcomes within populations. Therefore, theories guide the selection of key constructs and concepts thought to have influence on health and to explain the multiple social and biological processes that result in embodiment and consequent manifestation in the disease burden and health outcomes of populations and individuals. This new thinking conforms to the ideals of modern social epidemiology—a branch of epidemiology concerned with the way that social experiences, social structures, institutions, and relationships influence health [59]. Current dimensions of health call for integrative and cross-disciplinary research strategies or systems, each directed towards some aspects (including land tenure security) of the complex relationship of health and diseases to society and the individuals [57]. Theories underpin such research and provides the basis for explaining the connections between specified phenomena within and across specified domains by using interrelated sets of ideas whose plausibility can be tested by human action and thought [60]. Before delving into what theory is relevant for advancing the cause of land tenure security and health, it is important to understand what the two concepts denote.

2.1. Demystifying the Concepts of Land Tenure Security and Health

Like many concepts, the concepts of land tenure security and health are fluid and, as expected in intellectual discourse, connote different meanings to different individuals and groups, including that it is referred to as housing tenure in some jurisdictions. While the semantics of the concept may be argued differently, we hold the view that land tenure security encompasses housing tenure security. Thus, variants of housing including mortgage, ownership, or rental housing as classified in some jurisdictions, are variants of tenure that come with varying levels of tenure security. Our use of land tenure security is to underscore the fact that land is the basic building block of housing and as such land tenure security and housing tenure security cannot be separated. Particularly in developing country context where real estate markets are dominated by private housing development, tenure in relation to land is mostly precedent to housing tenure. Notwithstanding varying conceptualizations of land tenure [61–66], it fundamentally denotes a rights-based social relationship, whether legal or illegal, between people, land, and society and the basis upon which land is held, used, or owned. The concept of tenure security has evolved in response to efforts to clarify investment incentives of property holders [67]. Varying views on the concept of land tenure security has led to its definition from two broad schools of thought—the assurance and the substance of rights school of thoughts. On the one hand, proponents of the assurance school of thought define security of tenure in terms of the uncertainty of rights, the probability of loss of all or part of the rights held, the expected time of residence before eviction, the uncertainty of changes in government policy, and the impacts of changes in policy on tenure attributes [62,64,68–71]. On the other hand, the substance of right school of thought defines land tenure security in terms of the duration of rights, legal title to land, renewability of rights, and the right to sell or transfer land [72,73]. This study shares the views of Bruce and Migot-Adholia [74] that land tenure security exists when an individual perceives that he or she has legitimate claims to a piece of land on a continuous basis, free from imposition or interference from outside sources, as well as the ability to reap the benefits of labor and capital invested in that land, either in use or upon transfer to another holder. Thus, land tenure security exists when one's land tenure is legally recognized through formal registration or documentation; socially recognized through informal, customary, and undocumented tenure arrangements; and/or symbolically recognized through anti-eviction, adverse possession, social legitimacy, and extra-legal means, as well as when such tenure is also perceived to be secure by the tenure rights holder. We acknowledge that tenure is socially constructed and that there are different lenses for looking at land tenure—either via the lens of degree of formality (continuum of land rights) or legality (legal titles to land) both of which confer varying social obligations of land ownership and use—rights, restrictions, and responsibilities. A focus on legality often excludes legitimate and intermediary tenures that proffer enough security of tenure that are commensurate with people's circumstances of life. We hold the view that the legality of tenure by way of titles is not necessarily a precondition for land tenure security [75]. Instead, the individual's perception anchored on the degree of formality of tenure and the rights, restrictions, and responsibilities conferred by this formality, is the defining feature of land tenure security. The argument here is not that titles do not guarantee secure tenure, rather that there are multiple ways to achieving land tenure security that is not exclusive to land titles. Hence, we view land tenure security from a degree of formality perspective that enable us to capture the variants of tenure on the tenure continuum including but not limited to private rent, public rent, freehold, co-housing, and squatting.

The concept of health includes the traditional biomedical, ecological, holistic, and other universal concepts [16]. From the traditional point of view, health is the absence of disease and illness. That is a person is healthy if only all organs of the person are functioning normally [16]. From an ecological perspective, health is a relative concept that is influenced by the individual's quality of life and surroundings and depends on the individual's ability to adapt to a dynamic environment and society. Thus, health according to the ecological school of thought is the state of equilibrium between human beings and their environment. The holistic concept of health combines the traditional and ecological concepts by conceptualizing health as multi-dimensional concept that seeks to understand human health within the context of environment, which encompasses the physical and social environment.

Adopting a holistic approach to health, the World Health Organization (WHO) in 1986 defined health as "a state of complete physical, mental, and social well-being and not merely the absence of disease or infirmity". Critics [16,76,77] argue that WHO's definition of health lack specificity in defining well-being, for equating well-being to health, and for assuming health as a perfect state of being, which is practically not achievable. However, the various viewpoints of health articulated by the critics agree that the concept of health is multidimensional, involves the absence of disease, and is not influenced by a single criterion. Therefore, this study subscribes to the WHO's definition of health and multidimensionality of health, based on which the study envisions four dimensions of health—environmental, physical, social, and psychological, which can potentially be affected by land tenure security. The abstract question then is, what has a social science construct such as land tenure security got to do with a biomedical construct of health? A premise for answering this question is identifying a framework that allows for identifying and connecting the two interdisciplinary concepts.

2.2. Framing Land Tenure Security as a Social Determinant of Health

Although vigorously speculated in intellectual and policy discourses, land tenure security and health are two distinct concepts that are hardly investigated together. There is literature evidence that hints at linkages between land tenure security and health [50–53,78,79]. The connections between land tenure security and health are mostly investigated without mono-disciplinary theoretical and conceptual frameworks for testing the associations or causal pathways that link land tenure security to health outcomes. On one hand, there are health frameworks [80,81] that link health outcomes with environmental factors, but land tenure security is hardly one of such factors that are linked to health outcomes. On the other hand, there are land management frameworks including sustainable land management framework [82,83], responsible and smart land management framework [84], land governance assessment framework [85], land tenure security and spatial justice framework [86], and tenure responsive land use planning [87] which address land and land tenure security issues but rarely include health matters. Evidence from preliminary reviews of land tenure, urban and environmental health literature, and daily urban life and urbanization experiences suggests that there must be an inter-relation between land tenure security and health diseases. The manifestation of this inter-relation is that land tenure patterns dictate housing patterns and neighborhood conditions which directly affect health outcomes. Similarly, land management interventions such as slum upgrades, resettlement, and tenure formalization have delivered improved health and well-being as outcomes through improved tenure, housing, and environmental conditions [56]. Therefore, if land tenure influences housing and environmental conditions and housing and environmental conditions influence health outcomes, there must be a relation between land tenure security and health outcomes. In fact, this relationship exists and manifests in urban informal settlements, but has been given little attention, which has led to some implications on the health burden of urban areas. This makes it compelling to revisit this relationship. However, an understanding of this relation requires an inter-disciplinary integration of theories that combine society, environments, and health. One domain that presents an opportunity for such integration is social production of health and disease theories which combines ecology, biology, and social context to understand distal social and environmental determinants of health.

Three social epidemiology theories which are relevant for understanding health outcomes and socio-environmental context are psychosocial theory, social production of disease theory, and eco-social theory [18]. The three theories agree that health outcomes are results of an interplay between society and biology but differ in their respective emphasis on different aspects of social and biological conditions and how they integrate to shape population health and disease patterns.

The psychosocial theory hypothesizes that the social environment changes host susceptibility by affecting neuroendocrine function [88,89]. It emphasizes endogenous biological responses to human interactions by focusing on responses to stress and on stressed people in need of psychosocial resources but ignores who and what generates psychosocial insults and buffers as well as how their distribution is shaped by social, political, and economic policies [60]. In relation to this study, the theory offers insights for investigating the social and psychological stresses imposed on people by the presence or absence of land tenure security, and the consequences on health outcomes.

The social production of disease theory [90] posits that economic and political institutions and decisions that facilitate economic and social privilege and inequality are the "fundamental causes of inequalities" in health [91]. It focuses on the health impacts of state policies but offers few principles for identifying and investigating what the actual social determinants of health are [60]. Hence, the theory resonates with policy thinking, and in the case of this study, how land and tenure policy set broad social contexts that influence health outcomes.

The eco-social theory embraces the social production of disease approach while engaging a comparable analysis of biology and ecology [60]. The eco-social theory of disease distribution seeks to answer the principal question—what is responsible for population patterns of health, disease, and well-being, as manifested in the present, past, and changing social inequalities in health? [60]. It integrates social and biologic reasoning along with a dynamic, historical, and ecologic perspective to develop new insights into determinants of population distributions of disease and inequalities in health [92]. The key construct of the eco-social theory is embodiment—how humans biologically incorporate the material and social world in which we live, from conception to death.

Two propositions of the eco-social theory are key for relating land tenure and tenure security to health outcomes:

1. Societies' epidemiological profiles are shaped by ways of living which are created by current and changing societal arrangements of power, property, and the production and reproduction of both social and biological life, involving people, other species, and the biophysical world in which we live.
2. In societies exhibiting social divisions based on property and power, and in which those with the most power and resources constitute a small percentage of the population, the more prevalent the health outcome, the greater the absolute burden (and potentially the relative burden) on those with less power and fewer resources, because they constitute the majority of the population.

In the context of this study, these propositions are relevant for understanding the role of land tenure in health, particularly how variations of land tenure security stratify society, impose differential access to resources, rights, and power that shape health burdens and outcomes of populations and individuals.

Together, the three theories above emphasize the concept of "social position" which underscores social determinants of health [18]. Within this purview, we framed land tenure (security) as a social position, which shapes people's living conditions and health outcomes. In other words, viewed as a social determinant of health, land tenure creates a social structure or relations in society; variations in the nature of tenure and security of tenure create social stratification and assigns individuals to different social positions and circumstances that affect their health.

3. Materials and Methods

3.1. Study Design

Methodologically, this study adopted a narrative review approach [93]. Narrative reviews are less discriminatory in the identification, assessment, and inclusion of studies, which allows for pooling ideas and evidence from studies of varying scientific quality [94]. Unlike systematic reviews, a narrative approach enables searching more broadly across disciplines [95], which allowed us to search interdisciplinary trends in land tenure, urban health, environmental management, urban planning, land management, and health issues. In addition, new topics such as that under consideration in this study do not have enough primary research data upon which to base conclusions, which makes a narrative review appropriate for approaching both scientific and gray literature to derive new knowledge. Admitting that narrative reviews suffer the criticism of potential bias in literature selection, such bias is not spared in an individual-conducted systematic review. Hence, given the relatively new nature of this study, its interdisciplinarity, and the breadth of related literature across disciplines, we find a narrative review approach more appropriate for this study. Therefore, following Green et al.'s [93] methodological framework for writing narrative reviews, we conducted a narrative review of urban planning, health, and land tenure literature to understand the role of land tenure security in health outcomes. According to Green et al. [93], a narrative review comprises four stages—identification of sources of information, setting parameters and delimiting search terms, defining selection criteria, and summary and synthesis of results.

3.2. Identification of Sources

According to Green et al. [93], this stage comprises searching electronic databases, references of known articles and authoritative texts, including expert recommendations and unpublished research. Here, we conducted a search in PubMed, a major public health database, and Web of Science, an interdisciplinary research database. We also hand-searched expert recommended scientific and gray literature in Google and Google Scholar which could elude our search parameters due to different indexing in research databases. The inclusion of gray literature was necessary, as research on land tenure security and health linkages are scanty and mostly talked about in policy briefs and organizational reports such as the World Health Organization, UN-Habitat, and Global Land Outlook.

3.3. Setting of Search Parameters and Terms

Setting search terms and parameters is required to set comprehensive boundaries that make it possible for the authors to retrieve all relevant studies, while focusing the effort as it is not feasible and practicable to review every single study that has some relation to the topic of interest. Therefore, we conducted a systematic search in PubMed, Web of Science, Google Scholar, and Google in October 2019 (and updated in November 2020) using specific combinations of keywords and terms related to land tenure security, urban health, health, land use, built environment, and urban planning. Due to paucity of literature on the topic, the search was not limited geographically and temporally. However, there was a preference for more studies which were relatively recent (<10 years old) without compromising foundational ideas, facts, and logical arguments for recent publications. See Appendix A for the search terms and combinations that were used to identify relevant literature.

3.4. Definition of Selection Criteria

This stage of the review process involves defining search and selection criteria enabling the categorization of diverse studies into relevant and non-relevant literature. The selection involved a three-stage screening process and a final inclusion criterion. The literature search in PubMed, Web of Science, and specific searches in Google and Google Scholar yielded 15,371 records. Using Citavi, duplicates were removed during importation of citation text data, which brought the number of records to 10,178. First, the records were screened by titles to eliminate literature that, by their titles, were either irrelevant or unrelated to land tenure security and health. For instance, titles such as "Where Did They Come from-Multi-Drug Resistant Pathogenic Escherichia coli in a Cemetery Environment?" and "Similarities and Differences Between Yoruba Traditional Healers (YTH) and Native American and Canadian Healers (NACH)" were removed. This step excluded 9733 records. The remaining records were screened by reading their abstracts and selecting those which referred at least to a context in which either land tenure and/or health were discussed. This yielded 95 documents, which were selected for full review (see Figure 1 for flowchart of literature search process). A study was considered relevant literature for inclusion in this review, only if their abstracts suggested a focus on land tenure (in)security and variants of the term (including housing tenure, homeownership, tenure regularization, homelessness, housing security, land security, housing instability, etc.) and health variables including, sanitation, pollution, diseases, and illnesses. By full review, we mean reading and extracting data from the abstracts, introduction, findings, and conclusion of each study.

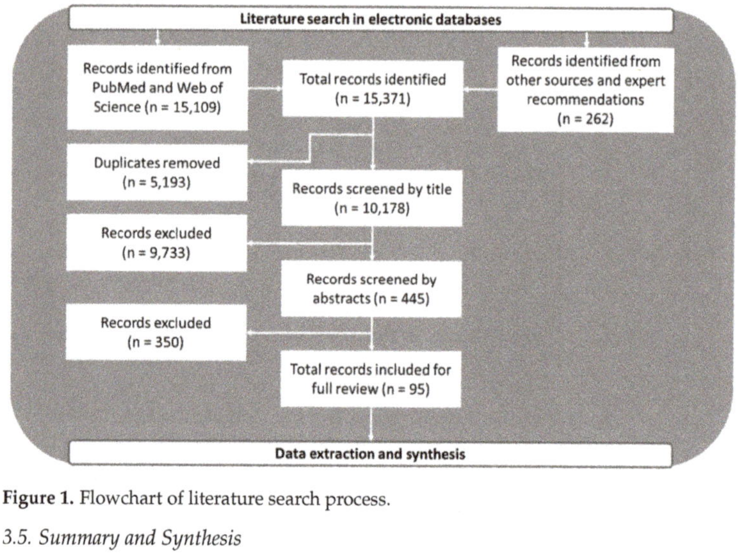

Figure 1. Flowchart of literature search process.

3.5. Summary and Synthesis

Green et al. [93] recommends taking notes on each of the literature works reviewed and constructing a short overview of the research topic based on the literature. Accordingly, we designed a data extraction sheet in Excel, with which we extracted data and took notes from each of the studies reviewed. The data extraction sheet was designed to retrieve data on the objective of each study, the methods used, the variables that were investigated, the findings relating to land tenure security and health, and a section for author note and comments. Relying on thematic analysis, we explored themes within the findings by color-coding the data extraction sheet according to themes, grouping and synthesizing the findings of each study to derive themes that address the connection between land tenure security and health. From this literature synthesis, we present our findings in the next section.

4. The Land Tenure Security and Health Nexus: Pathways of Association

The review identified four pathways through which land tenure security and health are connected—physical space and basic infrastructure, environmental conditions and stewardship, psycho-ontological security, and social cohesion. Our findings demonstrate that land tenure security, or the deprivation of it, fosters or hinders the above pathways and influences the severity or mildness of health burdens and health outcomes of individuals and urban neighborhoods.

4.1. Land Tenure Security, Physical Space, and Basic Infrastructure

One way to improve the living conditions and health of urban dwellers is the provision of affordable and secure shelter with access to basic services and amenities. Yet, literature shows that over one billion people are living in urban (mostly informal) areas that lack access to space and basic infrastructure and services [38]. Several scholars have simply attributed health outcomes in these areas to the provision of water, sanitation, and hygiene facilities through physical planning [96,97]. On the contrary, the solutions to poor health, material deprivation, lack of access to health care, clean water, and sanitation are not merely in the provision of these resources but in the access to these resources and the social milieu within which they are provided [98]. One such context to view the provision of infrastructure and basic services is land tenure. There is a connection between a person's land tenure status and health, which is mediated by the link between tenure and access to infrastructure and basic services [99]. Lack of secure tenure either leaves people with no other options but to settle on hazardous lands, deprives them access to basic amenities in cities, or disincentivizes people from undertaking private investment in infrastructure to improve their living conditions. Unsatisfactory land tenure relations and insecurity can impede investment in housing improvement, water, sanitation, and other services [100]. Moreover, viewed differently, the provision of these infrastructure by central government can lead to dispossessions and exposure to harsh living conditions where there is no security of tenure [101]. Merely providing infrastructure in urban neighborhoods is not enough to improve the living conditions and health of the inhabitants. Instead, people need secure tenure to access and enjoy these infrastructure and services. The lack of land tenure security means that people end up occupying high-risk land on which infrastructure and service provision are less attractive or not feasible at all. Another way in which insecure tenure affects the provision of infrastructure and services is the revenue leakages that are also associated with areas characterized by undefined and secure property rights. Low government revenues from these areas also mean low budgetary allocations for the provision of health and other infrastructure and services [102]. From a data perspective, the instability of persons and neighborhoods without secure tenure also presents a data problem, which is required for undertaking infrastructural and service interventions to improve health or reduce health burdens [103]. Beyond governments providing infrastructure and services, self-help provision, investment, and improvement of basic infrastructure and services are inhibited by the absence of secure tenure. This makes a difference in livelihood and health outcomes as it, for example, limits private provision of sanitation in deprived urban neighborhoods [100]. Neighborhoods with ample physical infrastructure and amenities such as housing, water supply, sanitation, basic drainage, and public space are less prone to poor health and disease contagion. Bhardwaj et al. [104] debunked claims of urban density rendering cities vulnerable to diseases and posit that density is not the problem but infrastructure. To leverage the health benefit of infrastructure however requires that people have secure tenure to access physical infrastructure and amenities to lower their health risks. Giving people secure tenure rights not only enables access to state provided infrastructure but can unleash investment in private infrastructure [105–107], which is an incremental and pro-poor approach to meeting infrastructure needs of urban neighborhoods, given that it is expensive to implement all the infrastructure needs of urban areas. Thus, when people have secure tenure, they can improve their living conditions and health through undertaking self-supportive basic infrastructural investment and improvement. According to Gomes et al. [108], land tenure insecurity inhibits the construction of definitive houses, which in their view makes homes vulnerable to Chagas diseases.

4.2. Land Tenure Security, Environmental Conditions, and Stewardwardship

The link between land tenure security and health is manifest in the complex connection between land tenure security, environmental justice, pro-environmental behaviors, and investment to improve environmental conditions. Urban areas are generators of polluting emissions, vast quantities of solid and human waste that contribute to health risks, and can have serious impacts on public health [99]. People without secure tenure who tend to live near environmentally hazardous facilities feel these impacts disproportionately. This is because either the lack of access to land pushed them to settle for whatever land is available regardless of dangers, or the absence of secure rights made them unable to defend their environmental rights and demand justice from other users of the environment. The insecurity of tenure also means that people are vulnerable to climate shocks and disaster [109,110]. Precarious land rights or land tenure insecurity make people vulnerable to environmental injustice. Without strong or secure land rights, people are unable to resist injustice. When tenure is secure, individuals can push back environmental hazards in defense of their health and environmental rights [111]. Security of tenure is therefore a precondition for environmental justice, the lack of which inhibits the ability of people to take transformative action that can either improve environmental justice or reduce environmental injustices that affect their health [112]. Transformative actions stimulate individuals to undertake investments that improve their environments, to undertake environmental management practices, or to participate in collective environmental activities to reduce their health burden. Secure tenure confers on people environmental rights, responsibility, and restrictions, i.e., rights to enjoy environmental benefits and defend their environments from pollution by others, responsibility to be environmentally accountable to others, and restrictions to act within the confines of environmental laws and other people's environmental rights. The sense of ownership, control, and responsibility that comes with land tenure security is the driving motivation for investing in environmental improvement which fosters good health [52,106,107,113]. Secure tenure also ensures participation in collective activities for better urban environmental conditions, which affects health. Compared to their counterparts who have secured tenure, people without secure tenure lack attachment with their surroundings and do not feel responsible for maintaining healthy environmental practices [114]. Therefore, land tenure security holds an important place in environmental health. First, it provides rights to resist unfair distribution of environmental benefits and burdens. Secondly, it confers responsibility on people to invest in their environments and act within environmentally permissible boundaries.

4.3. Land Tenure Security and Psycho-Ontological Security

A third pathway through which land tenure security connects with health is the psychological security afforded by security over land. Land tenure security contributes to improved health and well-being through two mechanisms: first, by allowing people to have a secure home, it allows some degree of ontological security. Second, it reduces stress associated with insecure tenure and frequent relocations [115]. Land tenure insecurity undermines the ontological security pertaining to the deep psychological need that all human beings have for a home or other locale to operate as a site of constancy, routine, and control in their lives [116]. Thus, the confidence of a person in the continuity of their self-identity and the constancy of the surrounding social and material environments of action [117]. On the contrary, secure tenure commands prestige, autonomy, and protection, which affect the individual's mental state and health. Mental health effects of tenure are not only related to the patterns of stress and anxiety associated with insecurity; the instability of tenure over a reasonable period also means that there is a regular fracturing of social ties and networks which affects a person's psychological being [47]. Insecurity of tenure is detrimental to the sense of stability and belonging of individuals. People who make multiple transitions due to instability of tenure are observed to show worse mental health outcomes as psychological distress increases with the number of transitions [118]. Literature generally supports the notion that homeownership is associated with better psychological health as it offers greater feelings of security compared with other forms of tenure [78,119–122] which are characterized by lack of control and threats of eviction. Owning a home reduces depressive symptoms. Park et al. [55] posit that compared with homeowners, renters are more likely to have depressive symptoms and poor self-rated health. This is mediated by the health effects of housing unaffordability which was observed among renters. Gentrification and residential displacements due to high rents exerts psychological stressors, which affect a person's mental health. A study showed that women who reported depressive symptoms attributed their health circumstances to high mobility, eviction, and problems paying rent [123]. While these studies limit tenure to ownership and leasehold, they agree in their stance that the security and control offered by ownership and the instability associated with rental housing is the distinguishing feature of tenure that affects the psychological health of owners and renters differently. As observed in the city of Belo Horizonte in Brazil, a tenure regularization project led to a downward trend in stress-related homicides in regularized slums, compared to the non-regularized slum settlements [124]. Similarly, a study of migrant populations in urban China indicated that persons with informal tenure have the highest level of perceived stress and worse mental health compared to their counterparts with formal tenure [125]. All these findings suggest that land tenure insecurity predispose holders of such volatile rights to psychosocial distress and diminishing ontological security. Although the literature has dominantly emphasized on homeownership and rental which are forms of tenure [78,119,120,122], it shrouds the nuances of tenure especially in developing country contexts where focus is put on individual land access rather than mortgage and investor supplied housing. Hence, it is relevant to investigate these linkages in different contexts where socio-cultural dynamics present multiple forms of tenure along the tenure continuum.

4.4. Land Tenure Security and Social Cohesion

Land tenure security gives people a sense of autonomy, social identity, and control over their living environment resulting in increased residential stability and reduced stress [115]. The increased residential stability and length of residence contributes to social cohesion, which influences the individual's social health. Social cohesion, social capital, and social networks are important to the material, psychosocial, and political aspects of empowerment that underpin social well-being and equitable health in urban settings. Observation and empirical data, mainly from high-income countries, indicate improvement in health outcomes following improvements in social support and social network [6]. By reducing residential mobility, land tenure security ensures that social ties and networks are maintained. Tenure instabilities deprive people from resources, and from opportunities for socializing. Moreover, the possibility of eviction at any time evokes anxiety, which makes it difficult to retain social ties. On the contrary, guaranteed tenure security creates a foundation for social engagement and strong friendships [126]. This contributes to individual self-esteem, sense of belonging, and community participation. Social mechanisms rely on social interactions between groups to influence behaviors and attitudes, formation of social networks that transmit information, resources, and social support which influences health outcomes [127]. Neighborhoods characterized by informality and insecurity of tenure suffer blemish of place and poor neighborhood reputation [128]. This affects individuals' sense of identity and social status, which leaves a feeling of social exclusion and social stigma. The way these neighborhoods are viewed by different groups including residents and policy makers affects social service provision which has an influence on health [128]. More formal and secure forms of tenure are associated with enhanced neighborhood reputation which has led many scholars to argue that reducing concentrations of informal and less secure tenures and increasing more mixed communities with secure rights reduces social stigma and exclusion which improves people social well-being and health [127,128]. Thus, secure tenure fosters social health through building strong social networks, sense of community, and participation such as in communal work and clean up campaigns for the social good. Where there is no secure tenure owing to gentrification and residential displacements, households lose the essential security and sense of knowing they can call a place home [116], and this affects a person's social and psychological well-being.

5. Conceptual Framework for Land Tenure Security and Health Nexus

Drawing evidence from the preceding section, we propose that looking into concerns and variations of land tenure security has the potential to address, promote, make resilient, or render susceptible multiple dimensions of neighborhood and individual health through four pathways described in the conceptual framework as displayed in Figure 2. Traditional biomedical frameworks for addressing health and diseases assume that health outcomes are a result of deviations from measurable biological factors. Hence, they leave little room for the social, environmental, psychological, and behavioral dimensions of health [129]. In contrast, social epidemiological and biopsychosocial frameworks acknowledge that a basis for understanding the determinants of health and arriving at appropriate interventions for addressing health patterns is that the social context in which a person lives is taken into account [18,57,129]. Land tenure security represents both a social context and social position that influence health through multiple dimensions and outcomes. From a health promotion perspective, land tenure security can be viewed as a multidimensional, behaviorally based socio-environmental intervention that has the potential to modify factors that affect health outcomes.

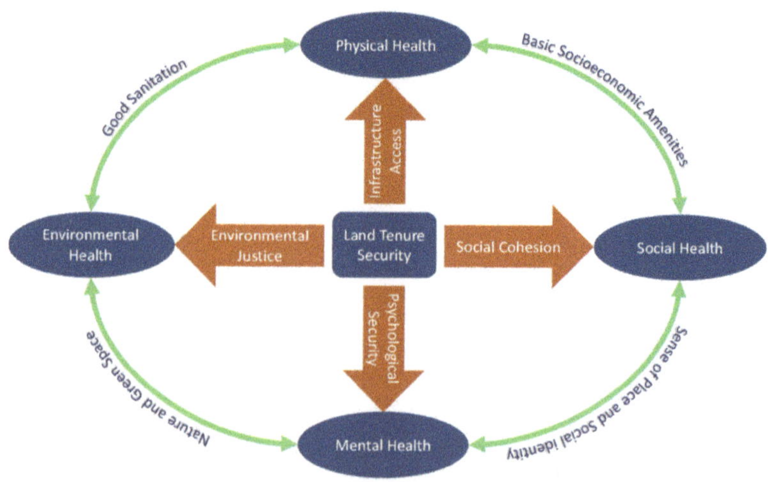

Figure 2. Conceptual framework for land tenure security and health nexus.

Land tenure security has been shown to provide psycho-ontological security which reduces the stress and anxiety associated with tenure insecurity to deliver positive mental health outcomes [47,55,115,116,122]. Reduced residential mobility and sustained social ties associated with tenure security is shown to promote social cohesion, keep families together, reduce place blemishes and social stigmatization, and enable households to focus on long-term goals such as personal relationships, career development, child health, and human capital development which collectively impact social health [115]. Similarly, secure tenure has been demonstrated to facilitate access to and investment in space improvement, basic infrastructure, and services such as water, sanitation, and hygiene facilities that can promote people's physical health [56,99,101,102,104,130–132]. Finally, land tenure security has been shown to be a socially empowering intervention that empowers people with environmental rights, instills environmental responsibility, and imposes environmental restrictions, which collectively enable individuals to lessen their environmental burdens or improve resilience against environmental ills and injustices. Thus, land tenure security empowers people to protect their living environments, while using their own environs responsibly and within permissible limits to improve environmental health [109–112,114]. Land tenure security influences health outcomes by tapping into the socio-physical environment and distal factors such as environmental justice, psychological security, social cohesion, and infrastructure access to modify health outcomes of neighborhoods and individuals. Our conceptual framework linking land tenure security and health was informed by social determinants of health thinking, underscored by three theories of social production of health—psychosocial theory [88,89], social production of disease theory [90], and eco-social theory [60].

From Figure 2, land tenure security, a socio-environmental intervention, guarantees psychological security to promote mental health. It does this by reducing psychosocial stresses and anxieties associated with tenure insecurities and threats of evictions, be it by state authorities through expropriation, planned land uses, development controls, building codes/standards, others claiming rights to land or natural disaster-induced tenure insecurity. Land tenure security ensures environmental justice, enabling a rights-based empowerment to individuals and neighborhoods to take transformative environmental actions against environmental injustices, such as pollution and insanitary conditions, to promote environmental health. Similarly, land tenure security promotes physical health through an infrastructure and service pathway by providing legitimacy and entitlement to use existing state-provided infrastructure or incentivizing individuals and households to undertake private investment in basic infrastructure and life-sustaining services. Finally, land tenure security promotes social health through social cohesion, leveraging on the sustainable social ties and networks afforded by residential stability. Improvements in any dimension of health is improvement in other dimensions of health. With appropriate tenure responsive interventions and policy, land tenure security will ensure good sanitation, access to basic socio-economic amenities, create a sense of place and social identity, and access to nature and green space. Good sanitation impacts environmental and physical health, basic socio-economic amenities impact physical and social health, a sense of place and social identity affects social and mental health, and access to nature and green space affects mental and environmental health.

6. Conclusions, Implications, and Recommendations for Further Research

To tackle the health burden of urban areas, there is the need to address the health risks, challenges of urbanization which define the social context within which people live, and the stratified social positions assigned to individuals and neighborhoods by these risks, challenges, and opportunities. Poor health and health inequalities can be addressed by focusing attention on the creation, management, and improvement of the socio-environmental settings in which people live [48]. In this paper, we set out to explore land tenure security both as a socio-environmental setting and a social position and its connection to health. Drawing upon results of 95 gray and peer-reviewed articles, this paper demonstrated that land tenure security influences health via four pathways—providing psycho-ontological security, enabling environmental justice, infrastructure access and fostering social cohesion. These four pathways promote four dimensions of health namely mental, environmental, physical, and social health. Underscored by social production of health theories, we framed land tenure security as a social determinant of health and accordingly proposed a conceptual framework for understanding the nexus between land tenure security and health.

The paper addressed a critical gap in land management, urban and global health literature, where the health impacts of land tenure (in)security are less understood and understudied, despite policy and intellectual discourses in the wake of coronavirus disease 2019 (COVID-19) which points to greater health burden of areas characterized by smacks of land tenure insecurity. The connection between land tenure security and health is a reality we live with, manifest in our daily encounter and interaction with informal settlements, and unsightly scenes of filth, and continuous (re)emergence of communicable and non-communicable diseases in our cities. This paper and subsequent studies advance a course for achieving the new urban agenda and key sustainable development goals including no poverty, zero hunger, life on land, sustainable cities, and clean water and sanitation, all of which feed into health and well-being.

Advancing the course of land tenure security and health nexus implies abandoning the mono-disciplinary silo approaches and opt for integrative inter-disciplinary approaches, which connect land tenure security and health. Land tenure policies and interventions must be designed to target and achieve health specific goals and outcomes. Similarly, health policies must integrate tenure security in the design of health interventions in a bid to deliver health equity. This means in practice that health and land professionals and institutions must collaborate and work together to deliver a common goal of good health. Thus, cities must work towards adopting health-in-all-policies with the ultimate objective of redressing social structures and factors such as variations in land tenure that generate inequalities in health and health outcomes.

The findings from the current study also open up further research avenues. First, despite the emerging evidence of a link between land tenure security and health, there are still no studies that empirically measure this connection. Hence, there is the need to go further to investigate empirically if there is a connection between land tenure and health and if tenure variations explain variations in health outcomes and disease outbreaks. Secondly, the findings open avenues for exploring tenure security and environmental health justice in cities. Land tenure security is a rights-based approach to justice. Therefore, wherever land tenure insecurities are served, environmental injustices are served as well. A third research avenue is the prospect of this study to inspire research into the mediating role of urban redevelopment, urban greening, and health in cities. Whereas urban greening and redevelopments have health justifications and benefits, they also carry the risk of gentrification and inequities in access, where there are insecure tenure rights. Thus, research is needed to understand how secure tenure can help mitigate the unintended health effects of urban redevelopments and associated gentrification. The findings also open up a fourth avenue for investigating variations in land tenure security in connection to health inequalities in cities. To this extent, neighborhood land tenure security patterns could be studied and analyzed spatially to understand health and disease patterns and distribution in cities. A key driver for activating the link between land tenure security and health is policy. Thus, the study also opens a fifth research dimension that would seek to understand the interactions of urban land tenure policy, urban planning policy and health policy to deliver equitable health outcomes in cities.

This study is only the beginning of a new research agenda to rethink urban health from a land tenure security lens. Moving beyond theory, a further step of this study is to develop a tenure security "health insults" index based on which an empirical investigation of the connection between land tenure security and health outcomes is possible.

Author Contributions: This manuscript is a part of ongoing PhD. research. The PhD. candidate (W.D.) and the supervisor (W.T.d.V.) made their respective contributions to the manuscript as follows: Conceptualization, W.D.; methodology, W.D.; validation, W.T.d.V.; formal analysis, W.D.; investigation, W.D.; resources, W.D. and W.T.d.V.; data curation, W.D.; writing—original draft preparation, W.D.; writing—review and editing, W.T.d.V.; visualization, W.D.; supervision, W.T.d.V. All authors have read and agreed to the published version of the manuscript.

Funding: This research received no external funding.

Institutional Review Board Statement: Not applicable.

Informed Consent Statement: Not applicable.

Data Availability Statement: Not applicable.

Acknowledgments: This study was carried out while undertaking a Ph.D. research program at the Chair of Land Management, Technical University of Munich (TUM). We wish to express our appreciation to the Konrad Adenauer Stiftung (KAS) for funding the doctoral studies. We would also like to thank the mentor of the Ph.D. candidate (U.E. Chigbu) for his mentorship, E.D. Kuusaana, fellow doctoral candidates, and reviewers whose intellectual discussions and constructive comments helped to improve the paper.

Conflicts of Interest: The authors declare no conflict of interest.

Appendix A

Table A1. Search terms.

Search Terms & Strategy	PubMed Records	Web of Science Records	Google and Google Scholar
"Land tenure" and "urban health"	42	33	-
"Land tenure" and "health"	0	64	-
"Land rights" and "health"	26	0	-
"Land rights" and "urban health"	2	0	-
"Built environment" and "diseases"	339	589	-
"Built environment" and "urban health"	183	82	-
"Land ownership" and "health"	84	77	-
"Land security" and "health outcomes"	4	0	-
"Land tenure" and "health vulnerability"	22	0	-
"Housing tenure" and "health"	382	187	-
"Built environment" and "health"	-	3650	-
"Built environment" and "health outcomes"	921	364	-
"Urban health vulnerability"	3022	1	-
"Built environment" and "health vulnerability"	299	0	-
Built environment and diseases	1135	0	-
"Land" and "health nexus"	2	8	-
"Land tenure" and "disease"	22	16	-
"Land rights" and "diseases"	4	0	-
Land tenure health nexus	165	10	-
Land tenure disease nexus	47	0	-
Land access health nexus	0	0	-
"Land tenure" and "health"	122	87	-
"Land tenure security" and "environment"	1	8	-
"Land tenure security" and "environmental"	1	24	-
"land tenure" and "environmental health"	2	4	-
"Land tenure" and "environment"	119	251	-
"Land tenure" and "diseases"	24	22	-
"Housing tenure" and "mental health"	46	63	-
"Land tenure" and "mental health"	1	2	-
"Land ownership" and "environmental health"	4	1	-

Table A1. *Cont.*

Search Terms & Strategy	PubMed Records	Web of Science Records	Google and Google Scholar
"Land tenure" and "social health"	3	2	-
"Housing tenure" and "environmental health"	3	2	-
"Land" and "social health	26	36	-
"Housing tenure" and "physical health"	8	11	-
"Homelessness" and "physical health"	151	193	-
"Homelessness" and "social health"	11	15	-
"Homelessness" and "environmental health"	26	4	-
"Informal tenure" and "health"	338	0	-
"Informal tenure" and "environmental health"	8	0	-
"Slums" and "social inclusion"	3	10	-
"Slums" and "social exclusion"	9	26	-
"Housing tenure" and "social exclusion"	3	11	-
"Informal settlements" and "health"	365	414	-
"Informal settlements" and "environmental health"	25	11	-
"Informal settlements" and "physical health"	5	7	-
"Informal settlements" and "social health"	3	3	-
"Informal settlements" and "mental health"	22	33	-
"Sense of place" and "health"	89	146	-
"Informal settlements" and "sense of place"	3	0	-
"Informal settlements" and "social exclusion"	3	11	-
"Informal settlements" and "social inclusion"	3	8	-
"Informal settlements" and "diseases"	67	55	-
"Land tenure" and "social identity"	0	5	-
"Land tenure" and "social exclusion"	0	6	-
"Land tenure" and "social inclusion"	0	6	-
"Land tenure security" and "sense of place"	0	1	-
"Land tenure" and "sense of place"	0	2	-
"Land tenure" and "cultural identity"	0	4	-
"Housing tenure" and "social inclusion"	0	4	-
Specific hand searches in Google and Google Scholar	-	-	262

References

1. Bodo, T. Rapid urbanisation: Theories, causes, consequences and coping strategies. *Ann. Geogr. Stud.* **2019**, *2*, 32–45.
2. Castells-Quintana, D.; Wenban-Smith, H. Population Dynamics, Urbanisation without Growth, and the Rise of Megacities. *J. Dev. Stud.* **2020**, *56*, 1663–1682. [CrossRef]
3. Connolly, C.; Keil, R.; Ali, S.H. Extended urbanisation and the spatialities of infectious disease: Demographic change, infrastructure and governance. *Urban. Stud. (Edinb. Scotl.)* **2021**, *58*, 245–263. [CrossRef]
4. United Nations. *World Urbanization Prospects: The 2018 Revision*; United Nations: New York, NY, USA, 2019.
5. Vlahov, D.; Freudenberg, N.; Proietti, F.; Ompad, D.; Quinn, A.; Nandi, V.; Galea, S. Urban as a determinant of health. *J. Urban Health Bull. N.Y. Acad. Med.* **2007**, *84*, i16–i26. [CrossRef] [PubMed]
6. De Snyder, V.N.S.; Friel, S.; Fotso, J.C.; Khadr, Z.; Meresman, S.; Monge, P.; Patil-Deshmukh, A. Social conditions and urban health inequities: Realities, challenges and opportunities to transform the urban landscape through research and action. *J. Urban Health Bull. N.Y. Acad. Med.* **2011**, *88*, 1183–1193. [CrossRef] [PubMed]

7. Godfrey, R.; Julien, M. Urbanisation and health. *Clin. Med.* **2005**, *5*, 137–141. [CrossRef]
8. Vearey, J.; Palmary, I.; Thomas, L.; Nunez, L.; Drimie, S. Urban health in Johannesburg: The importance of place in understanding intra-urban inequalities in a context of migration and HIV. *Health Place* **2010**, *16*, 694–702. [CrossRef] [PubMed]
9. Sverdlik, A. Ill-health and poverty: A literature review on health in informal settlements. *Environ. Urban.* **2011**, *23*, 123–155. [CrossRef]
10. WHO. *The Economics of Social Determinants of Health and Health Inequalities: A Resource Book*; World Health Organization: Geneva, Switzerland, 2013.
11. Chigbu, U.E. e-Tracking COVID-19 Disruptions to the Global Development Agenda on Land. *Int. J. Environ. Sci. Nat. Resour.* **2020**, *26*, 1–9.
12. Corburn, J.; Vlahov, D.; Mberu, B.; Riley, L.; Caiaffa, W.T.; Rashid, S.F.; Ko, A.; Patel, S.; Jukur, S.; Martínez-Herrera, E.; et al. Slum Health: Arresting COVID-19 and Improving Well-Being in Urban Informal Settlements. *J. Urban Health Bull. N.Y. Acad. Med.* **2020**, *97*, 348–357. [CrossRef]
13. Chigbu, U.E.; Onyebueke, V.U. The COVID-19 pandemic in informal settlements: (re)considering urban planning interventions. *Town Plan. Rev.* **2021**, *92*, 115–121. [CrossRef]
14. Barton, H. Land use planning and health and well-being. *Land Use Policy* **2009**, *26*, S115–S123. [CrossRef]
15. Barton, H.; Grant, M. A health map for the local human habitat. *J. R. Soc. Promot. Health* **2006**, *126*, 252–253. [CrossRef] [PubMed]
16. Grover, A.; Singh, R.B. *Urban. Health and Wellbeing: Indian Case Studies/Aakriti Grover, R.B. Singh*; Springer: Singapore, 2020.
17. Corburn, J. Urban Place and Health Equity: Critical Issues and Practices. *Int. J. Environ. Res. Public Health* **2017**, *14*, 117. [CrossRef] [PubMed]
18. World Health Organization (WHO). *A Conceptual Framework for Action on the Social Determinants of Health*; WHO: Geneva, Switzerland, 2010.
19. Corburn, J. City planning as preventive medicine. *Prev. Med.* **2015**, *77*, 48–51. [CrossRef]
20. Corburn, J. *Toward the Healthy City: People, Places, and the Politics of Urban Planning*; MIT Press: Cambridge, UK, 2009.
21. Arthurson, K.; Lawless, A.; Hammet, K. Urban Planning and Health: Revitalising the Alliance. *Urban. Policy Res.* **2016**, *34*, 4–16. [CrossRef]
22. Nawrath, M.; Guenat, S.; Elsey, H.; Dallimer, M. Exploring uncharted territory: Do urban greenspaces support mental health in low- and middle-income countries? *Environ. Res.* **2020**, *194*, 110625. [CrossRef]
23. Kruize, H.; van der Vliet, N.; Staatsen, B.; Bell, R.; Chiabai, A.; Muiños, G.; Higgins, S.; Quiroga, S.; Martinez-Juarez, P.; Aberg Yngwe, M.; et al. Urban Green Space: Creating a Triple Win for Environmental Sustainability, Health, and Health Equity through Behavior Change. *Int. J. Environ. Res. Public Health* **2019**, *16*, 4403. [CrossRef]
24. Labib, S.M.; Lindley, S.; Huck, J.J. Spatial dimensions of the influence of urban green-blue spaces on human health: A systematic review. *Environ. Res.* **2020**, *180*, 108869. [CrossRef] [PubMed]
25. Kim, M.; Rupprecht, C.; Furuya, K. Residents' Perception of Informal Green Space—A Case Study of Ichikawa City, Japan. *Land* **2018**, *7*, 102. [CrossRef]
26. Semeraro, T.; Scarano, A.; Buccolieri, R.; Santino, A.; Aarrevaara, E. Planning of Urban Green Spaces: An Ecological Perspective on Human Benefits. *Land* **2021**, *10*, 105. [CrossRef]
27. Gibson, L.; Rush, D. Novel Coronavirus in Cape Town Informal Settlements: Feasibility of Using Informal Dwelling Outlines to Identify High Risk Areas for COVID-19 Transmission from A Social Distancing Perspective. *Jmir Public Health Surveill.* **2020**, *6*, e18844. [CrossRef] [PubMed]
28. Seidlein, G.; von Alabaster, L.; Deen, J.; Knudsen, J. Crowding has consequences: Prevention and management of COVID-19 in informal urban settlements. *Build. Environ.* **2021**, *188*, 107472. [CrossRef]
29. Matan, A.; Newman, P.; Trubka, R.; Beattie, C.; Selvey, L.A. Health, Transport and Urban Planning: Quantifying the Links between Urban Assessment Models and Human Health. *Urban. Policy Res.* **2015**, *33*, 145–159. [CrossRef]
30. Nieuwenhuijsen, M.J. Urban and transport planning pathways to carbon neutral, liveable and healthy cities; A review of the current evidence. *Environ. Int.* **2020**, *140*, 105661. [CrossRef] [PubMed]
31. Gariazzo, C.; Carlino, G.; Silibello, C.; Renzi, M.; Finardi, S.; Pepe, N.; Radice, P.; Forastiere, F.; Michelozzi, P.; Viegi, G.; et al. A multi-city air pollution population exposure study: Combined use of chemical-transport and random-Forest models with dynamic population data. *Sci. Total Environ.* **2020**, *724*, 138102. [CrossRef]
32. Chen, S.-Y.; Chu, D.-C.; Lee, J.-H.; Yang, Y.-R.; Chan, C.-C. Traffic-related air pollution associated with chronic kidney disease among elderly residents in Taipei City. *Environ. Pollut. (Barking, Essex: 1987)* **2018**, *234*, 838–845. [CrossRef] [PubMed]
33. Opisa, S.; Odiere, M.R.; Jura, Walter, G.Z.O.; Karanja, D.M.S.; Mwinzi, P.N.M. Faecal contamination of public water sources in informal settlements of Kisumu City, western Kenya. *Water Sci. Technol. A J. Int. Assoc. Water Pollut. Res.* **2012**, *66*, 2674–2681. [CrossRef] [PubMed]
34. Corburn, J. Urban land use, air toxics and public health: Assessing hazardous exposures at the neighborhood scale. *Environ. Impact Assess. Rev.* **2007**, *27*, 145–160. [CrossRef]
35. Winkworth, C.L. Land-use change and emerging public health risks in New Zealand: Assessing Giardia risks. *New Zealand Med. J.* **2010**, *123*, 55–66. [PubMed]
36. Kim, H. Land Use Impacts on Particulate Matter Levels in Seoul, South Korea: Comparing High and Low Seasons. *Land* **2020**, *9*, 142. [CrossRef]

37. Durand-Lasserve, A.; Royston, L. *Holding Their Ground: Secure Land Tenure for the Urban Poor in Developing Countries*; Earthscan Publications: London, UK, 2002.
38. UN-Habitat. *WORLD CITIES REPORT 2020: The Value of Sustainable Urbanization*; United Nations Human Settlements Programme: Nairobi, Kenya, 2020.
39. Dachaga, W.; Chigbu, E.U. Understanding tenure security dynamics in resettlement towns: Evidence from the Bui Resettlement Project in Ghana. *JPLM* **2020**, *1*, 38–49.
40. Chigbu, U.E. Tenure Responsive Land-Use Planning as a Tool for Improving Quality of Life: The Perspective of Sub-Saharan Africa. In *Handbook of Quality of Life and Sustainability*, 1st ed.; Martinez, J., Mikkelsen, C.A., Phillips, R., Eds.; Springer International Publishing: Cham, Germany, 2021; pp. 17–33.
41. Chigbu, U.; Alemayehu, Z.; Dachaga, W. Uncovering land tenure insecurities: Tips for tenure responsive land-use planning in Ethiopia. *Dev. Pract.* **2019**, *29*, 371–383. [CrossRef]
42. Schnake-Mahl, A.S.; Jahn, J.L.; Subramanian, S.V.; Waters, M.C.; Arcaya, M. Gentrification, Neighborhood Change, and Population Health: A Systematic Review. *J. Urban Health Bull. N.Y. Acad. Med.* **2020**, *97*, 1–25. [CrossRef] [PubMed]
43. Cole, H.V.S. A call to engage: Considering the role of gentrification in public health research. *Cities Health* **2020**, *4*, 278–287. [CrossRef]
44. Cole, H.V.S.; Anguelovski, I.; Baró, F.; García-Lamarca, M.; Kotsila, P.; Del Pérez Pulgar, C.; Shokry, G.; Triguero-Mas, M. The COVID-19 pandemic: Power and privilege, gentrification, and urban environmental justice in the global north. *Cities Health* **2020**, 1–5. [CrossRef]
45. Boamah, E.F.; Amoako, C. Planning by (mis)rule of laws: The idiom and dilemma of planning within Ghana's dual legal land systems. *Environ. Plan. C Politics Space* **2020**, *38*, 97–115. [CrossRef]
46. Watson, N.L. Implications of land rights reform for Indigenous health. *Med J. Aust.* **2007**, *186*, 534–536. [CrossRef]
47. Baker, E.; Bentley, R.; Mason, K. The Mental Health Effects of Housing Tenure: Causal or Compositional? *Urban. Stud. (Edinb. Scotl.)* **2013**, *50*, 426–442. [CrossRef]
48. UN-Habitat; World Health Organisation. Integrating Health in Urban and Territorial Planning: A Sourcebook. Available online: https://unhabitat.org/integrating-health-in-urban-and-territorial-planning-a-sourcebook-for-urban-leaders-health-and (accessed on 23 December 2020).
49. Sena, A. Land Under Pressure—Health Under Stress. Available online: https://knowledge.unccd.int/publication/land-under-pressure-health-under-stress (accessed on 8 February 2021).
50. Choudhury, P.; Ghosh, R.K.; Sindhi, S. *COVID-19 Crisis, Pandemic Resilience and Linkages to Land: An Exposition*; Working Papers id:13058; eSocialSciences: New Mumbai, India, 2020.
51. Tseng, T.-W.J.; Robinson, B.E.; Bellemare, M.F.; BenYishay, A.; Blackman, A.; Boucher, T.; Childress, M.; Holland, M.B.; Kroeger, T.; Linkow, B.; et al. Influence of land tenure interventions on human well-being and environmental outcomes. *Nat. Sustain.* **2021**, *4*, 242–251.
52. Nyametso, J.K. Resettlement of Slum Dwellers, Land Tenure Security and Improved Housing, Living and Environmental Conditions at Madina Estate, Accra, Ghana. *Urban. Forum* **2011**, *23*, 343–365. [CrossRef]
53. Allendorf, K. Do Women's Land Rights Promote Empowerment and Child Health in Nepal? *World Dev.* **2007**, *35*, 1975–1988. [CrossRef]
54. Victora, C.G.; Vaughan, J.P. Land ownership and infant health in Rio Grande do Sul: Relationship between agricultural production, malnutrition and mortality. *Rev. Bras. De Estud. De Popul.* **1987**, *4*, 127–151.
55. Park, G.-R.; Jung, Y. Housing insecurity and health among people in South Korea: Focusing on tenure and affordability. *Public Health* **2019**, *171*, 116–122. [CrossRef]
56. Nyametso, J.K. Improvement of Squatter Settlements: The link between tenure security, access to housing, and improved living and environmental conditions: Improvement of Squatter Settlements: The Link between Tenure Security, access to Housing, and Improved Living and Environmental Conditions. Ph.D. Thesis, University of Otago, Dunedin, New Zealand, 2011.
57. Ansari, Z.; Carson, N.J.; Ackland, M.J.; Vaughan, L.; Serraglio, A. A public health model of the social determinants of health. *Soz. Und. Prav.* **2003**, *48*, 242–251. [CrossRef] [PubMed]
58. DiClemente, R.J.; Crosby, R.A.; Kegler, M.C. *Emerging Theories in Health Promotion Practice and Research*, 2nd ed.; Jossey-Bass: San Francisco, CA, USA, 2009.
59. Berkman, L.F.; Kawachi, I.; Glymour, M.M. *Social Epidemiology*; Oxford University Press: Oxford, UK, 2014.
60. Krieger, N. Theories for social epidemiology in the 21st century: An ecosocial perspective. *Int. J. Epidemiol.* **2001**, *30*, 668–677. [CrossRef] [PubMed]
61. Adler, M.A. Land Tenure, Archaeology, and the Ancestral Pueblo Social Landscape. *J. Anthropol. Archaeol.* **1996**, *15*, 337–371. [CrossRef]
62. Payne, G. Urban land tenure policy options: Titles or rights? *Habitat Int.* **2001**, *25*, 415–429. [CrossRef]
63. Pierce, J. Reinvigorating the Concept of Land Tenure for American Urban Geography. *Geogr. Compass* **2010**, *4*, 1747–1757. [CrossRef]
64. Food and Agriculture Organization of the United Nations (FAO). *Land Tenure and Rural Development*; Food and Agriculture Organization of the United Nations: Rome, Italy, 2002.

65. Kuhnen, F. *Man and Land: An Introduction into the Problems of Agrarian Structure and Agrarian Reform*; Breitenbach: Saarbrucken, Germany, 1982.
66. Dekker, H. *Pursuit of Land Tenure Security*; Leiden University Press Imprint; Amsterdam University Press: Amsterdam, The Netherlands, 2006.
67. Arnot, C.D.; Luckert, M.K.; Boxall, P.C. What Is Tenure Security? Conceptual Implications for Empirical Analysis. *Land Econ.* **2011**, *87*, 297–311. [CrossRef]
68. Augustinus, C. *Handbook on Best Practices, Security of Tenure, and Access to Land Implementation of the Habitat Agenda*; United Nations Human Settlements Programme: Nairobi, Kenya, 2003.
69. Van Gelder, J.-L. What tenure security? The case for a tripartite view. *Land Use Policy* **2010**, *27*, 449–456. [CrossRef]
70. Barry, M.; Augustinus, C. *Framework for Evaluating Continuum of Land Rights Scenarios*; UN-Habitat: Nairobi, Kenya, 2016.
71. Whittal, J. A New Conceptual Model for the Continuum of Land Rights. *South Afr. J. Geomat.* **2014**, 3, 13–32.
72. de Soto, H. *The Mystery of Capital: Why Capitalism Triumphs in the West. and Fails Everywhere Else*, 1st ed.; Basic Books: New York, NY, USA, 2006; p. 2000.
73. Deininger, K.W. *Land policies for growth and poverty reduction*; World Bank: Washington, DC, USA; Oxford University Press: Oxford, UK, 2003.
74. Bruce, J.W.; Migot-Adholla, S.E. *Searching for Land Tenure Security in Africa*; Kendall/Hunt: Dubuque, IA, USA, 1994.
75. Van Gelder, J.-L. Feeling and thinking: Quantifying the relationship between perceived tenure security and housing improvement in an informal neighbourhood in Buenos Aires. *Habitat Int.* **2007**, *31*, 219–231. [CrossRef]
76. Bircher, J. Towards a dynamic definition of health and disease. *Med. Health Care Philos* **2005**, *8*, 335–341. [CrossRef] [PubMed]
77. Meade, M.S.; Emch, M. *Medical Geography*, 3rd ed.; Guilford: New York, NY, USA; London, UK, 2010.
78. Ellaway, A.; Macdonald, L.; Kearns, A. Are housing tenure and car access still associated with health? A repeat cross-sectional study of UK adults over a 13-year period. *BMJ Open* **2016**, *6*, e012268. [CrossRef] [PubMed]
79. Victora, C.G.; Vaughan, J.P. Land tenure patterns and child health in southern Brazil: The relationship between agricultural production, malnutrition and child mortality. *Int. J. Health Serv. Plan. Adm. Eval.* **1985**, *15*, 253–274. [CrossRef]
80. Hambling, T.; Weinstein, P.; Slaney, D. A review of frameworks for developing environmental health indicators for climate change and health. *Int. J. Environ. Res. Public Health* **2011**, *8*, 2854–2875. [CrossRef] [PubMed]
81. Garchitorena, A.; Sokolow, S.H.; Roche, B.; Ngonghala, C.N.; Jocque, M.; Lund, A.; Barry, M.; Mordecai, E.A.; Daily, G.C.; Jones, J.H.; et al. Disease ecology, health and the environment: A framework to account for ecological and socio-economic drivers in the control of neglected tropical diseases. *Philos. Trans. R. Soc. B Biol. Sci.* **2017**, *372*. [CrossRef] [PubMed]
82. Smyth, A.J.; Dumanski, J. A framework for evaluating sustainable land management. *Can. J. Soil. Sci.* **1995**, *75*, 401–406. [CrossRef]
83. Thomas, R.; Reed, M.; Clifton, K.; Appadurai, N.; Mills, A.; Zucca, C.; Kodsi, E.; Sircely, J.; Haddad, F.; Hagen, C.; et al. A framework for scaling sustainable land management options. *Land Degrad. Dev.* **2018**, *29*, 3272–3284. [CrossRef]
84. DeVries, W.T.; Bugri, J.T.; Mandhu, F. *Responsible and Smart Land Management Interventions: An. African Context/Edited by Walter Timo de Vries, John Tiah Bugri, and Fatima Mandhu*, 1st ed.; CRC Press: Boca Raton, FL, USA, 2020.
85. Deininger, K.W.; Selod, H.; Burns, A. *The Land Governance Assessment Framework: Identifying and Monitoring Good Practice in the Land Sector*; World Bank: Washington, DC, USA, 2012.
86. Uwayezu, E.; de Vries, W.T. Exploring the connection between spatial justice and land tenure security: Insights from inclusive urban (re)development schemes in Recife, Brazil. *Geo. J.* **2020**, 1–22. [CrossRef]
87. Chigbu, U.E.; Schopf, A.; de Vries, W.T.; Masum, F.; Mabikke, S.; Antonio, D.; Espinoza, J. Combining land-use planning and tenure security: A tenure responsive land-use planning approach for developing countries. *J. Environ. Plan. Manag.* **2017**, *60*, 1622–1639. [CrossRef]
88. Cassel, J. The contribution of the social environment to host resistance. The Fourth Wade Hampton Frost Lecture. 1976. *Am. J. Epidemiol.* **1995**, *141*, 798–814, discussion 797. [CrossRef]
89. Marmot, M.; Wilkinson, R.G. Psychosocial and material pathways in the relation between income and health: A response to Lynch et al. *BMJ (Clin. Res. Ed.)* **2001**, *322*, 1233–1236. [CrossRef]
90. Conrad, P.E. *Sociology of Health and Illness: Critical Perspectives*; St Martin's Press: New York, NY, USA, 1981.
91. Phelan, J.C.; Link, B.G.; Tehranifar, P. Social conditions as fundamental causes of health inequalities: Theory, evidence, and policy implications. *J. Health Soc. Behav.* **2010**, *51*, S28–S40. [CrossRef] [PubMed]
92. Krieger, N.; Gruskin, S. Frameworks matter: Ecosocial and health and human rights perspectives on disparities in women's health–the case of tuberculosis. *J. Am. Med. Women's Assoc. (1972)* **2001**, *56*, 137–142.
93. Green, B.N.; Johnson, C.D.; Adams, A. Writing narrative literature reviews for peer-reviewed journals: Secrets of the trade. *J. Chiropr. Med.* **2006**, *5*, 101–117. [CrossRef]
94. Alaazi, D.A.; Aganah, G.A.M. Understanding the slum-health conundrum in sub-Saharan Africa: A proposal for a rights-based approach to health promotion in slums. *Glob. Health Promot.* **2020**, *27*, 65–72. [CrossRef]
95. Adams, E.A.; Stoler, J.; Adams, Y. Water insecurity and urban poverty in the Global South: Implications for health and human biology. *Am. J. Hum. Biol. Off. J. Hum. Biol. Counc.* **2020**, *32*, e23368. [CrossRef]

96. Ekumah, B.; Armah, F.A.; Yawson, D.O.; Quansah, R.; Nyieku, F.E.; Owusu, S.A.; Odoi, J.O.; Afitiri, A.-R. Disparate on-site access to water, sanitation, and food storage heighten the risk of COVID-19 spread in Sub-Saharan Africa. *Environ. Res.* **2020**, *189*, 109936. [CrossRef]
97. Brown, R.; Leder, K.; Wong, T.; French, M.; Ramirez-Lovering, D.; Chown, S.L.; Luby, S.; Clasen, T.; Reidpath, D.; El Sioufi, M.; et al. Improving human and environmental health in urban informal settlements: The Revitalising Informal Settlements and their Environments (RISE) programme. *Lancet Planet. Health* **2018**, *2*, S29. [CrossRef]
98. Ompad, D.C.; Galea, S.; Caiaffa, W.T.; Vlahov, D. Social determinants of the health of urban populations: Methodologic considerations. *J. Urban Health Bull. N.Y. Acad. Med.* **2007**, *84*, i42–i53. [CrossRef] [PubMed]
99. Payne, G.; Piaskowy, T.; Kuritz, L. Land Tenure in Urban Environments. Available online: https://www.land-links.org/issue-brief/land-tenure-in-urban-environments/ (accessed on 8 February 2021).
100. Cromwell, E. Land Tenure—Key Sheets for Sustainable Livelihoods Pro-Poor Infrastructure Provision—3. Available online: https://www.odi.org/sites/odi.org.uk/files/odi-assets/publications-opinion-files/3199.pdf (accessed on 1 January 2021).
101. Murthy, S.L. Land security and the challenges of realizing the human right to water and sanitation in the slums of Mumbai, India. *Health Hum. Rights* **2012**, *14*, 61–73. [PubMed]
102. Durand-Lasserve, A. Informal settlements and the millennium development goals: Global policy debates on property ownership and security of tenure. *Glob. Urban. Dev.* **2006**, *2*, 1–15.
103. Wilkinson, A. Local response in health emergencies: Key considerations for addressing the COVID-19 pandemic in informal urban settlements. *Environ. Urban.* **2020**, *32*, 503–522. [CrossRef]
104. Bhardwaj, G.; Esch, T.; Lall, S.V.; Marconcini, M.; Soppelsa, M.E.; Wahba, S. Cities, Crowding, and the Coronavirus. **2020**.
105. Nakamura, S. Revealing invisible rules in slums: The nexus between perceived tenure security and housing investment. *Habitat Int.* **2016**, *53*, 151–162. [CrossRef]
106. Scott, P.; Cotton, A.; Sohail Khan, M. Tenure security and household investment decisions for urban sanitation: The case of Dakar, Senegal. *Habitat Int.* **2013**, *40*, 58–64. [CrossRef]
107. Awunyo-Akaba, Y.; Awunyo-Akaba, J.; Gyapong, M.; Senah, K.; Konradsen, F.; Rheinländer, T. Sanitation investments in Ghana: An ethnographic investigation of the role of tenure security, land ownership and livelihoods. *BMC Public Health* **2016**, *16*, 594. [CrossRef] [PubMed]
108. Gomes, T.F.; Freitas, F.S.S.; Bezerra, C.M.; Lima, M.M.; Carvalho-Costa, F.A. Reasons for persistence of dwelling vulnerability to Chagas disease (American trypanosomiasis): A qualitative study in northeastern Brazil. *World Health Popul.* **2013**, *14*, 14–21. [CrossRef]
109. Lokonon, B.O.K. Land Tenure and Communities' Vulnerability to Climate Shocks: Insights from the Niger Basin of Benin. In *Building a Resilient and Sustainable Agriculture in Sub-Saharan Africa.*, 1st ed.; Shimeles, A., Verdier-Chouchane, A., Boly, A., Eds.; Springer International Publishing AG: Cham, Switzerland; Palgrave Macmillan: London, UK, 2018; pp. 147–180.
110. Ma, C.; Smith, T. Vulnerability of Renters and Low-Income Households to Storm Damage: Evidence from Hurricane Maria in Puerto Rico. *Am. J. Public Health* **2020**, *110*, 196–202. [CrossRef] [PubMed]
111. Ashwood, L.; Diamond, D.; Walker, F. Property rights and rural justice: A study of US right-to-farm laws. *J. Rural Stud.* **2019**, *67*, 120–129. [CrossRef]
112. Busscher, N.; Parra, C.; Vanclay, F. Environmental justice implications of land grabbing for industrial agriculture and forestry in Argentina. *J. Environ. Plan. Manag.* **2020**, *63*, 500–522. [CrossRef]
113. Twerefou, D.K.; Osei-Assibey, E.; Agyire-Tetteh, F. Land Tenure Security, Investment and the Environment in Ghana. *J. Dev. Agric. Econ.* **2011**, *36*, 261–273.
114. Arifwidodo, S.D.; Chandrasiri, O. The relationship between housing tenure, sense of place and environmental management practices: A case study of two private land rental communities in Bangkok, Thailand. *Sustain. Cities Soc.* **2013**, *8*, 16–23. [CrossRef]
115. Lewis, Jeanette: How does Security of Tenure Impact on Public Housing Careers? Available online: https://www.ahuri.edu.au/__data/assets/pdf_file/0012/3054/AHURI_RAP_Issue_78_Security_of_tenure_Synthesis.pdf (accessed on 1 January 2021).
116. Clair, A.; Hughes, A. Housing and health: New evidence using biomarker data. *J. Epidemiol. Community Health* **2019**, *73*, 256–262. [CrossRef]
117. Bentley, R.; Baker, E.; Simons, K.; Simpson, J.A.; Blakely, T. The impact of social housing on mental health: Longitudinal analyses using marginal structural models and machine learning-generated weights. *Int. J. Epidemiol.* **2018**, *47*, 1414–1422. [CrossRef]
118. Cairney, J. Housing tenure and psychological well-being during adolescence. *Environ. Behav.* **2005**, *37*, 552–564. [CrossRef]
119. Courtin, E.; Dowd, J.B.; Avendano, M. The Mental Health Benefits of Acquiring a Home in Older Age: A Fixed-Effects Analysis of Older US Adults. *Am. J. Epidemiol.* **2018**, *187*, 465–473. [CrossRef] [PubMed]
120. Howden-Chapman, P.L.; Chandola, T.; Stafford, M.; Marmot, M. The effect of housing on the mental health of older people: The impact of lifetime housing history in Whitehall II. *Bmc Public Health* **2011**, *11*, 682. [CrossRef] [PubMed]
121. Luginaah, I.; Arku, G.; Baiden, P. Housing and health in Ghana: The psychosocial impacts of renting a home. *Int. J. Environ. Res. Public Health* **2010**, *7*, 528–545. [CrossRef]
122. Daoud, N.; Matheson, F.I.; Pedersen, C.; Hamilton-Wright, S.; Minh, A.; Zhang, J.; O'Campo, P. Pathways and trajectories linking housing instability and poor health among low-income women experiencing intimate partner violence (IPV): Toward a conceptual framework. *Women Health* **2015**, *56*, 208–225. [CrossRef] [PubMed]

123. De Salles, D.; Maria, A.; de Lima, F.; Amélia, A.; Mingoti, S.A.; da Silva, C.D.Á.; de Rio, A.; Souza, A.; Amanda, C.; Freire, F.M.; et al. Mortality from Homicides in Slums in the City of Belo Horizonte, Brazil: An Evaluation of the Impact of a Re-Urbanization Project. *Int. J. Environ. Res. Public Health* **2019**, *16*, 154. [CrossRef] [PubMed]
124. Li, J.; Liu, Z. Housing stress and mental health of migrant populations in urban China. *Cities* **2018**, *81*, 172–179. [CrossRef]
125. Morris, A. 'I really have thought this can't go on': Housing tenure and health. *Aust. Dream Hous. Exp. Older Aust.* **2016**, 147–178.
126. Lawder, R.; Walsh, D.; Kearns, A.; Livingston, M. Healthy Mixing? Investigating the Associations between Neighbourhood Housing Tenure Mix and Health Outcomes for Urban Residents. *Urban. Stud.* **2014**, *51*, 264–283. [CrossRef]
127. Arthurson, K. Mixed tenure communities and the effects on neighbourhood reputation and stigma: Residents' experiences from within. *Cities* **2013**, *35*, 432–438. [CrossRef]
128. Farre, A.; Rapley, T. The New Old (and Old New) Medical Model: Four Decades Navigating the Biomedical and Psychosocial Understandings of Health and Illness. *Healthcare* **2017**, *5*, 88. [CrossRef] [PubMed]
129. Lall, S.V.; Freire, M.; Yuen, B.; Rajack, R.; Helluin, J.-J. *Urban. Land Markets: Improving Land Management for Successful Urbanization*; Lall, S.V., Ed.; Springer: Dordrecht, The Netherlands, 2009.
130. Sjöstedt, M. The impact of secure land tenure on water access levels in sub-Saharan Africa: The case of Botswana and Zambia. *Habitat Int.* **2011**, *35*, 133–140. [CrossRef]
131. Uwayezu, E.; de Vries, W.T. Scoping land tenure security for the poor and low-income urban dwellers from a spatial justice lens. *Habitat Int.* **2019**, *91*, 102016. [CrossRef]
132. Fitzpatrick, S.; Watts, B. Competing visions: Security of tenure and the welfarisation of English social housing. *Hous. Stud.* **2017**, *32*, 1021–1038. [CrossRef]

land

Concept Paper

Global Markets, Local Issues: The Hegemonic Process of Agri-Food Construction to Present Challenges

Guilherme Silva Fracarolli [1,2]

[1] SOCIUS—Research Centre in Economic and Organizational Sociology, ISEG—Lisbon School of Economics and Management, University of Lisbon, 1249-078 Lisbon, Portugal; guilherme.fracarolli@phd.iseg.ulisboa.pt or guilherme.fracarolli@agricultura.gov.br

[2] Rural Development Division, Federal Superintendence of Agriculture, Livestock and Food Supply of São Paulo, Ministry of Agriculture, Livestock and Food Supply, São Paulo 01327-002, Brazil

Abstract: The social construction of the agri-food market has undergone revolutionary changes throughout history since the Anthropocene. This conceptual paper discusses the embeddedness of institutions in this market construction. To do so, this work analyses the geographical indication (GI) of agri-food market formation through the lens of critical theory. Through dialectics, it analyzes the historical process of agrarian systems' shape according to their complexity, and the origins and effects of hegemonic interests in the construction of agri-food markets. Furthermore, this work shows how the market has evolved from different trade types as the capitalist system also evolved, changing the mechanics of trade and functions of food production. The results indicate that as agrarian systems evolved, food became more homogeneous and standardized in order to meet the demands of urban masses in capitalist economies. Regions where less complex systems predominate tend to hinder the creation, maintenance, and perpetuation of products such as GI, which may compromise their existence in the long run. Moreover, nations reproduce ideologically oriented interests according to the formation of dominant groups in each place, as also expressed in the agri-food market. This paper aims to provide new conceptual and theoretical insights into the institutional mechanisms and historical processes of agri-food market construction in terms of power interests.

Keywords: geographical indication; agrarian systems; economic sociology; cultural hegemony; agri-food complexity; critical theory

Citation: Fracarolli, G.S. Global Markets, Local Issues: The Hegemonic Process of Agri-Food Construction to Present Challenges. *Land* **2021**, *10*, 1182. https://doi.org/10.3390/land10111182

Academic Editors: Uchendu Eugene Chigbu, Ruishan Chen and Chao Ye

Received: 5 October 2021
Accepted: 31 October 2021
Published: 3 November 2021

1. Introduction

The relationship between humans, agriculture, and food production and consumption has been an issue since around ten thousand years ago [1]. The Neolithic saw the establishment of the first permanent human settlements, made possible by the domestication of animals and plant cultivation. Nevertheless, this relationship has not been the same in every location, nor has it developed the same way. As a complex combination of a cultivated ecosystem and productive social system [1], agriculture evolved heterogeneously across time and space.

The comprehension of this complex arrangement is crucial for understanding the dynamics of the diversity of food and agri-food production. The development of a wide variety of agrarian systems unrolled into a number of embedded constructions of particular realities involving food, from systems with high labor employment, small areas, and family use, to large tracts of land, highly mechanized and owned by multinational companies.

As these multiple agri-food systems evolved, societies became more complex in terms of production and consumption. As societies progressively started to transition from rural communities to urban and densely populated areas, food demand also changed. Some massive agricultural changes boosted production to provide food for the new boroughs near castles in the Middle Ages; however, the population remained predominantly rural until the 18th and 19th centuries, which brought an unprecedented agricultural boom. This period's output significantly increased production per area in English farms compared to others [1,2]. This period characterized the first agricultural revolution of modern times.

A second moment between the end of the 19th century and the mid-20th century brought motorization, mechanization, mineral fertilization, and specialization, changing the function of food in societies permanently, reducing it from its cultural role to the mere role of feeding the growing urban masses. The new characteristics of functions and agrarian structures put food in a global commodity perspective, changing the agri-food market.

The food market has also changed over time. If, centuries ago, food was more related to social and cultural construction, in modern times, food is either a necessary input for the great mass of workers or a luxury embedded by the allegory of three-star chefs as an item for the benefit of a few.

The necessity of satisfying the hunger of massive urban crowds in the wake of industrialization led to a standardization of agri-food products, which lost their identities and cultural significance over the years. Later, in the 20th century, the improvement of this centuries-old process would find a basis in Taylorism, affecting food production and leading to Fordism in the agri-food market [3].

Therefore, the modern agri-food market, dissociated from embedded cultural aspects, is simplified by the intrinsic matters related to commodity production, which could be more easily explained by orthodox economics. However, as a counterpart, there is an increasing demand for other food sources, such as those related to culture, geography, and historical meaning—for instance, geographical indication (GI) products.

As with the ways in which people have fed themselves over time, the market has also changed, from community trading of goods, followed by the regional exchange of food, to modern commerce, with commodities such as coffee and soybeans on stock markets. Furthermore, the comprehension of how institutions influence markets is crucial to understanding their operation.

Much work has been done on economic issues pertaining to specific crops, on the political economy of agricultural goods trading, and even approaches relating to niche markets. However, little has been debated with regard to sociological approaches to economic matters. A single approach to study such markets cannot explain the richness and complexity of their diversity.

The core theoretical question of this paper concerns the GI agri-food market's construction, and how institutions shape it according to power interests. However, markets are not all the same; therefore, they cannot all be analyzed using the same paradigms. Here, the proposition is supported by Allaire [4–6] that markets—especially food markets—are complex institutions, and are shaped by social construction. Thus, their analysis requires a more in-depth approach to their functioning, such as economic sociology to examine the complexity and diversity of different agricultural systems.

Critical theory in agriculture has been applied to specific and important themes concerning the rural environment. Its use is relevant in pedagogical processes in order to demonstrate the relevance of the hegemonic discourse [7,8], in the study of the behavior of social movements in face of the inherent accumulation of capital in capitalist societies [9,10], or even in dynamics of agrarian changes in local contexts [11]. However, it is in the context of the agrarian question, peasants, and the impacts of globalization in the 21st century that CT is most vital in the development of research [12–16].

The question is: what drives GI agri-food markets? Or, in other words, how do institutions act towards agri-food markets? Although grounded theory supports most of the works concerning agri-food markets, developing concepts and deepening the approaches from a theoretical spectrum is necessary. In this sense, this work addresses the development of such markets based on agrarian and critical theory (CT). Thus, supported by Gramsci's and other critical theorists' arguments about market institutions, we aim to explain how some regions developed differently from others in this niche.

Therefore, the purpose of this work is not to present new data or empirical analyses but, through dialectics, to present new insights and perspectives on the construction and development of agri-food markets—specifically GI. The design of these products and their market has been debated for some time. However, there is also a need to rethink and discuss the forms of production, as well as the causes and consequences of production models. Thus, the option of theoretical confrontation, addressing the global history of food construction, should add depth to the debate and raise new questions regarding the direction of global food production.

The debate over forms of production and consumption has been gaining attention and prominence from a sustainability perspective. The main challenge is to unravel the origins of the problems in order to propose action to achieve fundamental changes. However, few studies have been devoted to deepening the understanding of how food markets are created, with due consideration of the complexity of their object of study. Therefore, it is necessary to provide possible ways to make this approach more assertive and accurate. Furthermore, this work seeks to provide a debate over existing paths of discussion of the agri-food environment.

The concept of food and consumption as social constructions is a central theme of this paper. Thus, as social constructions, this work considers different economic approaches to investigate the agri-food markets' economic problems. To do so, this work analyses GI agri-food market formation through the lens of critical theory, explaining the development of markets based on the construction of institutions favoring regional elites.

Therefore, this paper aims to provide new conceptual and theoretical insights on the institutional mechanisms and historical processes of agri-food market construction towards power interests. Supported by critical theory, it aims to contribute with studies dedicated both to those interested in agricultural markets and to policymakers and practitioners in the conduct of policies aimed at rural development.

Firstly, this work presents agri-food markets in terms of agricultural systems, cultural aspects, and compelling implications for markets. The following section presents theoretical and empirical tools used by researchers to try to explain markets. Tools such as critical theory and the institutionalism perspective proposed by economic sociology enhance the discussion of the logic of market operations and complexities.

Later, this work discusses the findings between the convergences of the theories used and the different existing markets resulting from productive diversity. These meeting points are subsequently adapted to the concepts of hegemony, elites, and institutions.

Lastly, this work summarizes the findings, pointing to future pathways, and giving perspective for applications and possible usage by market actors in the pursuit of market improvement.

2. Methods

Humanity has always discussed food's origins, as well as its market and intrinsic human relations. However, all scientific approaches to debates regarding food are significantly recent. Many works have been conducted to reveal the engines of markets and how they operate, from seeding, to commerce and, lastly, consuming habits. Nevertheless, few studies have sought to understand the sociological issues of agri-food market construction. Thus, this work aims at providing explanations of agri-food market construction based on critical theory.

To achieve the proposed objectives, this paper dialectically analyzes the formation of agri-food systems. The analysis was carried out through the temporal reconstruction of the forms used by humanity in agriculture, exposing the sociological and economic concepts for these phases, and confronting them with the involvement of institutions in shaping these systems. From the perspective of the critical theory of sociology, this work shows the forms of construction of existing systems, and points to the reasons for such conformation. The choice of critical theory as an analytical path is based on the explanatory capacity of historical materialism.

For this work to face and fulfill the task of conceptually discussing the institutional mechanisms that exert influence and establish hegemonic visions of agrarian development, it is necessary to set the adopted parameters. Since this is not a literature review, the references do not follow a specific timeframe or database. Thus, the work is divided into the following themes: agri-food systems, the economics of agriculture, rural and economic sociology, globalization impacts, and agri-food markets. This thematic separation is crucial for selecting the works consulted. The different aspects that influence the deep analytical perspectives that permeate the hegemonic common sense are noticeable and differentiable. In each subsection, the basic concepts of the subject are presented, followed by its agrarian application and a discussion of how CT can be applied to it. Thus, this work must be understood as not just the sum of its parts, but the exponential result of its embeddedness. Therefore, each subject's key works, precursors, or transformers were approached for conceptual discussion, touching on the necessary essence of each subject, without losing the significance of the dialectics.

The investigation of markets without due consideration of their complexity can result in risky mistakes. It is crucial, then, to consider the peculiarities of production systems, such as the historical and cultural consumption aspects that influence this market. This, centered on critical arguments, demonstrates how agricultural systems of food production and commerce develop from different agrarian systems according to hegemonic interests and the embeddedness of institutions and elites.

In order to properly approach this issue and understand the functioning of modern agri-food markets, the main aspects needed to be discussed. Thus, this paper is divided into two main sections:

The first section regards the agricultural systems of production; it starts by detailing the cultivated system's influences and the social productive system, as well as their consequences for the products. Then, by structuring the systems, the work goes through the history of agri-food production, establishing a notion of intrinsic societal relations and food functions. Subsequently, with a historical construction of changes in productivity logic, the paper works through the transformation of food's cultural and functional roles in society.

The second section regards the main aspects of critical theory and institutions. The approaches of specific literature to the matter are vital to a thorough comprehension of the subject. As markets are social constructs [17–20], they must not be understood only as a matter of supply and demand. Moreover, the more complex the relations of the social networks and the actors involved in the construction of the market, the more factors must be taken into account.

As such, within the second section, this work presents an economic approach based on classical economics, political economy, and economic sociology. In terms of classical economics, this work presents the main concepts of competition—vital for market comprehension. Subsequently, this paper considers political economy as a tool for agricultural market analysis; it includes other elements in a macro-level approach, allowing a broader vision of reality, such as supply chains [21], value chains [22–24], and political factors [25,26]. In the last part of the section, this paper addresses economic sociology and discusses agri-food markets. This is an approach that attempts to understand the multifactorial influence of performance [27–29], networks [17,30,31], and institutional perspectives [32–35].

In the last part of the section, agricultural systems are confronted by critical and institutional theory. In this way, each approach gives a more explicit meaning to analysis according to agri-food production's evolution. By doing so, this paper aims to analyze how markets' development represents the interests of elites in each region.

3. Results and Discussion

3.1. Agri-Food Systems

Agriculture has been crucial for the development of humankind. Since the Neolithic period, when Homo sapiens learned to domesticate species of plants and animals, they began to transform themselves into cultivator societies. This transformation of the environment from the original represents the Neolithic agricultural revolution. Such change leads authors such as Childe [36] to affirm that this was the first revolution to transform the human economy.

The possibility of settling down in a particular place and producing one's own food was, in fact, revolutionary, changing the dynamics of the whole world. No species had done it before, and it allowed small communities to be born and become societies. The spread of this model of life happened differently across the world. The cultivated species and animals were different and adapted differently in each place. This difference implies two main subsystems of agriculture: a cultivated ecosystem, and a productive social system [1,37–40].

These two subsystems define the characteristics of food, as well as its social dynamics and economic environment. It is crucial to understand the concept of what food represents in all of its meanings. Food is not just what one eats; it is a product of thousands of years of interactions between humans and the environment. Furthermore, it has consequences for the formation of societies, economic dynamics, and culture. The following section of this work details both subsystems.

According to the authors of [1], the cultivated ecosystem deals with a set of practices and land use—direct and indirect—that affect soil fertility, its physical aspects, diversity, and any type of practice that changes its conditions of use and/or its surroundings. Factors such as the intensity of use and the techniques employed have a variable influence on both plant and animal production. This comprehension is set by the sum and interrelation between subsystems that do not exist independently.

On the other hand, the productive social system, according to the authors of [1], represents three main aspects of production: men and women, including labor force, knowledge, and savoir-faire; inert means, such as equipment and tools; and living matter, such as plant and animal species. The geographically localized combination of these factors, along with the type of usage, technical, economic, and social definitions, allows a theoretical construction of agrarian systems' dynamics, as Thom [41] pointed out to the value of systems' taxonomy.

Therefore, the development of a theory of agrarian systems is crucial for analysis. Despite the conceptual differences, these two are the core of these complex, structured, well-established subsystems that sustain a theoretical structure. Agrarian systems develop as a result of general changes in the form of production, the employed labor, the adopted techniques, the productive capacity, or the purpose. They can develop in an unequal, contradictory manner, or even go into crisis. According to Mazoyer and Roudart [1], when the changes in those factors engender a new cultivated ecosystem, an agricultural revolution takes place.

The theory of agrarian systems differs on some level from spatial production theory [42], in that the former relies on a view that the rural regions are organisms characterized by resources, rights, society, and capital, or the three-dimensional rural space system based on culture, society, and material [43], for example. Nevertheless, both theories are complementary and based on the work of Ren [44]. This work adopts Mazoyer and Roudart's [1] viewpoint, based on the agrarian system theory, which builds multilayer factors that influence the agrarian scenario, adopting a sociological perspective that characterizes rural spaces during space and time, considering all determinant factors.

In ancient agrarian systems, the concern was only to produce enough food to meet the family's own caloric needs, and perhaps those of the community in which the individuals were inserted. The development of agri-food production evolved differently worldwide. Soil, climate, availability of plants, and cultural habits influenced the necessary practices to feed the people better. Some of them, even in the Neolithic period, stood out, whether they were more irradiating—such as the center of the Near East, the Central American, Chinese, and New Guinean focuses—or those less irradiating centers such as the South American, North American, and Thai [1]. Such dispersion of agricultural development throughout history is so relevant that some researchers have theorized its relationship with civilizations' linguistic and cultural development [45].

The close relationship between cultural development and food has led to the characterization of certain products in specific regions. In addition, ancient records attribute qualitative aspects to the origins of some products, such as Lebanon cedar, Corinthian wine, or Brindisi oysters. Thus, historically, peoples' production, preparation, and consumption of food have been related to agrarian systems specific to each region. Some even gained notoriety for having specific characteristics that differentiated them from others.

However, the world has changed over the centuries. Civilizations have become less rural and more urban. Throughout history, some agricultural revolutions were crucial to multiply food production and favor urban centers. These revolutions allowed more outstanding production in the same areas due to new technologies. Therefore, less labor force was needed.

Revolutions such as the Neolithic, with sedentism, were significant. Even in antiquity, agrarian systems with the use of fallow and light traction allowed exponential gains, followed by the use of heavy traction, which brought about a new revolution in the Middle Ages. However, at the end of this last period, there were differences between regions—especially in Europe. With the use of new equipment more suitable for cultivation, productivity gains were accentuated, which led to changes in the social relations of the land and the structure of the domain. This resulted in the concentration of areas in the possession of dominant classes, characterized by an individualistic logic, starting to employ labor in place of land possession by the working class. This practice reduced the need for territorial expansion for agricultural production, thus replacing slash-and-burn methods.

A portion of the population started to concentrate on other functions in urban centers related to rural activities, such as boroughs. Thus, with this new urban mass, after introducing liberal ideas and the allowance of broader practices of production, trade, and circulation of goods and people, a new era was taking place in Europe. This new era began with a new agricultural revolution—the first of modern times. The substitution of systems without fallow for rotating crops with forage and grain provided new impetus in food production. This gain provided the industrializing cities with the necessary supply and, consequently, also provided more appropriate implements to increase production.

The growing urban concentration sustained by the supply of these new productive forces boosted the industrialization of large centers until the end of the 19th century. This achievement enabled the development of productive chains linked to the land, such as textiles, beer, sugar, and alcohol, which was only possible with abundant productive surpluses. The limitations of the production system were concentrated on the tools, since the properties used were mainly private, and labor was also employed, despite some regions remaining focused on family farming, lacking technology and tools. This all changed substantially with the Industrial Revolution.

With the Industrial Revolution in the 19th and early 20th centuries, not only industry and commerce were affected. New agricultural implements such as plows, seeders, and harvesters entered the agricultural equation, transforming the European and North American scenarios through the new steel technology available. However, at this point, another parallel phenomenon also occurred: Fordism, which also impacted the agrarian sector and changed the logic of food production. From that time, the world started to adopt a Weberian perspective of rationalization, which also applied to agriculture and food, although lacking a general theory for Fordist agriculture [46].

Commodities are defined generically as any goods that can be traded. However, the term is commonly used to refer to raw materials or goods with aspects and characteristics so uniform and abundant that they do not depend on the place of production. Therefore, they are traded in large volumes, with a constant flow and circulation throughout the planet. This also implies that, due to both the quantity available on the market and the intense standardization, their producers are only price-takers. In the agricultural case, coffee, soybeans, wheat, sugar, alcohol, corn, and orange juice can be categorized as commodities.

Beyond Gramsci's [47] concept of Fordism—of extreme rationalization of production and consumption through cultural and political means, based on Taylorism—this new logic of conceiving the global system also affected agrarian systems. The main argument relies on the transformation of the production and consumption paradigm towards massification. The process of commoditization changed the perspective of food around the globe.

Authors such as Kenney et al. [48] argue for the contribution of American agriculture to the production and consumption markets of undifferentiated commodities, while Potter and Tilzey [49] point to neoliberalism, neomercantilism, and multifunctionality on the European side. Both works see the succession of events from Fordism, post-Fordism, and globalization as a global phenomenon of standardization and homogenization of food, endorsed by institutions and states that affect the market unevenly.

This sequence of phenomena that affected rural regions, agricultural production, and food consumption resulted in reducing product diversity and local factors' influence on the product. The effects of massification and standardization are based on efforts to minimize differences between products so that they can be commercialized on a global scale. With the minimization of differences between products globally, there has also been a consequent change in agrarian systems to meet the productive demand. Due to the new logic of overproduction, there is homogenization, and consequent decreases in the number of species consumed, varieties produced, and differences in production. Thus, the forms of production are reduced to as few varieties of agrarian subsystems as possible. Mechanization becomes predominant, reducing differences in cultural treatment, the influence of edaphoclimatic conditions, the need for uniformity, and the social factors inherent in the attribution of value, reducing the relative workforce in the system.

However, according to the theory of agrarian systems, changes can occur as a natural part of development, despite the agrarian systems' consistency. These systemic changes have triggered shifts in trade blocs, globalization, liberalization, agro-technologies, societal demands, and climate change [50–54]. The more specific the system, the more complex it is. For example, agri-food producers with geographical indication (GI) registration base their products' differentiation centrally on natural, human, and historical factors [55].

The first item—the natural factors—is the concept of an "essential link between the location in which a food or beverage is produced, and its quality or other consumer attributes" [24]. The second, known as savoir-faire, refers to the techniques, materials, and production methods used. The last considers immeasurable elements attributable to the producing region's culture or history, which are applied to the product, making it notorious. Finally, the sum of the three composes what Allaire [6], based on the work of Goodman [56], qualifies as "the immaterialization of food and the institutionalization of quality"—a concept that considers environmental aspects such as soil and climate, but also cultural and human factors, characterizing intrinsic characteristics of agri-food products, and capable of providing specific regional qualities.

Therefore, such elements are central to the creation of more flourishing and more complex agrarian systems. Each of them directly influences the product, and provides a myriad of combinations that generate unique products.

The natural factors are the environmental aspects that qualitatively influence the products; this is what Josling [24] refers to as terroir. However, this is not entirely accurate; it represents a set of environmental factors—such as soil, climate, light, altitude, physical elements, and others—that yield specific characteristics to the products grown there. It is the foundation that gives uniqueness to each and every product coming from the field. It is so crucial and particular to the characterization that it cannot be reproduced elsewhere.

Savoir-faire relates to the labor put into practice. The concept refers to the human factors that can produce "typicity", or unique, traditional character [55]. It is the work of cultural bias in a geographic location that implies a historical process of knowledge construction over time, as endorsed by Guy [57].

The last item regards the cultural aspects that are embedded in the construction of such products. Sometimes referred to as "history" [55,58], it is more appropriately called culture, since history is part of cultural construction. Despite disagreements on a definition of the concept, this work adopts the understanding of Tylor [59], addressed by Abdel-Hadi [60]

(p. 12) as "that complex which includes knowledge, belief, art, morals, law, custom and any other capabilities and habits acquired by man as a member of society."

Therefore, in spaces where the simplification of systems predominates, agri-food production is based on commodities in large portions of land, tending not to have areas for production imbued with the necessary conditions for more complex products, such as those with GI. Thus, there is a tendency to widen the scope of prevailing and profit-maximizing systems in capitalist societies. Hence, with the decrease in the capacity for interaction between individuals in the rural space due to the increase in crops, decrease in the number of farmers, reduced exchange of experiences, and suppression of environmental factors in food, there is less terroir variability and, consequently, a smaller market for these products. In the long run, this fact tends to compromise the variability of these products and their markets.

Consequently, the direction of these markets is profoundly affected by the guidelines and regulations promoted by local, national, and regional policies. In this way, the development of agricultural markets centered on the production of commodities, or of products with culturally added value, is constructed according to the predominant ideas of the institutions. Thus, institutions that are socially oriented towards consolidating the maintenance of hegemonic systems tend to suppress systems that favor differentiated products and markets in the long run. Therefore, the maintenance or alteration of these systems must, necessarily, go through the institutions' composition to represent the intended interests.

There are discussions in academia about the concept of terroir containing other factors. Terroir is a traditional concept widely used to describe the particularities of GI products; it is widely used in the wine market, although it can be applied to all agri-food products [61–63]. It regards the relationship between the product's quality or taste and its geographic origin [64]. However, the concept of terroir addressed by this work is close to that used by Barham [55], which considers not only environmental aspects reflecting the product's quality, but also human and cultural factors that give particular attributes to agri-food products or wines. Therefore, dividing the concept into three parts allows for a better understanding of all influencing factors, broadening the comprehension of the complexity embedded in these products. Indeed, environment, savoir-faire, and culture are embedded concepts that comprise the GI market. Such products go beyond Marx's comprehension of homo faber, due to a complex, embedded paradigm of equally complex agrarian systems, needing a consideration of their roots in order to fully comprehend the issues of this market. To summarize the understanding of the differences in the formation of both types of agri-food market, Figure 1 illustrates the conceptual chain of both paths.

3.2. A View from Economics

Economics has been on the agenda as a science since the 18th century, sometimes associated with sociology, and sometimes dissociated from it. Initially, classical economics was treated as a pure and liberal science by figures such as Adam Smith [65]. Later, it was treated differently, via more critical views on its function by Marx [66] and Keynes [67], which brought new thoughts on employment matters in macroeconomics.

Stein [68] and Allaire [5] argue that the tools provided by mainstream economics are limited, and cannot provide sufficient elements to support development. Despite the charge of these elements being more related to institutions and structures of concepts, it brings an essential element of reflection on the role and consistency in using these tools.

Since Adam Smith's *The Wealth of Nations*, in 1776 [65], economics has been considered an independent subject. The aforementioned book is a landmark text on economics, and by discussing the issues of the division of labor, productivity, and free markets, is still vital reading on the subject. Moreover, the book was written during the Scottish agricultural revolution (therefore embedded in this context), seeking to form an economic theory opposing the theories of mercantilist foundations that could no longer respond to problems arising from new realities, such as protectionist tariffs on precious metal reserves.

As approached by Say [69], Ricardo [70], and Mill [71,72], classical economics is based on liberal perspectives of the market. Significant matters that ground such thoughts rely on self-regulating systems, in which external and state interference is not only unnecessary, but unwanted. Classical economists comprehend that such entities and measures limit and disrupt the market's perfect function, and that the market is governed by its own independent production laws and trade, needing no other external factors, and reaching its natural optimum by itself. Such understanding is best summarized by Smith's most famous analogy developing the invisible hand concept, central to the *laissez-fare* philosophy—the premise of neoclassical economics.

Figure 1. Opposing chains of agri-food market construction.

Regarding agrarian issues, Smith [65] understands that this activity is less prone to a division of labor than manufacturing, concluding that it does not result in significant differences in development between countries. However, he understands that such activity is more desirable than industrial work in the context of North America, due to land availability and owner control over the process.

Smith's view reveals a singular comprehension of the agrarian system, despite giving due importance to agriculture. He understood this system as complete land control and ownership along with total separation of labor between the urban and rural environments. This view shares the worldview that characterized the time; it endorses a utilitarian conceptualization of land use maximization. Moreover, by pointing to complete control as a positive asset, along with the large availability of land, it converges with the capitalist-based global system in formation at that time, based on profit maximization. As such, Smith understands that subsistence is of primary importance to long-term industrial-based economic growth; however, he also understands that a utilitarian view of the land function manifests that land should be comprehended as an asset for profitable use only by owners discharged from food sovereignty. Furthermore, by endorsing the use of large portions of land towards maximization, the author expresses the thought of land owners and elites of the time.

On the other hand, neoclassical economics was first quoted by Veblen [61] to set new perspectives based on new ideas of granting value based on the relationship between the material desire to acquire a specific good and the costs of production. These new ideas were based on the thought of maximization of utility and profit, based on rational choice theory, best defined by Arnsperger and Varoufakis [73] under three axioms (despite their observations on the development of this school of economic thought): methodological individualism, methodological instrumentalism, and methodological equilibration. In sum, based on the lack of pluralism, neoclassical economics reduces the analysis of reality into previously squared theory, and does not fully consider the concreteness of all social facts. However, it prevails both in academia and in public and private institutions.

The liberal view proposed by classical and neoclassical economics supposes that the market works more adequately when there is no or minimal regulation, since it develops naturally towards an optimal equilibrium, provided by the free competition of economic agents. Therefore, it is centered on a utilitarian view towards maximizing gains and specialization of functions through division of labor. The premise of self-determination of individuals towards their own gains, although legitimate, does not aim at social or collective gains as primary intentions.

Much of economics stems from Marshall's "perfect markets" model [74]. Based on the ideas of classical economics, Marshall believed that with an abundance of buyers and sellers, the market tends towards equilibrium. Despite providing good didactic models, such as the formulation of the model created by Pareto [75], and facilitating the understanding of economic concepts, their ideas were criticized both by Hayek [76], who understood competition as a process in constant change, and Granovetter [31], who understood that markets are determined by multiple factors, making such perfection impossible. From both sides, one can conclude that markets are not perfect, and that models can illustrate ideal but unrealistic situations. However, this model advanced a positivist, economic liberal ideal that less regulated markets tend to function better. In practice, they are determined by institutions that aim to maintain the status quo, creating maintenance tools for their holders.

The determining conditions of these perfect markets have been known for a long time. Firstly, that there are many buyers and sellers, so there is no personal influence in the market (atomization). Secondly, that there is a perfect substitute for the good on the producers' side (homogeneity). Thirdly, that there is free movement of goods and productive facilities for any party (mobility). Fourthly, that there is no barrier to entry into the market (permeability). Fifthly, that there is no imposition of any part of price holding, which results from the market itself (free price flow). Sixthly, that no social actor has information different from the others (transparency).

Within agri-food markets, commodities configure the nature of goods that are closest to the ideal model. These goods, as previously stated, considering their intrinsic characteristics, contemplate simplified realities in which ceteris paribus is best applied. These sorts of goods reduce the complexity of food, fitting more adequately with analysis that does not consider elements beyond the surface, and deepens through the causes, reasons, and hidden elements of the social factors. However, commodity production systems are designed only to maintain the hegemony of agrarian elites, in detriment to the production of food imbued with culture and destined for food sovereignty.

Since the first appearance of the term "political economy" in Montchrétien's work [77], the embeddedness of the state, economy, and society have become clearer, as noted by Mayntz [78] (p. 5). The Weberian concept of the term is rooted in government participation and intervention in employment and growth. Balaam and Veseth [79], for example, argue that the conceptual difference between political economy and economics lies in international trade. However, rooted along with economics, the political economy also centers the causes of social actions on self-maximization of benefits, rooted in utilitarianism, limiting its explanatory capacity.

Nevertheless, in the 18th century, François Quesnay—one of the pioneers in the field—reflected on the importance of agricultural production. He attributed value to it due to the multiplication of the farmer's effort and resources, while manufacturing, services, commerce, and trade would be "sterile" [80]. Years later, Theodore Schultz, upon receiving the Nobel Prize in economics in 1979, recognized the relevance of agriculture, placing it at the center of world economic development [81].

The relevance of agri-food matters to the field results from globalization. It is not by chance that it coincides with a new perspective on the state's role in the economy. In the agricultural sector, globalization resulted in the acceleration of the rationalization and mass production of goods, later adapted and optimized through Fordism. This process resulted in the massification and standardization of consumer goods. Gramsci [47] argues that the stability and maintenance of such a production system are integral to the performance and influence of the state.

Thus, under this interpretation of the system, the state acts towards the standardization and homogenization of agri-food products, while maintaining elites' status. Therefore, the globalization processes of massification, standardization, and transformation of the food sector sustain this logic of food and fiber for the industry. Thus, there is an evident loss of authenticity and diversification, consequently reducing the complexity of the agrarian system. This results in loss to the consumers, and simplifies the offer of agri-food products and political economy to answer the state's political and economic influence in the relations of production and consumption. As such, Benjamin [82] points to authenticity (uniqueness) and locale (physical and cultural) as attributes that embed an irreproducible character in goods and objects. Such a concept can be easily attributable to GI agri-food in order to sum value and work as a counter-movement to a mass-culture society.

In short, industrialization also plays a role in the construction of rural development—that is, not only in agro-industries, but also in the construction of combat spaces aimed at the greater potential for maximizing profits. In industrialized societies, these appear as the main battleground. However, in nation states where agrarian elites prevail, industrialization does not develop. Thus, extensive rural estates become the most significant source of power, and their owners constitute institutions, structuring the domination of these agrarian elites and strengthening their hegemony. In this way, the production of commodities in monoculture systems tends to suppress industrial development and stifle the growth of other agricultural systems.

3.3. *The Sociological Perspective*

In order for high-quality and meaningful agri-food products to be consumed by everyone, and not just seen as "Veblen goods", they need to be affordable and plentiful. For this to occur, systems capable of producing them must be possible and desirable and, therefore, have a favorable environment for them to flourish. Such environments are the institutions.

North, in 1991, stated that "Institutions are the humanly devised constraints that structure political, economic, and social interaction. They consist of both informal constraints (sanctions, taboos, customs, traditions, and codes of conduct), and formal rules (constitutions, laws, property rights)" [83]. The author understands that institutions are created to promote trust in trade, and work as an economy's incentive structure.

In other words, formal or informal institutions act in such a manner that allows or constrains economic and social development. The direction of the force exerted by institutions strengthens or weakens the performance of each sector. In any case, institutions play a decisive role in the economy's performance or, more specifically, the performance of markets. Additionally, the proper functioning of the involved institutions in societies is decisive with regard to the performance of the markets. Oriented institutions towards specific segments determine the success or failure of an economic sector. This concept is crucial for further conclusions.

In a complementary manner, Acemoğlu, Johnson, and Robinson also address the relevance of institutions in economic and social development, by studying the purposes of institutions based on the colonization process [84]. The authors argued that the development of European-colonized countries performs differently due to the home country's interests. Colonization of regions for purely exploratory purposes or to enrich certain states restrains local development, as seen in countries in Africa or South America. However, in regions where colonization took place for the purpose of permanence, the institutions allowed for development, such as in Australia and the USA. The argument is based on the types of colonial policies, the feasibility of settlements, and those institutions' persistence [84].

Additionally, Acemoğlu notes the influence of institutions on economic reforms. The author argues that in order to maintain privileges, interested and powerful groups act on their own behalf. However, the changes must reach not only formal but also informal institutions in order to achieve results [85–87].

The new economic sociology (NES) also brought essential elements that contribute to understanding of the influence of institutions on markets. Based on Weberian thinking, Swedberg [88] (p. 7) defines NES as "the application of sociological ideas, concepts and methods to economic phenomena—markets, companies, stores, unions, and so on". The author, supported by Weber, describes that its object studies "both economic phenomena, as well as how these phenomena influence the rest of society and how the rest of society influences them (economically relevant phenomena)".

In 1985, Granovetter used anthropological factors, such as those of Polanyi, to develop a critical theory of the relationships between individuals and institutions in a correlated manner [17]. With such relationships between them, embeddedness occurs, changing the market's characteristics. The main reason for this is that marketing behavior is based on trust and bad faith involved in the relationships between agents. Therefore, the connection of actors and institutions acts over the core fundaments of the market's functioning.

The theory of embeddedness in markets starts from the negation of classical and neoclassical utilitarian thinking, as well as from under and over-socialized views of consumer choices. The author was concerned about the atomization of human actions. This theory is based on the behavior of humans as a result of both the social web in which they are inserted, and their own initiative, avoiding theoretical extremes.

For the development of the NSE, Granovetter relies on three pillars: that economic action is a form of social action, that economic action is socially situated, and that economic institutions are social constructions [17,31,89]. Under any of them, the importance of complex analysis of a concrete economic issue from a non-isolationist perspective, but as part of a social context, is noteworthy. Alternatively, with regard to Granovetter's work, Raud-Mattedi [90] (p. 65) highlights this understanding, stating that "The market, therefore, does not consist of a free play of abstract forces, supply and demand, between actors atomized and anonymous, but in a set of actions closely intertwined in concrete networks of social relations". However, even within markets, authors differ as to the most appropriate approach for each market. This is typically divided into three theoretical groups [18,91]—(a) networks [17,30,31,92,93], (b) institutions [32,94–96], or (c) performativity [28,29,97,98]—as explanatory mechanisms for the emergence and dynamics of the markets.

Granovetter [17] (p. 488) describes a fundamental sense of validation and market gaps in his article on embeddedness. The author, without detailing explanations of over- or under-socialization, argues that "what has eroded this confidence in recent years has been increased attention to the micro-level details of imperfectly competitive markets, characterized by small number of participants with sunk costs and 'specific human capital' investments". However, in response to this work, Raud-Mattedi [90] (pp. 63–64) refers to this understanding, stating that "The market, therefore, does not consist of a free play of abstract forces, supply and demand, between atomized and anonymous actors, but in a set of actions closely intertwined in concrete networks of social relations". Still, Stein [99] proposes that institutions (property rights included) are crucial to examining markets, and that biases can hinder even neo-institutionalist perspectives in the neoclassical economy.

Still, with regard to the NES, Fligstein and McAdam [100] take a more in-depth approach regarding markets and institutions. The authors theorized about how institutions emerge, become stable, and are transformed to remain alive in what they call a theory of fields. This theory looks at disputes that occur at the intermediate or meso levels of dispute, implying that actions occur within organized local groups. It is in these disputes that institutions are built, stabilized, and transformed. Thus, Fligstein and McAdam argue that institutions result from social interaction between actors that confront one another in arenas or fields, and distance themselves from rational choice theorists by attributing importance to the social construction of identities, interests, actions, and action structures. Furthermore, they argue that the groups with the most significant influence in a strategic field of action promote shared identities and meanings by appropriating material and existential resources to legitimize their privileged position in the field, forging rules that favor them and defending the maintenance of their superior resources and power [100].

In another work, Fligstein went further and developed a theory of markets [94]. Here, the author theorizes that firms, faced with competition scenarios, tend to build what he calls "conception of control", where the incumbents' business model dominates the market's way of acting [94]. In the same work, he states that the creation of markets seeks stability in order to reduce the harmful effects of competition. For Fligstein, the state is a fundamental actor in reducing these harmful effects and promoting stability by acting not only through bureaucratic and legal means [94]. Sometimes, even its performance takes place through regulatory means and active participation [94].

In a third work by Fligstein, the author uses the metaphor of "markets as politics" to discuss how markets and states are intimately embedded. By dissociating the market into three phases (creation, stabilization, and transformation), he maintains that market arrangements develop around property rights, governance structures, conceptions of control, and rules of exchange [34]. Among the various propositions made, two stand out for this work: The first argues that "Laws and accepted practices often reflect the interests of the most organized forces in society" [34]. The second states that "the entry of countries into capitalism pushes states to develop rules about property rights, governance structures, and rules of exchange in order to stabilize markets for the largest firms" [34]. Along with the other 14 propositions, he concludes that "Markets are social constructions that reflect the unique political-cultural construction of their firms and nations" [34]. His approach adds to the sociological approach to markets by deepening and consolidating some aspects of how states act to guarantee their interests. Thus, we affirm in this conceptual paper that for the specific context of agri-food markets, the state not only acts in the markets, but also acts in defense of the interests of the social segments that it hegemonically represents.

One last sociological aspect is crucial in this work. Gramsci's theorization of hegemony is the glue that unites and gives purpose to the forms of construction of markets, emphasizing the different constructions in the agri-food sector. This conceptual work does not intend to extract all of the concepts and critical approaches carried out by the author, but simply to enter into what is pertinent to developing an understanding of the functioning of modern agri-food markets—cultural hegemony.

At the beginning of the 20th century, Gramsci developed the concept of cultural hegemony in a European context of intense industrialization and, therefore, of large masses of the population migrating from the countryside to the cities. He argued that this domination is usually built up due to the prestige of more powerful groups, as a result of their positions and roles in the world of production [47]. Consequently, this process of domination over these groups takes place through the state (or political society) to maintain the status quo [47]. The author goes deeper by asserting that this consent is given through ideological formation. Thus, the culture and values of the bourgeoisie become "common sense" for everyone, making the masses identify with it and defend their causes as their own, containing popular revolts [47].

Gramsci still distinguishes intellectuals (in what would fit better today with a concept of authority) from the countryside and cities, arguing that there are significant differences between them [47]. The author's established concept of traditional and organic intellectuals (adapted by us to authorities) derives from his notion of forms of ideology. For Gramsci, the agents responsible for these forms of articulation receive these names. In their theoretical elaboration, traditional intellectuals are those most detached from the economic structure, without a necessary relationship with the social or political class. In contrast, organic intellectuals are the agents responsible for the ideological promotion of their function in the economic field, due to their similarities with the ruling classes in the political and social areas.

From the point of view of these authorities as traditional and organic categories, they present opposite influences in their relationships. While there is a vertical relationship with the exercise of power by these authorities in the field, the opposite occurs in industry [47]. This conclusion implies that the exertion of power and ideological dissemination in rural areas occurs via a top-down social structure. In other words, the domination through assimilation occurs through the manufacture of consent, where the ruling social classes exercise their worldview over the popular classes. Meanwhile, in the industrial sector, this takes place through peers, in a horizontal manner.

Both the construction of the concept through the theorization of cultural hegemony and the distinction of the relationship of influence between the authorities of the countryside and the cities inexorably compose the construction of markets. In this work, we argue that this construction of hegemony is even more present in the agri-food markets due to this relationship, and that there is still a change-resistant productive structure for the dominant groups to remain in a position of influence, privilege, and power.

3.4. Impacts of Globalization

As agricultural systems evolved into more specialized and simplified agri-food processes, the agricultural market also walked the same path—notably after the Second World War, through scaling up and diversification in the trade of agricultural goods [101]. However, these systems evolved unevenly on a global scale. Wallerstein points out how globalization affects countries differently; his theory rejects the conception of "first", "second", and "third world". Instead, he proposed a modern world system, classifying nation states in three possible positions—center, semi-periphery, and periphery [102]—derived from dependence theory.

Wallerstein's theory is adequately applied to the functioning of agri-food markets in the modern world. Nation states that play a leading role in the agricultural sector seek to do so through soy, corn, and sugar commodities. Such agricultural cultures are only possible with the simplification of agrarian systems, and are exercised in large portions of land—often in monocultures represented by high concentrations of rural properties.

The nation states situated in these positions are located in the periphery and semi-periphery of the world system, supplying the countries of the center with food and primary products, and acquiring these more industrialized products. On the other hand, due to industrialization, countries in the center of this system manage to buy primary products from countries on the periphery of the globe, and dedicate themselves to producing food in more complex agrarian systems and, consequently, in more complex agri-food markets. Hence, globalization conditions the geography of the production process and the market for agri-food products, consolidating the exchange relationship. This relationship reinforces the thesis that the central countries started to occupy distinct and privileged positions relative to the others in the world system, due to prioritizing the industrialization of their economies.

However, this intra-nation relationship reproduces asymmetric effects in the construction of agri-food markets. Due to the characteristics of the productive systems inherent in these markets, globalization impacts the world's production and food systems, where peripheral nations have the development of complex markets compromised due to the development of capitalism, causing an unequal agri-food development between nation states.

The construction of a global market guided by the center–periphery logic places agri-food production in an asymmetric perspective. On the one hand, countries in the periphery are characterized by the production of primary products, or commodities, to supply raw materials to the central nations. On the other hand, the more industrialized and central countries consume these goods and export value-added products. In this way, the international market is built with disproportionate weights in terms of values. Thus, considering these characteristics of production systems, countries on the periphery and semi-periphery of the global system tend to maintain this format, due to the hegemonic process that benefits the ruling classes. Therefore, in a world system of low mobility between nations for a commercial balance, the tendency towards alterations in the productive systems becomes equally reduced.

Roland Robertson states that the ideas, cultural forms, and goods reach the world. However, due to the cultural diversity of each place, those global forms are perceived differently and adapted to each reality. He calls this phenomenon "glocalization" [103]. In the GI case, this phenomenon is well observed in cheese, for example. The Parmigiano-Reggiano cheese is a protected designation of origin (PDO), made in the Italian regions of Emilia-Romagna and Lombardy. When Italian immigrants went to South America at the beginning of the 20th century, they brought their cheesemaking knowledge along with them. This resulted in analogous cheeses called Parmesão, in Brazil, or Parmesano, in Argentina. However, according to EU regulations on GI products, neither is recognized as the same as the original.

Boaventura de Sousa Santos, on the other hand, points to the idea of an uneven conflict between hegemonic states and ideologies on one side, and collective dominated groups on the other, as counter-hegemonic [104]. According to the author, this polarized position is due to several areas of knowledge based on epistemological exclusion. The unequal struggle pushes the dominant models and interests of the North towards the South of the globe via an unfair and hegemonic social hierarchy of knowledge, stretching social inequality from the perspective of Boaventura de Sousa Santos or, from Wallerstein's perspective, in the dominant models and interests from the center to the periphery of the globe. By disregarding and invalidating other forms of thought and cultures, a standard model for the construction of science and society is established, consolidating the body of knowledge and possibilities for building society. Thus, the author concludes that modern capitalism needs alternatives to eradicate inequalities, and that this would only be possible with what he calls "global cognitive justice". The thinking of Sousa Santos is consistent with Wallerstein's.

Thus, the causes and effects of the process of globalization in the market and the agri-food production structure become clear. The position of countries regarding the function and products in the world system affects how countries produce food, the type of food, for whom this market is constructed, and the biggest beneficiaries of the consolidation of this market.

According to Milton Santos [105], globalization is characterized by a hierarchically structured market articulated by hegemonic, national, and foreign firms, commanding the territory supported by the state. This is precisely how globalization impacts markets. The consolidation of agri-food systems aimed at maintaining the current status quo, both in the periphery and in the center of the world system, makes the dominant interests in all parts become hegemonic.

Thus, colonization also plays a role and generates consequences. Settlement- and permanence-oriented colonization is capable of promoting development in nations in a less predetermined and dispute-oriented way. In exploration-oriented colonization, the formation of the agrarian structure is previously established and divided for the elite construction. This fact supports the model that places these same nations in peripheral conditions.

3.5. *Agri-Food Markets*

Starting from Ilbery and Kneafsey's [106,107] studies regarding specialty agri-food markets, the authors concluded that this results from interactions between producers, customers, and institutions. This embeddedness does not occur only by chance; the involvement of these three aspects sustains a market that cannot sustain itself with customers and producers alone. As such, the present work is dedicated to discussing the relevance and influence of the third aspect of this market: the institutions.

From the mentioned definitions of perfect markets, it is utterly clear that there is a significant variation in production models among agricultural markets. In the first case, large agricultural markets that produce commodities—such as sugar, soybeans, or corn—are significantly closer to the definition of the perfect competition conditions. There is extreme homogeneity between products, mobility, volatile prices, a slight permeability of participating actors, and information about production, logistical, and stock conditions is known to any buyer or seller.

On the other hand, there is another relevant agricultural market. Local and regional products are part of agricultural product niches that do not fit into this market. Such niches are the definition of imperfectly competitive markets, full of details that need to be looked at in depth, with few participants with sunk costs and investments in specific capital through the terroir of each producing region. Therefore, the natural, evident, and unique path for these agricultural markets is to fit within the approach proposed by the NES.

Both economic and agrarian matters present a myriad of complex forms of approach, as seen previously. On the one hand, agrarian systems vary in their complexity, subject to environmental conditions, human influence, and cultural factors. The more complex the system, the more unique the resulting products, and the more complex the markets become. On the other hand, the distinction of how to classify the analysis is not a simple toolbox. However, the more subjective those analytic tools are, the more details can be perceived and, therefore, the more capable the tools are of in-depth analysis. Sociological tools, for example, allow for the examination of social actions through magnifying lenses, with more detailed visualization and understanding. This facilitates the extraction of information about the functioning and its causes, in addition to the results.

Agricultural practices developed over time; however, the development of these practices is asymmetrical in geographical and chronological terms. Thus, not only environmental conditions, but also historical events and cultural aspects, pushed specific regions towards specialization or diversification of technological advances, practices, production, changes, and the role of agriculture in each society. The reasons for such differences were previously discussed in this article. However, there are still some aspects of agricultural markets that require attention.

Thom [41] argued that adopting a taxonomy of systems is imperative for the proper analysis and development of a theory. Thus, the creation of a theory of agrarian systems by Mazoyer and Roudart [1] allowed for the deepening of the subject. Furthermore, the distinction of systems into the cultivated ecosystem and the productive social system supports their theory. These two components suggest that agrarian systems can only be altered if at least one of them is changed.

Both components are embedded. The cultivated ecosystem relates to a set of practices, land use intensity, and environmental relations, and varies as socio-political moments vary. For example, when times demanded more food production as the population grew more rapidly, there was a need to intensify the land use and apply techniques to extract more from the environment.

Productive social systems are no different. However, the need for the development of new tools and equipment, selection of animals and plants for growth of production, and the labor force dedicated to it also changed according to socio-political demand for more food. However, as the demand over time has changed chiefly in quantitative terms, the changes in these two components have also changed (cultivated ecosystem and the productive social system supports) in qualitative aspects. Nevertheless, the rise of a new agrarian system does not imply the demise of the existing ones; different agrarian systems can coexist. However, there is a tendency for specific systems to prevail. This depends on the combination of forces capable of exerting political pressure. According to Gramsci [47], within the world's capitalist system, such pressure is exerted by the holders of economic power through cultural hegemony. Therefore, any type of change in agrarian systems depends on the ability to exert political pressure to change them, since the institutions have the tools capable of influencing this process [108].

Although there are differences in conception at the start of the Anthropocene, this work chose to consider the dawn of agriculture as the first agrarian system constituted. However, even in discussions of origins, scholars consider the relevance of all the socio-ecological complexity of agri-food systems [109]. The complexity of the systems hinges on the impossibility of reproducing terroir-related quality, regardless of intrinsic regional characteristics. Therefore, standardization of agri-food products over a wide geographic area, reducing human, cultural, and environmental factors, implies a reduction in complexity.

Globalization started a process that improved Fordism [3]. This process resulted in a struggle between local and global agri-food systems, and pushed smallholders and communities towards niche formation [110,111]. The process of globalization is a result of modern capitalism, and has goals of standardization and homogenization at its core. Thus, the agri-food products that prevail in the current capitalist system are commodities that are only possible in low-complexity agrarian systems. On the other hand, complex agrarian systems result in non-reproductive agri-food products in other locations, due to characteristics arising from cultural, human, and environmental elements. The GI represents the complexity of such products from complex systems. The materialization of terroir is institutionalized through the granting of intellectual property rights. Therefore, through its institutionalization in GI, terroir constitutes a niche market, deserving a more in-depth approach [112].

As shown in the discussion in the previous section, economics enables and is of great analytical use to the agri-food market. However, as Smith [55] initially observed, economics endorses a utilitarian worldview by maximizing land use. The theory provided in *The Wealth of Nations* by the same author converges with the capitalist-based world system in formation at that time, based on profit maximization. Smith understands that individuals act for their own benefit by devoting efforts and resources to it. The denial of secondary interest of societies is crucial to understanding the principles of classical economics; it is sustained by individuals, lacking intent for the collective good. The core of classical and neoclassical economics is based on rational choice theory—a thesis that conflicts with the foundations of complex agrarian systems embedded with human, natural, and cultural factors.

As Arnsperger and Varoufakis [73] previously noted, the three axioms of the theory are methodological individualism, methodological instrumentalism, and methodological equilibration. Thus, these theoretical foundations properly match the conditions of perfect competition. Commodity conditions such as atomization, homogeneity, mobility, permeability, free price flow, and transparency, due to their characteristics, are less easy to influence and, therefore, more suited to classical and neoclassical tools.

Although also utilitarian, political economy has as its object of study the geopolitical and globalization factors in transnational trade. Its development, along with globalization and the post-Fordist society, added a political variable after an era of mercantilism. As such, the conception of an accelerating mobility of capital along with worldwide urbanization also developed based on everlasting development that progressed to the capitalism of mass production and consumption.

From a rural perspective, this production and consumption philosophy is no different. However, there are inherent differences in the means of production between urban and rural living and production. For example, agri-food production is not a mechanical process like industrial production. Soil, climate, and pests, among others, influence and interfere with production. Nevertheless, capitalism has pushed agri-food production in the same direction, aiming for standardization and homogenization of mass production. Moreover, Bonanno and Constance [3] point to the increase in the rationalization process, pushing massification and standardization as only being possible with the participation and influence of the state [47].

As addressed in the other perspectives, economic sociology also approaches the agri-food markets. Additionally, more complex agrarian systems are endowed with cultural, human, and environmental elements that influence the differentiation of their products. These elements provide characteristics capable of producing unique and irreproducible agri-foods in different areas. Thus, agri-food products of greater complexity reproduce this complexity in the markets in which they participate and, consequently, provide the conditions that keep them from perfect competition.

The best examples of these product markets arising from more complex agrarian systems that are so particular and require differentiated markets are the products labeled with GI. Such agri-food products are significantly different from commodities, since they are irreproducible in areas other than those for which they are registered, for cultural, human, and environmental reasons. However, due to the multiple factors derived from the three terroir builders, the markets for these differentiated products can only be adequately investigated through the lens of economic sociology.

Agri-food markets become more complex as their production systems add more elements and produce more complex foods. However, as previously discussed, these systems have developed asymmetrically around the world.

Some nations support more complex systems, enabling the development of a more significant number of products imbued with their local cultures, the exercise of human practices, and those influenced by the environment. Meanwhile, other nations maintain simpler agrarian systems aimed at producing agricultural commodities.

Given the market imperfections addressed by Granovetter and Hayek [17,31,76], it is clear that markets are not only formed by free trade relations between buyers and sellers. External influences are present, as are built-in institutions—formal or informal [83]. Such conformations sustained in the markets have built institutions with different purposes around the globe.

As discussed by Acemoğlu, the formation of institutions in the world is strongly influenced by colonization [84]. Thus, in countries where colonization was carried out in an exploratory manner, the construction of institutions was supported by similar agrarian and productive systems, as in Latin America and Africa. Thus, despite the existence of productive initiatives in more complex market niches, nations with this type of colonization mostly maintained commodity-producing systems. On the other hand, colonized nations with purposes of permanence developed similarly to those of their origin, industrialized, formed urban elites, and opened space for the construction of more complex markets in the food field. In addition to the form of colonization, regions where the dominant groups do not come from rural areas were able to implement formal institutions with greater capacity for the development of more complex agri-food markets.

Thus, countries colonized in an exploratory manner created agrarian elites that reproduce themselves in political power. Formal institutions, be they state structures or legal instruments, result from the political constructions into which they are inserted [18,34,94]. Thus, institutions represent the thinking and interests of dominant groups. In the case of nations that maintained exploratory agricultural systems and, consequently, less complex markets, their institutions became reflections of groups with greater power in the countryside.

Therefore, with the formation of agrarian elites resulting from exploratory colonization, structured groups that produce large-scale commodities are formed. On the one hand, such groups exert local power and political influence and, on the other hand, consolidate the common sense of the field's function according to the theory elaborated by Gramsci [47]. In the same vein, Michels states that the interests of those on top of organizations always come first, and oligarchies tend to sustain the elite's interests and suppress people's interests [113].

While consolidating the ideology and values of the dominant groups, institutions suppress the development of systems that could threaten their hegemony. Thus, the agrarian elites, upon establishing themselves as the dominant group in certain regions, build systems of political and ideological tools that make the development of other productive systems unfeasible. In these regions, by establishing the commodity production system as a model, they suppress the development of more complex agrarian systems and niche market products, such as products with GI.

The fact that institutions result from the embeddedness of the social actors involved allows for mutability in their construction. However, since institutions reflect the groups that influence them, in order for there to be transformations, it is necessary to change the groups that dominate the construction of these institutions. Thus, in order for new systems, products, and markets to flourish, it is necessary for new groups to become dominant over the construction of these institutions. In the same sense, Acemoğlu and Robinson argue that political institutions need to increase state capacity and distribution of power in a balanced way in order to be inclusive [86]. Therefore, modernization is achieved through inclusive and balanced institutions. Even more specifically, Allaire and Wolf point out the importance of hybridity in institutions in the qualification process of agri-food systems [114]; the authors' approach solidifies the importance of transforming institution-forming forces in order to objectify identity-based food systems.

In this way, the process of globalization consolidates the position of the nations concerning their commercial function in the world, allowing little or no mobility between them [102]. For this reason, Bonanno and Constance argue that global post-Fordism is a system that takes advantage of economic and social rigidity, seeing the labor market and local consumption as forces to be included or excluded according to their corporate interests [3]. Thus, the capitalist logic of serving the interests of hegemonic groups is maintained, to the detriment of the development of complex agri-food markets such as GI. This view of the disproportionate effects caused by globalization is endorsed by Friedland, who sees it as a phenomenon of heterogeneous effects and proportions across sectors, regions, and products, and proposes a neo-Fordist approach to cross-cutting commodities [115].

This concept paper indicates that the market (and especially the agri-food market) has different levels of embedded influences, via economic and agrarian analysis. Therefore, we can conclude that there is no such thing as an invisible hand. Economic issues and, more specifically, markets are always oriented by a power balance. This balance is a result of the embeddedness of social, political, and economic matters. The outcome of this struggle pushes the profits towards the most powerful actors in play. Furthermore, GI agri-food products arise from embedded agrarian systems resulting from terroir, as the fruition of the multiple hands acting towards creating and stabilizing a market.

Finally, institutions are built to consolidate the ideas of the ruling elites. If, in turn, these elites exercise power through domination over land, the tendency is for these institutions to be oriented towards perpetuating this form of power and maintaining the interests of dominant groups. Thus, the formation of these groups allows for divergent models of agri-food production: One, oligarchic and commoditized, where colonization was exploratory, and other, in productive niches where industrialization was able to emerge.

4. Conclusions

The present work sought to discuss the embeddedness of institutions in agri-food markets, based on critical theory. According to agrarian complexity, as well as the consequent formation of the market, the present concept paper sought to approach the differences in the construction of institutions by the dominance of interest groups. Much study has been devoted to agricultural markets. This work sought to present contemporary approaches to the theme and contextualize them in terms of their formation, central ideas, and analytical skills associated with different agrarian complexities and their products. In no way does this work aim to exhaust the debate; simply to present possible, viable, and assertive paths for the future discussion of these markets.

The first conclusion is that products such as GI, imbued with cultural values derived from environmental conditions and proper knowhow, are only possible in complex agrarian systems. In turn, such systems are reminiscent of practices in specific regions, carried over time by the cultural factors that allowed their current existence. Therefore, as the complexity of agrarian systems increases, the determining variables in the market for such products also increase. Thus, regions where less complex systems predominate tend to hinder the creation, maintenance, and perpetuation of such products, which may compromise their existence in the long run.

A clear conclusion is based on the principle of the formation of agrarian systems, with embedded relations with civilizations' cultural formation. Food and culture are part of the same matrix, and cannot be dissociated. Barham [55] and Allaire [4] suggest that the embeddedness perspective along with convention theory analysis can enlighten the discussion of origin-related food issues. Such a path could be a future avenue of research.

The second conclusion is that more or less complex agri-food markets develop due to the elite formation in each region. In regions where there is an agrarian elite sustained by the production of commodities, institutions tend to be built with their own interests in mind. Regions with industrialized economies tend to set the interest groups on this sector and open a window for dispute in the agri-food sector, allowing for the development of more complex products.

On the other hand, as a third conclusion, regions colonized through exploration, without goals of permanence, built institutions capable of maintaining this vision, as noted by Acemoğlu [84,86,87]. In commodity-oriented nations, these institutions are formed by agrarian elites who exercise power and influence over them. Furthermore, the theory developed by Wallerstein applies to the present case in terms of maintaining positions regarding their functions in the periphery and semi-periphery of the world.

The fourth conclusion is that the construction of institutions is carried out to promote the maintenance of dominant groups' interests through ideological means, as highlighted by Gramsci [47]. Thus, in agri-food markets, nations reproduce these interests according to the formation of dominant groups in each place: oligarchic elites where colonization was exploratory, and productive groups dedicated to niches in regions where industrialization was a driving force.

Finally, in response to the question presented at the beginning of this work as to what drives GI agri-food markets, it is clear that the construction of these markets does not result merely from productive capacity, from the number of individuals involved in agriculture, or from the diversity of the environment that influences the goods. The primordial and determining factor for the construction of these markets is the result of the social conformation and power struggle where dominant interests prevail, which exercise control through institutions, which is called hegemony. In other words, where agrarian elites from fundamentally exploratory colonizing processes predominate, they tend to perpetuate the dominance of low-complexity agrarian models, constraining more complex embedded systems such as those endowed with terroir, such as GI. Meanwhile, in regions where the dispute for power takes place in other fields, there is room for developing factors capable of producing agri-food products and more complex markets. For agri-food markets to be altered, it is necessary to break the hegemony of dominant interest groups over the structures that form institutions. New systems can be developed only by breaking the hegemony of these groups and expanding the base of influence.

Funding: This research received no external funding.

Institutional Review Board Statement: Not applicable.

Informed Consent Statement: Not applicable.

Data Availability Statement: Not applicable.

Acknowledgments: The author wishes to thank M.P.C. and R.J.S.D.M. for their support and assistance with the research.

Conflicts of Interest: The views and opinions expressed in this article are those of the author, and do not necessarily reflect the official position of the Brazilian Ministry of Agriculture, Livestock, and Food Supply.

References

1. Mazoyer, M.; Roudart, L. *História das Agriculturas no Mundo*, 1st ed.; Editora UNESP: São Paulo, Brazil, 2010; ISBN 978-85-7139-994-5.
2. Allen, R.C. Agriculture during the industrial revolution. In *The Economic History of Britain Since*; Floud, R., McCloskey, D.N., Eds.; Cambridge University Press: Cambridge, UK, 1994; Volume 1700, ISBN 978-0-521-42520-9.
3. Bonanno, A.; Constance, D.H. Globalization, Fordism, and Post-Fordism in Agriculture and Food: A Critical Review of the Literature. *Cult. Agric.* **2001**, *23*, 1–18. [CrossRef]
4. Allaire, G. Economics of Conventions and the New Economic Sociology and Our Understanding of Food Quality and New Food Markets and Trade Institutions: What Are Markets That Pure Economics Does Not Know? In Proceedings of the Plenary Session 5 "The New Landscape and the Need for an Extension to the Agricultural Economics Toolkit", Beijing, China, 16 August 2009.
5. Allaire, G. Applying Economic Sociology to Understand the Meaning of "Quality" in Food Markets. *Agric. Econ.* **2010**, *41*, 167–180. [CrossRef]
6. Allaire, G. Quality in economics: A cognitive perspective. In *Qualities of Food*; Manchester University Press: Manchester, UK, 2018; ISBN 9781526137609.
7. Freire, P. *Pedagogy of the Oppressed: 50th Anniversary Edition*; Bloomsbury Publishing: New York, NY, USA, 2018; ISBN 978-1-5013-1415-5.
8. Hartmann, K.; Martin, M. A Critical Pedagogy of Agriculture. *J. Agric. Educ.* **2021**, *62*, 51–71.
9. Hardnack, C. Gramsci and Goffman, Together at Last: Toward a Counterhegemonic Framing Approach to Movement Research. *Interface A J. Soc. Mov.* **2019**, *11*, 200–215.
10. Cini, L.; Chironi, D.; Drapalova, E.; Tomasello, F. Towards a Critical Theory of Social Movements: An Introduction. *Anthropol. Theory* **2017**, *17*, 429–452. [CrossRef]
11. Jacka, T. Translocal Family Reproduction and Agrarian Change in China: A New Analytical Framework. *J. Peasant. Stud.* **2018**, *45*, 1341–1359. [CrossRef]
12. Levien, M.; Watts, M.; Yan, H. Agrarian Marxism. *J. Peasant. Stud.* **2018**, *45*, 853–883. [CrossRef]
13. Byres, T.J. In Pursuit of Capitalist Agrarian Transition. *J. Agrar. Chang.* **2016**, *16*, 432–451. [CrossRef]
14. Bernstein, H. Historical Materialism and Agrarian History. *J. Agrar. Chang.* **2013**, *13*, 310–329. [CrossRef]
15. Bernstein, H. Is There an Agrarian Question in the 21st Century? *Can. J. Dev. Stud.* **2006**, *27*, 449–460. [CrossRef]
16. Borras, S.M. Agrarian Change and Peasant Studies: Changes, Continuities and Challenges—An Introduction. *J. Peasant. Stud.* **2009**, *36*, 5–31. [CrossRef]
17. Granovetter, M. Economic Action and Social Structure: The Problem of Embeddedness. *Am. J. Sociol.* **1985**, *91*, 481–510. [CrossRef]
18. Fligstein, N.; Dauter, L. The Sociology of Markets. *Annu. Rev. Sociol.* **2007**, *33*, 105–128. [CrossRef]
19. Smelser, N.J.; Swedberg, R. *The Handbook of Economic Sociology*; Princeton University Press: Princeton, NJ, USA, 2010; ISBN 9781400835584.
20. Steiner, P. A Plea for a Weberian Economic Sociology. *Arch. Eur. Sociol.* **2017**, *58*, 545. [CrossRef]
21. Swinnen, J.F. *Global Supply Chains, Standards and the Poor: How the Globalization of Food Systems and Standards Affects Rural Development and Poverty*; CABI: Cambridge, MA, USA, 2007; ISBN 1-84593-185-8.
22. Swinnen, J.; Kuijpers, R. Value Chain Innovations for Technology Transfer in Developing and Emerging Economies: Conceptual Issues, Typology, and Policy Implications. *Food Policy* **2019**, *83*, 298–309. [CrossRef]
23. Meloni, G.; Swinnen, J. Trade and Terroir. The Political Economy of the World's First Geographical Indications. *Food Policy* **2018**, *81*, 1–20. [CrossRef]
24. Josling, T. The War on Terroir: Geographical Indications as a Transatlantic Trade Conflict. *J. Agric. Econ.* **2006**, *57*, 337–363. [CrossRef]
25. Garrone, M.; Emmers, D.; Olper, A.; Swinnen, J. Jobs and Agricultural Policy: Impact of the Common Agricultural Policy on EU Agricultural Employment. *Food Policy* **2019**, *87*, 101744. [CrossRef]
26. Barjolle, D.; Quiñones-Ruiz, X.F.; Bagal, M.; Comoé, H. The Role of the State for Geographical Indications of Coffee: Case Studies from Colombia and Kenya. *World Dev.* **2017**, *98*, 105–119. [CrossRef]
27. Callon, M. Why Virtualism Paves the Way to Political Impotence: A Reply to Daniel Miller's Critique of "The Laws of the Market". *Econ. Sociol. Eur. Electron. Newsl.* **2005**, *6*, 3–20.
28. MacKenzie, D. *An Engine, Not a Camera: How Financial Models Shape Markets*; MIT Press: Cambridge, MA, USA, 2008; ISBN 978-0-262-25004-7.
29. MacKenzie, D.; Millo, Y. Constructing a Market, Performing Theory: The Historical Sociology of a Financial Derivatives Exchange. *Am. J. Sociol.* **2003**, *109*, 107–145. [CrossRef]
30. Burt, R.S. *Structural Holes*; Harvard University Press: Cambridge, MA, USA, 2021; ISBN 9780674029095.

31. Granovetter, M. *The Sociology of Economic Life*; Routledge: New York, NY, USA, 2019; ISBN 9780429494338.
32. Powell, W.W.; DiMaggio, P.J. *The New Institutionalism in Organizational Analysis*; University of Chicago Press: Chicago, IL, USA, 2012; ISBN 978-0-226-18594-1.
33. Fligstein, N. Social Skill and Institutional Theory. *Am. Behav. Sci.* **1997**, *40*, 397–405. [CrossRef]
34. Fligstein, N. Markets as Politics: A Political-Cultural Approach to Market Institutions. In *Readings in Economic Sociology*; John Wiley & Sons, Ltd.: Hoboken, NJ, USA, 2008; pp. 197–218, ISBN 978-0-470-75567-9.
35. Stone Sweet, A.; Sandholtz, W.; Fligstein, N. *The Institutionalization of Europe*; Oxford University Press: Oxford, UK, 2001; ISBN 0-19-924796-X.
36. Childe, V.G. *Man Makes Himself*; Watts: London, UK, 1936.
37. FAO (Food and Agriculture Organization). *Guidelines for Agrarian Systems Diagnosis*; FAO: Rome, Italy, 1999.
38. Vieira, R.M.S.P.; Tomasella, J.; Alvalá, R.C.S.; Sestini, M.F.; Affonso, A.G.; Rodriguez, D.A.; Barbosa, A.A.; Cunha, A.P.M.A.; Valles, G.F.; Crepani, E.; et al. Identifying Areas Susceptible to Desertification in the Brazilian Northeast. *Solid Earth* **2015**, *6*, 347–360. [CrossRef]
39. Xie, L.W.; Zhong, J.; Chen, F.F.; Cao, F.X.; Li, J.J.; Wu, L.C. Evaluation of Soil Fertility in the Succession of Karst Rocky Desertification Using Principal Component Analysis. *Solid Earth* **2015**, *6*, 515–524. [CrossRef]
40. Torres, L.; Abraham, E.M.; Rubio, C.; Barbero-Sierra, C.; Ruiz-Pérez, M. Desertification Research in Argentina. *Land Degrad. Dev.* **2015**, *26*, 433–440. [CrossRef]
41. Thom, R. La méthode expérimentale: Un mythe des épistémologues (et des savants?). *Le Débat* **1985**, *34*, 11–20. [CrossRef]
42. Lefebvre, H.; Nicholson-Smith, D. *The Production of Space*; Oxford Blackwell: Oxford, UK, 1991; Volume 142.
43. Halfacree, K.H. Locality and Social Representation: Space, Discourse and Alternative Definitions of the Rural. *J. Rural Stud.* **1993**, *9*, 23–37. [CrossRef]
44. Ren, K. Following Rural Functions to Classify Rural Sites: An Application in Jixi, Anhui Province, China. *Land* **2021**, *10*, 418. [CrossRef]
45. Bellwood, P.; Gamble, C.; Le Blanc, S.A.; Pluciennik, M.; Richards, M.; Terrell, J.E. First Farmers: The Origins of Agricultural Societies. *Camb. Archaeol. J.* **2007**, *17*, 87–109. [CrossRef]
46. Fine, B. Towards a Political Economy of Food. *Rev. Int. Political Econ.* **1994**, *1*, 519–545. [CrossRef]
47. Gramsci, A. *Selections from the Prison Notebooks*; Hoare, Q., Smith, G.N., Eds.; Reprint, 1989 edition; International Publishers Co.: New York, NY, USA, 1971; ISBN 978-0-7178-0397-2.
48. Kenney, M.; Lobao, L.M.; Curry, J.; Goe, W.R. Midwestern Agriculture in US Fordism: From the New Deal to Economic Restructuring. *Sociol. Rural.* **1989**, *29*, 131–148. [CrossRef]
49. Potter, C.; Tilzey, M. Agricultural Policy Discourses in the European Post-Fordist Transition: Neoliberalism, Neomercantilism and Multifunctionality. *Prog. Hum. Geogr.* **2005**, *29*, 581–600. [CrossRef]
50. Hanh, H.Q.; Azadi, H.; Dogot, T.; Ton, V.D.; Lebailly, P. Dynamics of Agrarian Systems and Land Use Change in North Vietnam. *Land Degrad. Dev.* **2017**, *28*, 799–810. [CrossRef]
51. Araya, A.; Stroosnijder, L.; Habtu, S.; Keesstra, S.D.; Berhe, M.; Hadgu, K.M. Risk Assessment by Sowing Date for Barley (Hordeum Vulgare) in Northern Ethiopia. *Agric. For. Meteorol.* **2012**, *154–155*, 30–37. [CrossRef]
52. Alexander, P.; Paustian, K.; Smith, P.; Moran, D. The Economics of Soil C Sequestration and Agricultural Emissions Abatement. *SOIL* **2015**, *1*, 331–339. [CrossRef]
53. Galati, A.; Crescimanno, M.; Gristina, L.; Keesstra, S.; Novara, A. Actual Provision as an Alternative Criterion to Improve the Efficiency of Payments for Ecosystem Services for C Sequestration in Semiarid Vineyards. *Agric. Syst.* **2016**, *144*, 58–64. [CrossRef]
54. Jlassi, W.; Romero, M.E.N.; Ruiz, J.M.G. Modernization of New Irrigated Lands in a Scenario of Increasing Water Scarcity: From Large Reservoirs to Small Ponds. *Cuad. Investig. Geográfica/Geogr. Res. Lett.* **2016**, *42*, 233–259. [CrossRef]
55. Barham, E. Translating Terroir: The Global Challenge of French AOC Labeling. *J. Rural Stud.* **2003**, *19*, 127–138. [CrossRef]
56. Goodman, D. Rethinking Food Production–Consumption: Integrative Perspectives. *Sociol. Rural.* **2002**, *42*, 271–277. [CrossRef]
57. Guy, K.M. Silence and Savoir-Faire in the Marketing of Products of the Terroir. *Mod. Contemp. Fr.* **2011**, *19*, 459–475. [CrossRef]
58. Barjolle, D.; Boisseaux, S.; Dufour, M. *Le Lien Au Terroir. Bilan Des Travaux de Recherche*; Ecole Polytechnique Fédérale de Zurich, Institut D'économie Rurale: Zurich, Switzerland, 1998. Available online: www.originfood.org/pdf/wp1/wp1-ch.Pdf (accessed on 12 May 2021).
59. Tylor, E.B. *Primitive Culture: Researches into the Development of Mythology, Philosophy, Religion, Art and Custom*; Gordon Press: New York, NY, USA, 1974; ISBN 978-0-87968-091-6.
60. Abdel-Hadi, A. Culture, Quality of Life, Globalization and Beyond. *Procedia-Soc. Behav. Sci.* **2012**, *50*, 11–19. [CrossRef]
61. Veblen, T. The Preconceptions of Economic Science. *Q. J. Econ.* **1900**, *14*, 240–269. [CrossRef]
62. Kyraleou, M.; Herb, D.; O'Reilly, G.; Conway, N.; Bryan, T.; Kilcawley, K.N. The Impact of Terroir on the Flavour of Single Malt Whisk(e)y New Make Spirit. *Foods* **2021**, *10*, 443. [CrossRef] [PubMed]
63. Millet, M.; Keast, V.; Gonano, S.; Casabianca, F. Product Qualification as a Means of Identifying Sustainability Pathways for Place-Based Agri-Food Systems: The Case of the GI Corsican Grapefruit (France). *Sustainability* **2020**, *12*, 7148. [CrossRef]
64. Van Leeuwen, C.; Seguin, G. The Concept of Terroir in Viticulture. *J. Wine Res.* **2006**, *17*, 1–10. [CrossRef]
65. Smith, A. *The Wealth of Nations*; Modern Library: New York, NY, USA, 1994; ISBN 978-0679424734.

66. Marx, K. *O Capital—Livro 1: Crítica da Economia Política. Livro 1: O Processo de Produção do Capital*; Boitempo Editorial: São Paulo, Brazil, 2015; ISBN 978-85-7559-321-9.
67. Keynes, J.M. *The General Theory of Employment, Interest, and Money*; Palgrave Macmillan: Cham, Switzerland, 2018; ISBN 978-3-319-70343-5. [CrossRef]
68. Stein, H. Institutionalizing Neoclassical Economics in Africa: Instruments, Ideology and Implications. *Econ. Soc.* **2021**, *50*, 120–147. [CrossRef]
69. Say, J.-B. *A Treatise on Political Economy*; Routledge: Boca Raton, FL, USA, 2017; ISBN 978-1-351-31568-5.
70. Ricardo, D. *The Principles of Political Economy and Taxation*; Cambridge University Press: Cambridge, UK, 2015; ISBN 9781107589421.
71. Mill, J.S. *On Liberty*; Cambridge University Press: Cambridge, UK, 2011; ISBN 978-1-108-04083-9.
72. Mill, J.S. *Utilitarianism*; Cambridge University Press: Cambridge, UK, 2015; ISBN 9781139923927.
73. Arnsperger, C.; Varoufakis, Y. What Is Neoclassical Economics? The Three Axioms Responsible for Its Theoretical Oeuvre, Practical Irrelevance and, Thus, Discursive Power. *Panoeconomicus* **2006**, *53*, 5–18. [CrossRef]
74. Marshall, A. *Principles of Economics*; Prometheus Books: Amherst, NY, USA, 1997; ISBN 978-1-57392-140-4.
75. Pareto, V.; Montesano, A.; Zanni, A.; Bruni, L.; Chipman, J.S.; McLure, M. *Manual of Political Economy: A Critical and Variorum Edition*; Illustrated edição; OUP: Oxford, UK, 2014.
76. Hayek, F.A.V. *Individualism and Economic Order*; 1a edição; University of Chicago Press: Chicago, IL, USA, 1996; ISBN 978-0-226-32093-9.
77. Montchrétien, A. *de Traité de L'oeconomie Politique*; Laudet, M., Ed.; Classiques Garnier: Paris, France, 2017; ISBN 978-2-406-06628-6.
78. Mayntz, R. Changing Perspectives in Political Economy. 2019. Available online: https://pure.mpg.de/rest/items/item_3149015_4/component/file_3149430/content.Pdf (accessed on 20 May 2021).
79. Balaam, D.N.; Veseth, M.A. Political Economy. Available online: https://www.britannica.com/topic/political-economy (accessed on 17 May 2021).
80. Eltis, W.A. Francois Quesnay: A Reinterpretation 1. The Tableau Economique. *Oxf. Econ. Pap.* **1975**, *27*, 167–200. [CrossRef]
81. Schultz, T.W. Nobel Lecture: The Economics of Being Poor. *J. Political Econ.* **1980**, *88*, 639–651. [CrossRef]
82. Benjamin, W. *The Work of Art in the Age of Mechanical Reproduction*; Penguim: London, UK, 2008; ISBN 9780141963426.
83. North, D.C. Institutions. *J. Econ. Perspect.* **1991**, *5*, 97–112. [CrossRef]
84. Acemoglu, D.; Johnson, S.; Robinson, J.A. The Colonial Origins of Comparative Development: An Empirical Investigation. *Am. Econ. Rev.* **2001**, *91*, 1369–1401. [CrossRef]
85. Acemoglu, D.; Robinson, J.A. *Economic Origins of Dictatorship and Democracy*; Cambridge University Press: Cambridge, UK, 2006; ISBN 978-0-521-85526-6.
86. Acemoğlu, D.; Robinson, J.A. Paths to inclusive political institutions. In *Economic History of Warfare and State Formation. Studies in Economic History*; Eloranta, J., Golson, E., Markevich, A., Wolf, N., Eds.; Springer: Singapore, 2016; pp. 3–50, ISBN 978-981-10-1605-9.
87. Acemoglu, D.; Ticchi, D.; Vindigni, A. Emergence and Persistence of Inefficient States. *J. Eur. Econ. Assoc.* **2011**, *9*, 177–208. [CrossRef]
88. Swedberg, R. Sociologia Econômica: Hoje e Amanhã. *Tempo Soc.* **2004**, *16*, 7–34. [CrossRef]
89. Granovetter, M. The Old and the New Economic Sociology: A History and an Agenda. In *Beyond the Marketplace: Rethinking Economy and Society*; Routledge: New York, NY, USA, 1990; pp. 89–112, ISBN 978-0-202-30371-0.
90. Raud-Mattedi, C.H.J. Análise crítica da Sociologia Econômica de Mark Granovetter: Os limites de uma leitura do mercado em termos de redes e imbricação. *Polit. Soc.* **2005**, *4*, 59–82. [CrossRef]
91. Fourcade, M. Theories of Markets and Theories of Society. *Am. Behav. Sci.* **2007**, *50*, 1015–1034. [CrossRef]
92. White, H.C. *Markets from Networks: Socioeconomic Models of Production*; Princeton University Press: Princeton, NJ, USA, 2018; ISBN 978-0-691-18762-4.
93. White, H.C. Where Do Markets Come From? *Am. J. Sociol.* **1981**, *87*, 517–547. [CrossRef]
94. Fligstein, N. *The Architecture of Markets: An Economic Sociology of Twenty-First-Century Capitalist Societies*; Princeton University Press: Princeton, NJ, USA, 2002; ISBN 0-691-10254-6.
95. Fligstein, N. *The Transformation of Corporate Control*; Harvard University Press: Cambridge, MA, USA, 1993; ISBN 0-674-90359-5.
96. Dobbin, F. *Forging Industrial Policy: The United States, Britain, and France in the Railway Age*; Cambridge University Press: Cambridge, UK, 1994; ISBN 978-0-521-62990-4.
97. Callon, M. Introduction: The Embeddedness of Economic Markets in Economics. *Sociol. Rev.* **1998**, *46*, 1–57. [CrossRef]
98. Callon, M. What does it mean to say that economics is performative? In *Do Economists Make Markets?* MacKenzie, D., Muniesa, F., Siu, L., Eds.; Princeton University Press: Princeton, NJ, USA, 2020.
99. Stein, H. Theories of Institutions and Economic Reform in Africa. *World Dev.* **1994**, *22*, 1833–1849. [CrossRef]
100. Fligstein, N.; McAdam, D. *A Theory of Fields*; Oxford University Press: Oxford, UK, 2012; ISBN 0-19-985995-7.
101. Friedland, W.H. Agrifood Globalisation and Commodity Systems. *Int. J. Sociol. Agric. Food Online* **2004**, *12*, 5–16. [CrossRef]
102. Wallerstein, I. *The Modern World-System I: Capitalist Agriculture and the Origins of the European World-Economy in the Sixteenth Century*; Univ of California Press: Berkeley, CA, USA, 2011; Volume 1, ISBN 0-520-94857-2.
103. Robertson, P.R. *Globalization: Social Theory and Global Culture*; SAGE: London, UK, 2000; ISBN 978-0803981874. [CrossRef]

104. Santos, B.d.S. *Epistemologies of the South: Justice Against Epistemicide*; Routledge: New York, NY, USA, 2014; ISBN 9781612055459. [CrossRef]
105. Santos, M. A Revolução Tecnológica e o Território: Realidades e Perspectivas. *TL.* 2015. Available online: https://publicacoes.agb.org.br/index.php/terralivre/article/view/101.Pdf (accessed on 25 May 2021).
106. Ilbery, B.; Kneafsey, M. Niche Markets and Regional Speciality Food Products in Europe: Towards a Research Agenda. *Environ. Plan. A* **1999**, *31*, 2207–2222. [CrossRef]
107. Ilbery, B.; Kneafsey, M.; Bamford, M. Protecting and Promoting Regional Speciality Food and Drink Products in the European Union. *Outlook Agric.* **2000**, *29*, 31–37. [CrossRef]
108. Fracarolli, G.S. The Effects of Institutional Measures: Geographical Indication in Mercosur and the EU. *Sustainability* **2021**, *13*, 3476. [CrossRef]
109. Reisman, E.; Fairbairn, M. Agri-Food Systems and the Anthropocene. *Ann. Am. Assoc. Geogr.* **2021**, *111*, 687–697. [CrossRef]
110. McMichael, P. Globalization: Myths and Realities. *Rural Sociol.* **1996**, *61*, 25–55. [CrossRef]
111. McMichael, P. Rethinking Globalization: The Agrarian Question Revisited. *Rev. Int. Political Econ.* **1997**, *4*, 630–662. [CrossRef]
112. Fracarolli, G.S. Mapping Online Geographical Indication: Agrifood Products on E-Commerce Shelves of Mercosur and the European Union. *Economies* **2021**, *9*, 84. [CrossRef]
113. Michels, R. *Political Parties*; Free Press: Hong Kong, China, 1966; ISBN 978-0029212509.
114. Allaire, G.; Wolf, S.A. Cognitive Representations and Institutional Hybridity in Agrofood Innovation. *Sci. Technol. Hum. Values* **2004**, *29*, 431–458. [CrossRef]
115. Friedland, W.H. Reprise on Commodity Systems Methodology. *Int. J. Sociol. Agric. Food Online* **2001**, *9*, 82–103. [CrossRef]

MDPI
St. Alban-Anlage 66
4052 Basel
Switzerland
Tel. +41 61 683 77 34
Fax +41 61 302 89 18
www.mdpi.com

Land Editorial Office
E-mail: land@mdpi.com
www.mdpi.com/journal/land

www.ingramcontent.com/pod-product-compliance
Lightning Source LLC
Chambersburg PA
CBHW041944260926
51850CB00018B/767
* 9 7 8 3 0 3 6 5 3 8 8 4 6 *